Bayesian Precision Medicine

Bayesian Precision Medicine presents modern Bayesian statistical models and methods for identifying treatments tailored to individual patients using their prognostic variables and predictive biomarkers. The process of evaluating and comparing treatments is explained and illustrated by practical examples, followed by a discussion of causal analysis and its relationship to statistical inference. A wide array of modern Bayesian clinical trial designs are presented, including applications to many oncology trials. The later chapters describe Bayesian nonparametric regression analyses of datasets arising from multistage chemotherapy for acute leukemia, allogeneic stem cell transplantation, and targeted agents for treating advanced breast cancer.

Features:

- Describes the connection between causal analysis and statistical inference
- Reviews modern personalized Bayesian clinical trial designs for dose-finding, treatment screening, basket trials, enrichment, incorporating historical data, and confirmatory treatment comparison, illustrated by real-world applications
- Presents adaptive methods for clustering similar patient subgroups to improve efficiency
- Describes Bayesian nonparametric regression analyses of real-world datasets from oncology
- Provides pointers to software for implementation

Bayesian Precision Medicine is primarily aimed at biostatisticians and medical researchers who desire to apply modern Bayesian methods to their own clinical trials and data analyses. It also might be used to teach a special topics course on precision medicine using a Bayesian approach to postgraduate biostatistics students. The main goal of the book is to show how Bayesian thinking can provide a practical scientific basis for tailoring treatments to individual patients.

Peter F. Thall is a global leader in the development and application of Bayesian methods in medical research, with over 300 publications in professional journals. His research interests include Bayesian statistics, clinical trial design, precision medicine, and dynamic treatment regimes.

Chapman & Hall/CRC Biostatistics Series

Series Editors
Shein-Chung Chow, Duke University School of Medicine, USA
Byron Jones, Novartis Pharma AG, Switzerland
Jen-pei Liu, National Taiwan University, Taiwan
Karl E. Peace, Georgia Southern University, USA
Bruce W. Turnbull, Cornell University, USA

Recently Published Titles

Quantitative Methods for Precision Medicine
Pharmacogenomics in Action
Rongling Wu

Drug Development for Rare Diseases
Edited by Bo Yang, Yang Song, and Yijie Zhou

Case Studies in Bayesian Methods for Biopharmaceutical CMC
Edited by Paul Faya and Tony Pourmohamad

Statistical Analytics for Health Data Science with SAS and R
Jeffrey Wilson, Ding-Geng Chen, and Karl E. Peace

Design and Analysis of Pragmatic Trials
Song Zhang, Chul Ahn, and Hong Zhu

ROC Analysis for Classification and Prediction in Practice
Christos Nakas, Leonidas Bantis, and Constantine Gatsonis

Controlled Epidemiological Studies
Marie Reilly

Statistical Methods in Health Disparity Research
J. Sunil Rao, Ph.D.

Case Studies in Innovative Clinical Trials
Edited by Binbing Yu and Kristine Broglio

Value of Information for Healthcare Decision Making
Edited by Anna Heath, Natalia Kunst, and Christopher Jackson

Probability Modeling and Statistical Inference in Cancer Screening
Dongfeng Wu

Development of Gene Therapies: Strategic, Scientific, Regulatory, and Access Considerations
Edited by Avery McIntosh and Oleksandr Sverdlov

Bayesian Precision Medicine
Peter F. Thall

Statistical Methods Dynamic Disease Screening Spatio-Temporal Disease Surveillance
Peihua Qiu

For more information about this series, please visit: https://www.routledge.com/Chapman--Hall-CRC-Biostatistics-Series/book-series/CHBIOSTATIS

Bayesian Precision Medicine

Peter F. Thall

CRC Press
Taylor & Francis Group
Boca Raton London New York

CRC Press is an imprint of the
Taylor & Francis Group, an **informa** business
A CHAPMAN & HALL BOOK

First edition published 2024
by CRC Press
2385 Executive Center Drive, Suite 320, Boca Raton, FL 33431, U.S.A.

and by CRC Press
4 Park Square, Milton Park, Abingdon, Oxon, OX14 4RN

CRC Press is an imprint of Taylor & Francis Group, LLC

ISBN: 978-1-032-75446-8 (hbk)
ISBN: 978-1-032-75496-3 (pbk)
ISBN: 978-1-003-47425-8 (ebk)

DOI 10.1201/9781003474258

Typeset in Nimbus Roman
by KnowledgeWorks Global Ltd.

Publisher's note: This book has been prepared from camera-ready copy provided by the authors.

Developing statistical solutions to biomedical research problems is a highly collaborative process. Over the years, it has been especially enjoyable to work with Richard Simon, Peter Mueller, Yanxun Xu, Juhee Lee, Abdus Wahed, Satoshi Morita, Kyle Wathen, Tom Murray, Lu Wang, Andrew Chapple, Ying Yuan, Ruitao Lin, Randy Millikan, Borje Andersson, and Pavlos Msaouel. They have taught me many interesting and useful things, and their fingerprints are all over the methods and data analyses described in this book.

Contents

Preface **xi**

1 Evaluating New Treatments **1**

1.1 Estimation and Testing 1
1.2 Simpson's Paradox 9
1.3 Linking Early and Late Outcomes 18
1.4 Elements of Practical Bayesian Statistics 20

2 Statistics and Causality **29**

2.1 Causal Relationships 29
2.2 Inverse Probability of Treatment Weighting 38
2.3 Generalized Computation Methods 43

3 Precision Dose Optimization **55**

3.1 Dose Finding Based on Time to Toxicity 55
3.2 Choosing Estimands 65
3.3 Utility Functions for Dose Finding 70
3.4 Dose Finding for Natural Killer Cells 79
3.5 Generalized Phase I-II Designs 95

4 Basket Trials **111**

4.1 Patient Heterogeneity and Baskets 111
4.2 A Phase II Design for Non-Exchangeable Subgroups 113
4.3 An Early Basket Trial 118
4.4 A Basket Trial Design with Bayesian Model Averaging 121
4.5 Monitoring Response and a Longitudinal Biomarker 126

5 Precision Randomized Phase II Designs **132**

5.1 A Two-Stage Phase II Design with Adaptive Matching 132
5.2 Precision Treatment Screening and Selection 137

6 Precision Randomized Phase III Designs **154**

6.1 Nutritional Prehabilitation and Post-Operative Morbidity 154
6.2 Precision Confirmatory Survival Time Comparisons 159

7 Enrichment Concepts and Methods **176**

7.1 Disease Heterogeneity and Targeted Agents 176
7.2 Enrichment Using a Vector of Biomarkers 178
7.3 Outcome Adaptive Randomization 181
7.4 Treatment-Biomarker Interactions 184
7.5 Estimation Bias 190

8 Adaptive Enrichment Designs **193**

8.1 Adaptive Signature Designs 193
8.2 Enrichment Designs with One Biomarker 195
8.3 A Hybrid Utility-Based Enrichment Design 198
8.4 A Phase II-III Enrichment Design 202
8.5 Biomarker Subgroups as Random Partitions 204
8.6 Enrichment Using Two Regressions on Biomarkers 211

9 Bayesian Nonparametric Models **226**

9.1 Regression Analysis in Medical Research 226
9.2 Dirichlet Process Priors 229
9.3 Dependent Dirichlet Processes 238

10 Evaluating Multistage Treatment Regimes for Leukemia **242**

10.1 Chemotherapy for Acute Leukemia 242
10.2 A Randomized Trial of Frontline Chemotherapies 243
10.3 Accounting for Multistage Therapy 245
10.4 Bayesian Regression Analysis of the Leukemia Data 253

11 Personalizing Dose in Stem Cell Transplantation **259**

11.1 Allogeneic Stem Cell Transplantation 259
11.2 A DDP with a Refined Gaussian Process Prior 263
11.3 Optimizing Personalized Dose Intervals 266
11.4 Simulation Study 267
11.5 Analyses of the Allosct Data 268

12 Choosing a Breast Cancer Treatment **271**

12.1 A Limitation of the Medical Literature 271
12.2 Evaluating Targeted Agents for Breast Cancer 273
12.3 Bayesian Regression Analysis of the Breast Cancer Data 277
12.4 A Precision Utility Function 281
12.5 Choosing a Best Personalized Treatment 284

Bibliography **289**

Index **309**

Preface

This book describes modern Bayesian statistical methods for using patient characteristics to identify optimal personalized treatments. The particular methods and examples that I have included are based on my experiences working as a biostatistician at M.D. Anderson Cancer Center. Chapter 1 discusses the process of evaluating a new treatment. Chapter 2 reviews elements of causal inference. Chapters 3–8 describe clinical trial designs that account for patient heterogeneity while addressing a wide variety of different goals. These include identifying optimal subgroup-specific doses, screening treatments for safety and efficacy, using latent variables to cluster similar subgroups, identifying a treatment sensitive subgroup using patient biomarkers, and doing confirmatory subgroup-specific treatment comparisons. Chapter 9 gives a brief review of Bayesian nonparametric models. Chapters 10–12 describe Bayesian nonparametric regression analyses of datasets arising from multistage chemotherapy for acute leukemia, allogeneic stem cell transplantation, and targeted agents for treating advanced breast cancer. My goal is to show how Bayesian thinking can provide a practical scientific basis for tailoring treatments to individual patients.

1

Evaluating New Treatments

1.1 Estimation and Testing

Practicing physicians routinely make treatment decisions for their patients using established prognostic variables such as age, disease severity, and performance status. A modern refinement in the treatment of cancers and other diseases also uses markers for biological targets of engineered molecular, cellular, or immunological agents. This raises the possibility that treatment may interact with a biomarker for the target that the agent is designed to attack. In this case, the treatment has different effects in biomarker-positive and biomarker-negative patients, and the biomarker is said to be a *predictive* covariate. Using a patient's biomarkers to guide treatment choice, known as *precision* or *personalized* medicine, is a modern technological elaboration of what physicians have been doing for thousands of years.

In recent decades, a powerful array of new statistical models, methods, and computational algorithms have emerged. This has been facilitated by the rapid development and wide availability of high-speed computers. Biomedical research has provided a particularly strong impetus for the development of new statistical methods for clinical trial design and data analysis. Statistical thinking can greatly facilitate the process of developing and evaluating new precision medicines. This is done by characterizing and exploiting relationships between treatments, prognostic variables, biomarkers, and clinical outcomes such as response to treatment, toxicity, quality of life, and survival time. From a statistical viewpoint, human beings are extremely noisy objects, and clinical outcomes following a given medical treatment are inherently random. This motivates the use of probability models to characterize randomness and uncertainty. These provide a basis for developing statistical methods for experimental design and data analysis. *Prognostic* variables affect the probability distributions of clinical outcomes in the same qualitative way regardless of what treatment is used, and consequently, they are included routinely as additive variables in the linear components of regression models used to evaluate outcomes as functions of treatments. For example, in general, older patients are less likely to have favorable outcomes with any treatment. Ideally, the presence of a biomarker for a cancer cell surface marker or signaling pathway targeted by a particular agent improves the clinical outcomes of a patient with that cancer treated with that agent. In both biological and statistical terms, the biomarker and the agent have an interactive effect on clinical outcomes. The distinction between prognostic and predictive covariates may be

DOI: 10.1201/9781003474258-1

very important in the process of identifying optimal personalized treatments. This issue will be discussed in detail below. The scientific process of treatment evaluation and optimization may involve designing clinical trials and statistical analyses of data obtained from randomized or observational studies. This process has led to many challenging statistical problems, including optimizing a new agent's dose or schedule, identifying treatment-sensitive subpopulations, making treatment choices based on risk-benefit trade-offs, and correcting for confounding between patient covariates and treatments in non-randomized, observational data. These are some of the topics that will be explored in this book.

When a new experimental treatment, E, is invented preclinically, it must be evaluated in patients with the disease that E has been designed to target before it can be established for routine use by physicians. This requires determining whether E is safe and effective. To do this, it may seem that one can simply ask physicians to use E to treat some of their patients who have been diagnosed with the disease, and see what happens. A bit of thought shows that this ignores many important issues. These include how one should choose the physicians, choose the patients, decide how many patients to treat, obtain informed consent from the patients, decide what patient characteristics to record, decide what clinical outcomes to evaluate, train people to assign treatments to patients following a particular statistical design, record patients' data as they are treated and their outcomes are observed, construct a computerized database to house the data, determine who will pay for this process, and make sure to follow all regulations and laws governing how new treatments may be evaluated in humans. This list actually is incomplete, and a federal regulator, database manager, nurse, or physician researcher each would describe things differently, reflecting their perspective.

Because the process of clinical treatment evaluation involves data, inevitably it requires statistical methods for experimental design, data analysis, and making inferences. These are what this book is about. To begin thinking about this, it helps to assume, temporarily, that patients are homogeneous so that it makes sense to talk about an average treatment effect. If there is an active control treatment, C, that has some anti-disease effect but is not curative, so that there is room for improvement, the question is whether E is better than C. The word "better" refers to the rates of one or more clinical outcomes that are used to compare the treatments. For life-threatening diseases, the two outcomes commonly used to establish efficacy are early response to treatment in terms of a specified anti-disease effect, and survival time. In practice, these may give contradictory results since, for example, E may be associated with a higher response rate but provide a shorter average survival time compared to C. Numerical examples of this apparent paradox will be given in the following text. Because treatment safety is always a concern, adverse events that may be caused by E also must be monitored, and considering safety and efficacy together may lead to evaluating risk-benefit trade-offs. For example, if E has a higher response rate than C but also a higher rate of severe toxicities, then it is not obvious how to decide whether one treatment is better than the other. Several of the clinical trial designs and data analyses in later chapters will discuss how utilities may be used to make decisions based on multiple outcomes by quantifying efficacy-toxicity trade-offs. Initially, I will focus on response probabilities, π_E for E and π_C for C, and assume that safety has been established. The main questions then are whether π_E is larger than π_C and, if so, whether the difference $\Delta = \pi_E - \pi_C$ is large enough to matter to a patient. Most of the methods discussed in this book will deal with elaborations of this simple setting that include patient covariates and more complex outcomes.

Even this simplified version of the treatment evaluation process can go wrong, in many ways. The parameters π_C and π_E are conceptual objects that are easy to understand, but they must be estimated statistically from data. Where the data on E and C come from is a major issue, since this determines what the estimates actually represent. To make valid statistical inferences, ideally, a sample of patients treated with E should be representative of the patent population that one wishes to study. The sample also should be large enough to provide a reliable estimate of π_E. Are 10 patients enough? What about 100? A reported response rate of 20% has very different interpretations for these two sample sizes.

To formalize this, suppose that Y is a binary indicator of an event, such as treatment response, that occurs with probability $\pi_E = \Pr(Y = 1)$ with treatment E. If a sample $Y_1, \cdots, Y_n$ of n such indicators is observed, then $X_n = \sum_{i=1}^{n} Y_i$ is the number of successes, and $X_n \mid \pi$ follows a binomial distribution with parameters n and π. written $X_n \sim binom(n, \pi)$. If, under a Bayesian model, one assumes that π_E follows a beta(.5, .5) prior distribution, which has mean .5 and effective sample size 1, then it is easy to compute a posterior 95% credible interval (CrI) for π_E. For example, if $Y = 2$ out of n =10, for a 20% response rate, a 95% CrI is .04 to .50. This implies that, given this small amount of observed data, all one really knows is that π_E probably is less than .50. If n =100, then $Y = 20$ gives the same empirical response rate of 20%, but 95% posterior CrI .13 to .29, which is much smaller and thus much more informative. The general point is that reporting a response rate or other parameter estimate requires also providing an index of variability or uncertainty.

If historical data on C are available but, for example, the E patients tend to be younger and have less severe disease than the C patients, then a conventional statistical analysis of the two samples ignoring patient prognosis will give an unfair, or *biased* treatment comparison. If historical data are used naively as a basis for treatment comparison, without accounting for the fact that patients were not randomized between E and C, an apparent advantage of E over C based on sample estimates may be due, in part or entirely, to the between-treatment difference in prognosis rather than a difference between treatment effects. Below, statistical methods that correct for such covariate imbalances will be discussed. While accounting for reliability and bias is pretty simple stuff, unfortunately, these basic scientific issues often are ignored by people who analyze medical data and report their results. So, even in this simple setting, evaluating E leads to the statistical concepts of patient heterogeneity, sampling, parameters, estimation, reliability, comparison, and bias.

It is not surprising that, for many members of the medical research community, it is not obvious how to go about obtaining data and making inferences about a comparative treatment effect like Δ. The above list of issues and potential complications suggests that performing a well-designed experiment is the best way to proceed. This leads to a clinical trial, which is a medical experiment with human subjects. Its two purposes are to treat the patients enrolled in the trial and to use the trial's data to make inferences about the treatments being studied that may benefit future patients. A clinical trial design must reconcile achieving the trial's scientific goals with the ethical requirement that physicians must treat the trial's patients in an ethically acceptable manner. In addition, a statistical design also must provide a practical basis for trial conduct.

The problem of how to design an experiment and use its results to do one or more comparisons has been discussed extensively in statistics, medical research, computer science, causal analysis, and many other fields. Long after Sir Ronald Fisher (1925) and his colleagues established fundamental structures of modern experimental design, motivated largely by problems in agriculture, the basic principles have been applied and extended to accommodate medical research settings. In oncology and other disease areas, a common practice is to estimate Δ using data from a single-arm trial of E and historical data on patients treated with C. As I will explain below, if ones wishes to compare treatments by estimating Δ, then using data from single-arm trials often is a bad way to proceed. A major statistical issue is that there may be study-versus-historical differences in patient characteristics that can lead to a flawed comparison. To elaborate the example given above, suppose that, in fact, E and C have identical population response rates $\pi_E = \pi_C$ but, due to the play of chance, the sample of patients treated with E had a better prognosis, on average, than the historical patients treated with C. Using the sample proportions $\hat{\pi}_E^{trial}$ from the trial of E and $\hat{\pi}_C^{hist}$ from the historical data on C, the naive estimator $\hat{\Delta}^{naive} = \hat{\pi}_E^{trial} - \hat{\pi}_C^{hist}$ may appear to show that E is superior to S. For example, the observed estimates may be $\hat{\pi}_E^{trial,obs} = .50$ and $\hat{\pi}_C^{hist,obs} = .25$. If the between-study difference in patient prognosis is ignored, it may appear that E is greatly superior because the estimates appear to show that E doubles the response rate of C. But because, in this example, the treatments have identical true response probabilities, the estimated difference of .25 must be due to a combination of prognostic variable effects, random variation, and possibly effects of unknown external variables. Of course, in practice one does not know what the response rates would be if every patient with the disease were treated with E, or if every patient with the disease were treated with C. All one has are data from the two samples. What may not be obvious is that misleading statistical estimates may be obtained from samples that, intentionally or not, were designed to be non-representative. To see why this may occur, it is useful to consider the treatment development process.

Developing a new molecular agent or cellular therapy engineered to attack a particular disease often takes years of intensive, costly laboratory research. Based on preclinical laboratory data showing that E kills cancer cells in culture or in tumors in xenografted mice, investigators may become convinced that E will be effective in humans. This mouse-to-human transportability assumption (Pearl and Bareinboim, 2011) is implicit when positive preclinical results are optimistically reported in professional journals or in the mass media. Because poor prognosis patients are less likely to respond to any therapy, investigators may feel that enrolling such patients in a single-arm phase II trial of E will cast it in a bad light, and possibly negate many years of work devoted to developing E. This belief may lead investigators to conduct a single-arm clinical trial of E with entry criteria that restrict enrollment to patients with good prognosis, who are more likely to respond to any treatment. This increases the likelihood of obtaining a higher empirical response rate with E, compared to what would be obtained if both good and poor prognosis patients were enrolled, and thus of concluding that E is "promising". But $\hat{\Delta}^{naive}$ is biased in favor of E over C, since the historical data on C include both good and poor prognosis patients, while the trial of E only includes good prognosis patients. The source of such problems is the original decision to conduct a single-arm trial of E, rather than a randomized trial of E versus C. This is an immensely important issue to which I will return repeatedly.

To formalize these ideas, index subgroups by $Z = G$ for good prognosis and $Z = P$ for poor prognosis, and denote the subgroup-specific response probabilities by $\pi_{X,Z}$ for $X = E$ or C. Accounting for both Z and X gives two subgroup-specific E-versus-C effects, $\Delta_G = \pi_{E,G} - \pi_{C,G}$ and $\Delta_P = \pi_{E,P} - \pi_{C,P}$. Denoting the prevalences of good and poor subgroups by λ and $1-\lambda$, the overall average response probability for each treatment X is the subgroup-weighted average

$$\bar{\pi}_X = \lambda \pi_{X,G} + (1-\lambda)\pi_{X,P}.$$

The top portion of Table 1.1 provides a numerical illustration of a setting where treatment and prognostic subgroup effects are additive. In the table, if the prevalence $\lambda = .20$ and the true subgroup-specific response probabilities are $\pi_{E,G} = .50$ and $\pi_{E,P} = .20$ for E, and $\pi_{C,G} = .60$ and $\pi_{C,P} = .30$ for C, then the overall average probabilities are $\bar{\pi}_E = .26$ and $\bar{\pi}_C = .36$. In this example, E has a lower response probability than C for good prognosis patients, for poor prognosis patients, and overall. So E is worse than C any way you look at it, and if this is known then E should not be used. But the practice described above restricts the trial of E to enroll only good prognosis patients, and compares the trial estimator $\hat{\pi}_E^{trial} = \hat{\pi}_{E,G}^{trial}$ to the historical data-based estimator,

TABLE 1.1
Response probabilities for standard treatment C and experimental treatment E within Poor and Good prognosis subgroups, and overall. For each treatment $X = E$ and C, the overall mean response probability is $\bar{\pi}_X = .80\pi_{X,P} + .20\pi_{X,G}$.

Treatment	Poor(80%)	Good(20%)	Overall
	Additive Treatment and Prognostic Subgroup Effects		
E	$\pi_{E,P} = .20$	$\pi_{E,G} = .50$	$\bar{\pi}_E = .26$
C	$\pi_{C,P} = .30$	$\pi_{C,G} = .60$	$\bar{\pi}_C = .36$
	Interactive Treatment and Prognostic Subgroup Effects		
E	$\pi_{E,P} = .20$	$\pi_{E,G} = .50$	$\bar{\pi}_E = .26$
C	$\pi_{C,P} = .40$	$\pi_{C,G} = .60$	$\bar{\pi}_C = .44$

which is the subgroup-weighted average

$$\hat{\pi}_C^{hist} = \lambda \hat{\pi}_{C,G}^{hist} + (1-\lambda)\hat{\pi}_{C,P}^{hist}.$$

In this example, $\hat{\pi}_E^{trial}$ has a mean of .50 while $\hat{\pi}_C^{hist}$ has a mean of .36, and the expected difference .50 - .36 = .14 favors E over C. But this difference is due entirely to the fact that the sample of E patients has a good prognosis, while the sample of C patients has 20% good and 80% poor prognosis. This sort of biased comparison may be done simply by reporting $\hat{\pi}_E^{trial}$ = .50, citing a published value $\hat{\pi}_C^{hist}$ = .36, and failing to notice that, due to selectively biased sampling in the trial of E, these two statistics are not comparable. If the prognostic difference between these sample estimates is made explicit, then it is obvious that the comparison is unfair and misleading. But if the results are presented in a way that obfuscates or ignores the difference in prognosis between the E and C samples, then it may appear that the inferior treatment E outperforms C. In statistical practice, this is not a small technical error. It is a large mistake that can lead to incorrect conclusions with potentially disastrous consequences for future patients, since they would be given the inferior treatment E.

To numerically illustrate the harm that can be done by making such a biased statistical comparison, suppose that, rather than representing the probability of response, $\pi_{X,Z}$ represents the probability of surviving one year, for a patient with prognosis Z treated with X. Suppose that the four true but unknown numerical probabilities are as given above, and based on a flawed statistical analysis it is concluded, incorrectly, that E provides an advance over C. Suppose that, based on this incorrect inference, E is used to treat 1000 future patients, 200 with good prognosis and 800 with poor prognosis. The expected number of one-year survivors with E then would be 200$\times$.5 + 800$\times$.2 = 260. If, instead, these 1000 future patients were treated with the superior standard treatment, C, the expected number of one-year survivors would be 200$\times$.6 + 800$\times$.3 = 360. Consequently, the effect of poor statistical practice that leads to the incorrect decision to treat future patients with E rather than C is an additional 100 deaths within one year among 1000 patients. This simple example can be elaborated in numerous ways, but the general point is quite important. Bad statistical practice can mislead physicians and harm or kill a large number of patients.

This sort of misleading treatment comparison based on biased parameter estimates also may be done using a test of hypotheses. If prognosis is ignored and the focus is on π_E and π_C, a common practice is to use the observed value of a statistic, such as an estimate $\hat{\pi}_C^{hist,obs}$ = .36 based on the historical data on C, to formulate the null hypothesis $H_0 : \theta_E = .36$ and alternative $H_1 : \theta_E > .36$ for a one-sided test. This practice is extremely common. A problem with these hypotheses is that it is not known that π_C = .36, but only that a sample estimate of π_C took on this value. To perform this one-sided test, a commonly used but incorrect test statistic is based on the estimator $\hat{\pi}_{E,G}^{trial}$ obtained from a single-arm trial of good prognosis patients treated with E and the estimate .36 from historical patients treated with C. This incorrect test statistic is

$$T^{(Incorrect)} = \frac{\hat{\pi}_{E,G}^{trial} - .36}{\{var(\hat{\pi}_{E,G}^{trial}\}^{1/2}}.$$

Conventionally, a large sample approximation, known as the *De Moivre-Laplace Theorem*, is used to determine the null distribution of $T^{(Incorrect)}$. This theorem is misapplied here because the statistic $\hat{\pi}_{E,G}^{trial}$ is used, rather than the sample proportion that would have been obtained if accrual in the trial of E had not been restricted to good prognosis patients. This leads to the incorrect assumption that $T^{(Incorrect)}$ follows a distribution under H_0 that is approximately N(0,1), that is, *standard normal.* But $T^{(Incorrect)}$ is not standard normal under H_0. Because this practice ignores patient heterogeneity, the biased sample in the trial of E, and the variability in $\hat{\pi}_C^{hist}$, using $T^{(Incorrect)}$ to test these one-sided hypotheses relies on two incorrect assumptions. The first incorrect assumption is that, like the historical data, the sample of patients treated with E included both good and poor prognosis patients. The second incorrect assumption is that $var(\hat{\pi}_C^{hist}) = 0$, which could only be true if the historical data consisted of the entire population of patients ever treated with C, in which case the statistic $\hat{\pi}_C^{hist}$ would be the population mean. In practice, this is impossible. Since $\hat{\pi}_C^{hist}$ was computed from whatever historical data were available, it has a distribution that varies around some unknown mean, provided that the sample was reasonably representative of historical experience with C in the patient population being considered. The numerical estimate .36 is an observed value of $\hat{\pi}_C^{hist}$, obtained from one particular dataset, and another dataset would give a different numerical estimate. That is, $\hat{\pi}_C^{hist}$ has a sampling distribution, and the numerator of the test statistic actually is $\hat{\pi}_{E,G}^{trial} - \hat{\pi}_C^{hist}$. The test is biased because it is more likely to reject $H_0 : \theta_E = .36$ than a test based on data from a trial of E that includes both poor and good prognosis patients. Because the test statistic actually is based on the difference between two statistics, $\hat{\pi}_{E,G}^{trial} - \hat{\pi}_C^{hist}$, the correct denominator should be $\{var(\hat{\pi}_{E,G}^{trial}) + var(\hat{\pi}_C^{hist})\}^{1/2}$. This is an example of a *Doubly Incorrect* statistical test of hypotheses, since the test statistic includes both (1) a biased estimate of Δ in the numerator and (2) the wrong variance in the denominator. Assuming that $T^{(Incorrect)}$ follows an approximately N(0,1) null distribution is just plain wrong. Doubtless, if either Abraham De Moivre or Pierre-Simon Laplace were alive to see such an egregious misuse of their theorem, they would be appalled.

While this common practice is deeply flawed, it may be made even worse by ignoring historical data on C entirely. A one-sided null hypothesis sometimes is formulated by using a conveniently small, arbitrary, fixed null value for θ_C, such as .20, that has no connection to historical experience with C. This gives the even worse test statistic

$$T^{(Even\ Worse)} = \frac{\hat{\pi}_{E,G}^{trial} - .20}{\{var(\hat{\pi}_{E,G}^{trial})\}^{1/2}}.$$

This is the sort of thing that may be done by someone who has learned how to run a canned program that computes a one-sample test statistic based on binary response data, with a null value of the success probability being one of the program's required inputs. Because it is more likely that $H_0 : \theta_E = .20$ will be rejected than $H_0 : \theta_E = .36$ for a given dataset, the test now has been rigged even further to be even more likely to reject H_0 and thus conclude that E provides a treatment advance over C. The test is likely to lead to the incorrect conclusion that an E for which $\pi_E < \pi_C$ is superior to

C. While $T^{(Even\ Worse)}$ may appear to be a valid test statistic, performing such a test is fake science. This is an example of a *Triply Incorrect* statistical test of hypotheses. This test statistic includes (1) a biased estimate of the between-treatment effect Δ, (2) an incorrect variance that is too small and thus inflates the value of the test statistic, and (3) an arbitrary fixed null value of π_E that is disconnected from historical experience with standard treatment. A practical consequence is that, if it leads to rejection of H_0, this test may harm future patients by leading physicians to give them a treatment E that actually is inferior to C. It also may lead a pharmaceutical company to invest a large amount of time, money, and human resources to conduct a future phase III trial of E that a valid statistical analysis of existing data would show is very unlikely to produce a positive result. This illustrates how bad statistical practice may not only kill patients, it also may waste millions of dollars and years in the lives of medical researchers, while causing patients to be enrolled in a clinical trial that never should have been conducted.

Comparing treatments based on nonrandomized data without correcting for bias can produce a variety of flawed inferences. Statistical techniques that use patient covariates to correct for bias when using observational data to make comparisons will be discussed below. Unfortunately, the fact that bias correction methods often are not used by medical researchers suggests that they are not widely understood. There appears to be a common belief in the medical research community that, if one fits nonrandomized data on two treatments similar to that described above using a conventional regression model that includes prognostic variable and treatment effects additively, then randomization is not needed. This is not true. Consider the linear term of a logistic regression model accounting for treatments $X = E$ or C and patient subgroups $Z = G$ or P in the above example with the form

$$\eta_{X,Z}^{additive} = \text{logit}\{\pi_{X,Z}^{additive}\} = \alpha + \beta\, I[X = E] + \gamma\, I[Z = G].$$

Under this model, β is the E-versus-C treatment effect, γ is the G-versus-P prognostic subgroup effect, and all three parameters (α, β, γ) can be estimated, despite the absence of any data on (E,P) patients. The assumption of additivity implies that the good-versus-poor prognostic subgroup effects are identical for E and C, and thus can be represented by the single parameter γ. Consequently, the three parameters (α, β, γ) determine the four subgroup-specific response probabilities $(\pi_{E,G}, \pi_{E,P}, \pi_{C,G}, \pi_{C,P})$. If the additivity assumption is incorrect, however, then this model is incorrect, the formulas for the subgroup-specific response probabilities are incorrect, and this can lead to incorrect inferences. For example, suppose that, in fact, $\pi_{C,P}^{true} = .40$, as in the bottom portion of Table 1.1, which illustrates a case with treatment-prognosis interaction. The true E-versus-C effects are $.20 - .40 = -.20$ in poor prognosis patients and $.50 - .60 = -.10$ in good prognosis patients, with average true response probabilities $\bar{\pi}_C^{true} = .44$ and $\bar{\pi}_E^{true} = .26$. In this case, the additive model is incorrect and there should be four model parameters rather than three, but the absence of data from (E,P) patients makes it impossible to determine this statistically. Fitting this incorrect additive regression model to the data will give estimates of parameters that do not exist.

These examples illustrate potential consequences of what many people actually do in practice. Single-arm clinical trials, often called "phase II trials", are quite common in oncology. The medical literature is full of papers reporting data from single-arm trials, including nonrepresentative samples and biased comparisons to historical data. Numerous published papers reporting the results of such trials include treatment comparisons based on non-randomized study data, observational data, or data from separate single-arm trials, with no correction for between-study bias. Such papers may include a flawed comparative statistical analysis by fitting a regression model under the naive, incorrect assumption that accounting for covariate effects additively will somehow correct for bias.

A model that accounts for treatment-subgroup interactions has the linear term

$$\text{logit}\{\pi_{X,Z}\} = \alpha^* + \beta^*\, I[X=E] + \gamma^*\, I[Z=G] + \xi^*\, I[Z=G]\, I[X=E].$$

The superscripted * is included so that, for example, the E-versus-C treatment effect β^* in poor prognosis patients in this interaction model is not confused with the main treatment effect β in the additive model, since they are different parameters. The interaction model specifies different subgroup-specific E-versus-C effects, $\beta^* + \xi^*$ in good prognosis patients and β^* in poor prognosis patients. This model re-expresses the four probabilities $(\pi_{E,G}, \pi_{E,P}, \pi_{C,G}, \pi_{C,P})$ as the four real-valued parameters $(\alpha^*, \beta^*, \gamma^*, \xi^*)$, since there is a one-to-one relationship between the two vectors. Under the assumption $\xi^* = 0$, that there is no treatment-subgroup interaction, the interaction model is reduced to the simpler additive form, and in this case $\alpha^* = \alpha$, $\beta^* = \beta$, and $\gamma^* = \gamma$. While it may be argued, on fundamental grounds, that a treatment should not interact with prognosis, in practice this may or may not be true. If the good and poor prognostic subgroups are replaced in the illustration by biomarker positive and biomarker negative subgroups, for a biomarker that E has been engineered to attack, then it may not make sense to assume an additive model, since by the design of E the two biomarker subgroups should have different E-versus-C effects.

1.2 Simpson's Paradox

In the examples given above, patients were classified as having either good or poor prognosis, with subgroup-specific response probabilities. If E-versus-C treatment effects $\Delta_G = \pi_{E,G} - \pi_{C,G}$ and $\Delta_P = \pi_{E,P} - \pi_{C,P}$ are not the same, then good and poor prognosis patients may have different optimal treatments. For example, if poor prognosis patients with a hematological malignancy are immunocompromised due to combined effects of the disease and previous treatments, then a new targeted cellular immunotherapy that is effective in good prognosis patients, who have stronger immune systems, may be ineffective in poor prognosis patients. Data showing this may lead to the conclusions that the new treatment is best in good prognosis patients, but standard therapy is best if a patient has a poor prognosis. As mentioned above,

TABLE 1.2
Example of Simpson's Paradox. Hypothetical data on 850 patients with a rapidly fatal disease, each either treated with targeted immunotherapy or relying on prayer, and living in either New York or California. The entry in each cell is [Number of Patients Surviving Six Months] / [Number of Patients] (%).

	Location of Patient		
Treatment	New York	California	Overall
Prayer	30 / 100 (30%)	210 / 300 (70%)	240 / 400 (60.0%)
Immunotherapy	160 / 400 (40%)	40 / 50 (80%)	200 / 450 (44.4%)

a structurally similar setting is one where patients are either biomarker positive or biomarker negative, so a treatment engineered to target the biomarker may be appropriate for biomarker-positive patients but not for biomarker-negative patients. In this case, a biomarker-treatment interaction is a consequence of the way that the targeted treatment was designed to behave biologically. But this belief may be overly simplistic. A designed molecule may have additional, unplanned anti-disease effects, not only the anti-disease effect that was engineered to attack the target represented by the biomarker. In this case, the agent may have unexpected anti-disease effects in biomarker-negative patients.

With observational medical data, which is what may become available when patients are not randomized between treatments, a phenomenon known as *Simpson's Paradox* may arise. This occurs when effects of a treatment or exposure variable seen within subgroups of a sample disagree with the overall effect seen when the subgroups are ignored. This counterintuitive statistical phenomenon can occur if the observed outcome distributions vary with patient subgroups or covariates, and the subgroup sample sizes are unbalanced. Table 1.2 provides a toy illustration, where 850 patients living in either New York or California were diagnosed with a rapidly fatal disease. Each patient either was treated with a targeted immunotherapy, or relied on prayer to treat their disease. Among 500 patients living in New York, 100 relied on prayer and achieved a 30% six-month survival rate. Among the 400 New Yorkers who received the immunotherapy, 40% survived six months. So, immunotherapy had a higher six-month survival rate than prayer among New Yorkers. For the Californians, 300 relied on prayer and achieved a 70% six month survival rate, while 50 accepted the immunotherapy and achieved an 80% survival rate. So, the immunotherapy also was better than prayer for the Californians. If the states where the patients live are ignored, the last column of the table shows that the six-month survival rate was 60.0% for prayer compared to only 44.4% for immunotherapy so, overall, prayer was better for achieving six-month survival. The paradox is that immunotherapy was better than prayer for both New Yorkers and Californians, but if the states are ignored then prayer was better than immunotherapy. Of course, this makes no sense. A possible conclusion is that, if you are diagnosed with this disease, you should reject the immunotherapy and instead pray to God to save you, but make sure not to tell God where you live.

This paradox can be resolved by applying the *Law of Total Probability*. Table 1.2 presents statistical estimates of Pr(Survive at least six months) for each of the four different combinations of (Treatment, State), and overall for each of the two treatment groups, Prayer and Targeted Immunotherapy (TI). While the first four empirical estimates are perfectly correct, the paradox arises from to the fact that each of the combined estimates in the table weights the two states incorrectly. To fix this, first note that the overall sample proportions are 500/850 = .588 for New York (NY) and 350/850 = .412 for California (Cal). Denoting T = survival time, the *Law of Total Probability* implies that

$$\Pr(T \geq 6 \mid Pray) = \Pr(T \geq 6 \mid Pray, NY)\Pr(NY) + \Pr(T \geq 6 \mid Pray, Cal)\Pr(Cal).$$

Plugging in the sample estimates to obtain an overall estimate gives

$$\widehat{\Pr}(T \geq 6 \mid Pray) = \left(\frac{30}{100} \times .588\right) + \left(\frac{210}{300} \times .412\right) = .465.$$

Similarly,

$$\widehat{\Pr}(T \geq 6 \mid TI) = \left(\frac{160}{400} \times .588\right) + \left(\frac{40}{50} \times .412\right) = .565.$$

Thus, if one obeys *The Law of Total Probability* and obtains correctly weighted estimates, they show that immunotherapy also has a higher overall six-month survival probability than Prayer. This may come as a relief for anyone worried about possible consequences of praying to God but not telling Him where you live.

To see why the values in the last column of Table 1.2 are incorrect estimates of the overall six-month survival probabilities, note that the New Yorkers were far more likely to choose immunotherapy, with rate 400/500 (80%), versus only 50/350 (14%) of Californians. This is hardly surprising. Additionally, the sample had 500 (59%) New Yorkers and 350 (41%) Californians. The data must be observational since, if they had come from a fairly randomized study, the probability that 500 New Yorkers would be randomized with 100 going to a Prayer arm and 400 to an immunotherapy arm is less than 10^{-43}. Of course, a randomized study comparing prayer and immunotherapy would be impossible to carry out, since many people do not believe in the efficacy of immunotherapy. If someone with the disease prayed to God and rejected the immunotherapy, He might tell them, "*You prayed to me, so I made the targeted immunotherapy available to you. Why did you ignore my answer to your prayer?*"

Taking this a bit further, there also is the possibility that the combination therapy consisting of both Prayer and TI might be more effective than either component alone. To see how this might work, the example may be elaborated by including an additional group of patients who received targeted immunotherapy and also prayed. An extended version of the example including 150 additional patients who received this combination therapy is given in Table 1.3. For this larger dataset with 1000

TABLE 1.3
Elaborated example of Simpson's Paradox. Hypothetical data on 1000 patients with a rapidly fatal disease, each treated with targeted immunotherapy (TI), relying on prayer (P), or both TI and P, living in either New York or California. The entry in each cell is [Number of Patients Surviving Six Months] / [Number of Patients] (%).

	Location of Patient		
Treatment	New York	California	Overall
Prayer	30 / 100 (30%)	210 / 300 (70%)	240 / 400 (60.0%)
Immunotherapy	160 / 400 (40%)	40 / 50 (80%)	200 / 450 (44.4%)
Prayer + Immunotherapy	57 / 142 (40.1%)	7 / 8 (87.5%)	64 / 150 (42.7%)

patients, the sample proportions are .642 for New York and .358 for California. Table 1.3 shows that, in each state, prayer was less effective than targeted immunotherapy, which in turn was less effective than the combination. But the overall rates go in the opposite direction, so Simpson's Paradox arises again. Applying *The Law of Total Probability* and weighting the probabilities correctly shows that, for Prayer alone,

$$\Pr(T \geq 6 \mid Pray) = \Pr(T \geq 6 \mid Pray, NY)\Pr(NY) + \Pr(T \geq 6 \mid Pray, Cal)\Pr(Cal)$$

$$= (.30 \times .642) + (.70 \times .358) = .4432,$$

for targeted Immunotherapy alone,

$$\Pr(T \geq 6 \mid TI) = (.40 \times .642) + (.80 \times .358) = .5432,$$

and for the combination

$$\Pr(T \geq 6 \mid TI + Pray) = (.4014 \times .642) + (.875 \times .358) = .5710.$$

Thus, correctly weighting the survival probabilities for each treatment in each state using the overall sample proportions of the states shows that the overall 6-month survival probabilities were lowest for prayer alone (44.3%), about 10% higher for targeted immunotherapy (54.3%), and slightly higher if a patient received TI and also prayed (57.1%). From a mechanistic viewpoint, whether this may be caused by an enhanced immune system due to a positive attitude from praying, or The Hand of God, may be a subject for debate. In any case, obeying *The Law of Total Probability* gives the opposite of the incorrect ordering in the last column of Table 1.3. These examples show that it is always best to obey *The Laws of Probability*.

To see how Simpson's Paradox may arise when a continuous variable is involved, consider a life-threatening disease with data that classifies patients into three prognostic subgroups, Good ($g = 1$), Intermediate ($g = 2$), or Poor ($g = 3$). Denote the parameter subvector $\boldsymbol{\theta}_g = (\alpha_g, \beta_g)$ for each g and $\boldsymbol{\theta} = (\boldsymbol{\theta}_1, \boldsymbol{\theta}_2, \boldsymbol{\theta}_3)$. Suppose that the conditional mean of Y = log survival time given d = dose of a treatment and prognostic subgroup g is

$$\mu_g(d,\boldsymbol{\theta}) = E\{Y \mid d,g,\boldsymbol{\theta}\} = \alpha_g + \beta_g d,$$

with each $\beta_g > 0$. This says that mean log survival time increases linearly with dose within each prognostic subgroup, and the subgroups may have different dose-mean log survival time lines. The fact that, for each dose d, expected log survival time decreases as prognosis worsens from Good to Intermediate to Poor is formalized by the inequalities $\mu_1(d,\boldsymbol{\theta}_g) > \mu_2(d,\boldsymbol{\theta}_g) > \mu_3(d,\boldsymbol{\theta}_g)$. Figure 1.1 illustrates the relationships among survival time, prognosis, and dose for two different assumed regression models. Denoting $\boldsymbol{\theta}^* = (\alpha^*,\beta^*)$ if the prognostic subgroups are ignored and the naive model does not include dose-subgroup interactions, with means given by

$$\mu(d,\boldsymbol{\theta}^*)^{naive} = E^{naive}\{Y \mid d\} = \alpha^* + \beta^* d$$

is fit to the data then, as shown in the left portion of Figure 1.1, the scattergram and fitted line show that Y decreases with d. The right side of the figure is based on the correct model accounting for prognostic subgroup-dose interactions. The three scattergrams of the (d,Y) data within the subgroups are represented by red for Good, green for Intermediate, and blue for Poor prognosis. This shows that, under the model correctly accounting for subgroup-dose interactions, each fitted line has a positive slope, leading to the correct inferences that $E(Y \mid d,g,\boldsymbol{\theta})$ increases with dose and

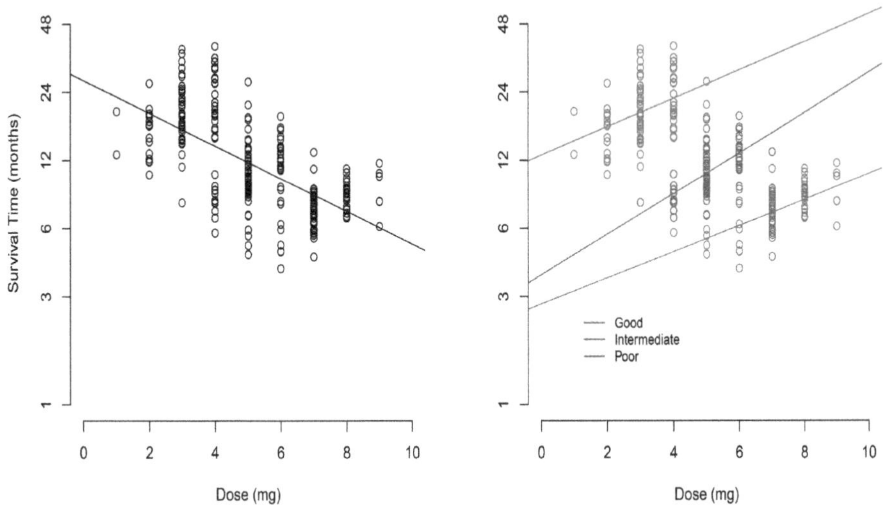

FIGURE 1.1
Illustration of Simpson's Paradox for survival time as a function of dose and prognostic subgroup. If subgroups are ignored (left side), the scattergram and fitted line imply that survival time decreases with dose. Accounting for subgroup effects (right side) shows that survival time decreases with worse prognosis but increases with dose within each prognostic subgroup.

decreases as prognostic level becomes worse. The paradox arises if the effect of dose on log survival time is estimated as one dose effect parameter β^* under the incorrect naive model. When fit to the data, this incorrect model gives a negative estimate of β^*. Together, the two different statistical analyses show that survival time increases with dose within each prognostic subgroup, but if prognosis is ignored then one is misled to infer that survival time decreases with dose. The key to the paradox is that ignoring the ordinal prognostic subgroups makes the subgroup effects appear to be a dose effect.

In general, Simpson's paradox arises from ignoring a variable that may affect the outcome in a multi-variable setting by disobeying *The Law of Total Probability*. This leads to an incorrect estimate of the effect of a variable of interest, such as treatment. This problem also may occur in more complex settings where more than three variables are involved and a model simplification, such as variable selection, has been done that removed one or more key variables. Another possibility is that a variable having a substantive effect on an outcome simply is not included in a dataset, possibly because it is unknown. Such a *lurking variable*, also known as an *external confounder*, can produce estimated effects of known variables that are very different from their true effects. Many statisticians, scientists, and causal effects experts are so terrified of external confounders that they ritualistically assume that they do not exist. VanderWeele and Onyebuchi (2011) described how to conduct sensitivity analyses to assess possible biasing effects of unmeasured confounders.

To summarize, it is useful to distinguish between the following different approaches to treatment evaluation and comparison.

Approach 1. Do not conduct a clinical trial. Instead, rely on preclinical *in vivo* mouse data to infer what E-versus-C effects should be in humans. This is the *mouse-to-human without humans* approach, which relies on the belief that treatment effects in mice and humans are qualitatively identical. This is an example of assuming perfect inter-species causal transportability.

Approach 2. Collect data from a single-arm clinical trial of E that enrolls only good prognosis patients, and compare the empirical response rate to that computed from historical data on C. Ignore patient prognosis when doing the comparison. Depending on one's perspective, this may be considered an example of either *Lying with Statistics* or *Inexcusable Stupidity*.

Approach 3. Collect data from a single-arm clinical trial of E enrolling only good prognosis patients, and compare the trial's data to good prognosis patients in historical data on C. This produces a more fair comparison. Since the comparison still may be biased due to the lack of randomization, use patient covariates to compute a bias-corrected estimator, using a method such as inverse probability of treatment weighting, pair matching, or G-estimation.

Approach 4. If observational data with response rates for E and C classified by subgroups are available, compute subgroup weighted average estimates to correct for bias. Compute the subgroup weights using the combined E and C data.

Approach 5. If you are diagnosed with the disease, pray to God for assistance, but make sure to take full advantage of whatever God provides, bearing in mind that He may work in mysterious ways.

Approach 6. Randomize patients between E and C. This yields unbiased estimates of $\pi_{E,Z} - \pi_{C,Z}$ for each subgroup Z, and of the overall average E-versus-C difference $\bar{\pi}_E - \bar{\pi}_C$. Randomization helps to avoid the sort of scientific and medical disasters, described above, that may result from conducting single-arm trials that lead to biased comparisons. Randomization may be refined by stratifying on prognosis to balance sample sizes within the subgroups G and P. As a basis for precision medicine, two subgroup-specific tests may be performed, of the null hypotheses

$$H_{0,P} : \theta_{E,P} = \theta_{C,P} \quad \text{and} \quad H_{0,G} : \theta_{E,G} = \theta_{C,G}.$$

The prevalence of single-arm clinical trials and biased treatment comparisons in oncology is, in large part, a sociological and cultural phenomenon. Among oncologists, it may be traced back to *activity trials*, which are conducted in patient subpopulations where no safe standard treatment with substantive anti-disease activity exists. This is the case if the disease is very severe at diagnosis or the patient population being considered has a very poor prognosis, such as cancer patients who were refractory to frontline therapy or relapsed after achieving disease remission with one or more previous lines of therapy. In this case, the question for a practicing physician is whether treating a patient with E is better than not treating them at all. In such settings, a randomized trial of a new treatment E cannot be conducted because there is no comparator. A reasonable approach is to conduct a single-arm activity trial of E with the aim to decide whether E has at least a small response rate that is meaningfully different from 0, such as .20 or .30. This rationale disappears, however, in settings where there is a comparator, in the form of an effective standard therapy C having a non-trivial estimated response rate $\hat{\pi}_C \geq .20$. In this case, randomizing patients between E and C is ethical and is the most scientifically valid approach for comparison because it provides an unbiased estimator of the treatment effect difference $\Delta = \pi_E - \pi_C$. Arguably, not randomizing and instead conducting a single-arm trial is less ethical, because the resulting data do not provide a simple basis for comparing E to C, which in this case requires some form of bias correction, such as pair matching. Provided that there is no empirical or rational basis for preferring one treatment over the other clinically, randomization is the most ethical approach, because an unbiased comparison may be done using only the trial's data, and it provides the best scientific basis for deciding which treatment is best for future patients.

In oncology, conducting non-randomized single-arm phase II trials is a pervasive convention, despite the problem that it often leads to flawed inferences. This dysfunctional practice was motivated, in large part, by arguments that many clinicians consider randomization to be undesirable and that the use of historical controls for comparing E to C may be scientifically preferable to randomization (Gehan and Freireich, 1974). An additional problem is that representatives of a pharmaceutical company with a new agent E, who provide funding for a trial, may not wish to have patients randomized between E and C in an early phase trial, because such a trial might show that E is not promising, or even inferior, compared to C. This is a false economy, however, since if E does not provide an improvement of C then a trial that finds E promising has made a false positive conclusion that will not be borne out by future use of E.

Applying valid statistical bias correction methods that use historical data to make comparisons of a new treatment to a historical control C is good scientific practice, given that one wishes to evaluate data from a single-arm trial of E and data on C are available. Rather than using bias correction methods, many medical papers reporting single-arm phase II trial results for a new agent E give estimates of response rates and survival time parameters, and cite previously published numerical estimates for one or more standard treatments. For example, it may be reported that estimated median survival with E was 36 months, with a 95% confidence interval of 28–44 months, and that the previously seen median survival with C was 30 months, with no attempt to correct for bias. The implication is that the reader should to make the obvious but inherently biased comparison of 36 to 30. It also is common to see this accompanied by fitted regression models for survival or progression-free survival time, or response probability, including patient covariates additively. The implication is that this somehow corrects for selection bias (Gehan and Freireich, 1974), which is not true. This problem has persisted for decades, despite the availability of practical, reliable statistical methods for making bias-corrected comparisons using historical data. Some of these methods will be discussed later in this chapter, and also in Chapters 2, 4, 7, and 9. Reporting biased comparisons is not merely a technical error, because practicing clinicians base their treatment decisions on the published medical literature.

Several of the above examples illustrate problems that may arise if treatment recommendations ignore individual patient prognostic variables or biomarker status. Most of the applications and methods described in this book aim to improve on the common practice of "one size fits all" statistical procedures that ignore patient heterogeneity. The central issue, which is both statistical and medical, is that comparative treatment effects may vary with patient covariates. Accounting for good and poor prognostic subgroups statistically requires estimating $\pi_{X,G}$ and $\pi_{X,P}$ for each treatment X and making subgroup-specific inferences, rather than ignoring prognosis and assuming that a single probability π_X adequately characterizes the response rate for each X. More generally, for a vector $\mathbf{Z} = (Z_1, \cdots, Z_q)$ of baseline covariates that may include some combination of prognostic variables and biomarkers, denoting $\mu(\mathbf{Z}, X) = E(Y \mid \mathbf{Z}, X)$ for a treatment X, one may compute an overall mean of Y for X by averaging over the covariate distribution. For a representative sample $\{(Y_i, \mathbf{Z}_i), i = 1, \cdots, n\}$ of patients treated with X, in practice, this may be the sample mean

$$\mu(X) = \frac{1}{n}\sum_{i=1}^{n} \mu(\mathbf{Z}_i, X)$$

that weights the observed sample covariate vectors $\mathbf{Z}_1, \cdots, \mathbf{Z}_n$ equally. To identify optimal precision treatments in a practical way when comparing E and C, it may be useful to construct a treatment assignment rule that classifies each possible $\mathbf{Z}$ vector as being either E-optimal or C-optimal.

An important problem in developmental therapeutics is that effects of a biomarker that is found to be predictive based on preclinical data in cells or mice may not turn out to be as anticipated in humans. A predictive biomarker identifies the presence (+) or absence (–) of a biological target that a designed molecule or immunotherapy E has been engineered to attack. By transporting preclinical knowledge based on mice to humans, one may believe that the biological mechanisms which motivated the design of E imply that biomarker + patients should be more likely to have a favorable outcome than biomarker – patients when treated with E, formally $\pi_{E,+} > \pi_{E,-}$. If this is validated statistically in humans, then E behaves as it was designed. However, it is not necessarily true that E provides a substantive improvement over C in biomarker + patients, where $\Delta_+ = \pi_{E,+} - \pi_{C,+}$ is large enough to be clinically meaningful. A different, not unlikely possibility is that a designed molecule or engineered cellular therapy may have multiple effects on a patient's disease, some of which are not anticipated. If E is safe, it may be the case that $\Delta_- = \pi_{E,-} - \pi_{C,-} > 0$, that is, E provides benefit in biomarker – patients. In this case, an established mechanistic biological explanation of how a new agent works may only provide a partial picture of its effects. In such settings, to compare E to C statistically requires reliably estimating both Δ_+ and Δ_-. This motivates conducting a randomized trial of E versus C that enrolls both biomarker positive and negative patients. Once trial results have been obtained and analyzed, in order to use its results to guide medical practice, the question is whether a putatively predictive biomarker should be used to decide whether to give a patient a new treatment.

In some settings, a covariate may be both prognostic and predictive. For example, in breast cancer, amplification of the HER2/neu gene leads to increased cancer cell proliferation and inhibition of apoptosis. Consequently, this amplification is associated with a worse prognosis regardless of what treatment is given, so it is prognostic. If a monoclonal antibody, E, is engineered to attack tumors that overexpress HER2/neu, and E behaves as it was engineered by providing a higher response rate than standard therapy in patients with this overexpression, then the presence of this abnormality is also predictive for E.

Safety is never a secondary concern in a clinical trial. It may turn out that a new targeted agent E has unacceptable, possibly unanticipated adverse effects in biomarker positive patients. Because adverse events are common in oncology, they play a major role in treatment evaluation and clinical practice. Thus, in addition to evaluating subgroup-specific or covariate-specific effects on response probabilities or expected survival times, an important consideration is that adverse event rates also may vary with both the treatments being studied and patient covariates. In a given subgroup, a treatment E may extend the expected survival time or PFS time compared to C, but E also may be more likely to carry a higher risk of severe adverse effects. For example, suppose that, among older patients who are biomarker +, a six-month improvement in expected survival time with E, from 30 to 36 months, is accompanied by an increase in the rate of grade 3 or 4 toxicities, from 10% with

C to 40% with E. The question then becomes whether an additional six months of expected survival time is a desirable trade-off for spending that additional time suffering from the toxicities. A decision analysis dealing with this type of problem, based on data from a clinical trial of targeted agents for advanced breast cancer, will be discussed in Chapter 12.

1.3 Linking Early and Late Outcomes

Figure 1.2 illustrates a setting with two outcome variables, early response and survival time, and treatment. This is an example of a directed acyclic graph (DAG). A DAG consists of nodes and edges, where each node is a covariate, treatment, or outcome variable, and each edge represents a causal effect from one node to another, subject to the requirement that no set of edge directions can form a closed loop, or cycle. A DAG, also known as a causal diagram, provides a visual guide for how to formulate a regression model that accurately reflects actual causal relationships. In oncology, typical early response variables are an indicator of $\geq 50\%$ shrinkage of a solid tumor, complete remission of acute leukemia, or an indicator of whether the level of a biological variable, such as a particular cytokine level or cell count, exceeds a fixed threshold. Examples in other disease areas include the resolution of an infection in an antibiotic study, dissolving a cranial blood clot when treating acute ischemic stroke, and lowering blood pressure by $\geq 10\%$ in an anti-hypertension treatment study. Each of these response variables is defined so that it is observed within a fixed early time period from the start of treatment. Response commonly is used as a surrogate outcome in oncology to speed up treatment evaluation because it is scored much sooner than PFS or overall survival time. Quite often, this common practice is a serious error, and it may lead to fallacious inferences. This is because, while early treatment response may be related to these long-term outcomes, it is never a perfect surrogate. Merino et al. (2023) discuss the disconnect between early response, PFS time, and OS time that often occurs in oncology, and illustrate this problem with six randomized trials in non-Hodgkin lymphomas (NHLs) or chronic lymphocytic leukemia (CLL).

A general illustration is provided by the following toy numerical example. In the causal diagram in Figure 1.2, treatment has direct effects on both early response and survival time, and early response has a direct effect on survival time. Making the incorrect assumption that response is a surrogate for survival or PFS time can lead quite easily to incorrect conclusions and poor decisions. Some illustrations will be given below. A practical approach is to characterize the probabilistic relationship between these variables, covariates, and treatments. Consider the goal of comparing an experimental treatment E $(X = 1)$ to a control C $(X = 0)$ based on an early response indicator Y and survival or PFS time T. Assume for simplicity that Y is evaluated very soon after treatment. To focus on the relationship between Y and T, also assume that patients are homogeneous. Denote the response probabilities by $\pi_x = \Pr(Y = 1 \mid X = x)$

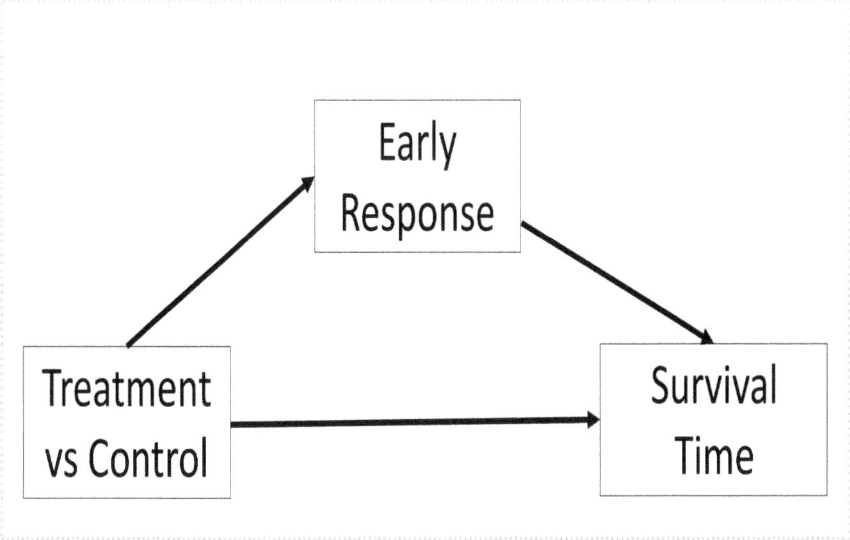

FIGURE 1.2
Causal diagram of treatment, early treatment response, and survival time.

and the conditional 12-month survival probabilities by

$$\overline{F}_x(y) = Pr(T > 12 \mid Y = y, X = x), \qquad \text{for} \quad y = 0,1 \;\; \text{and} \;\; x = 0,1.$$

Suppose that C has a much higher response probability, π_0 = .40, versus π_1 = .15 for E, and that the 12-month survival probabilities among responders and non-responders treated with E are

$$\overline{F}_1(1) = .95, \;\; \text{and} \;\; \overline{F}_1(0) = .90,$$

and for patients treated with the control are

$$\overline{F}_0(1) = .90, \;\; \text{and} \;\; \overline{F}_0(0) = .70.$$

Thus, achieving an early response increases the 12-month survival probabilities by .95 - .90 = .05 for E and .90 - .70 = .20 for C. This is in accordance with the commonly held belief that, because an early response increases survival time, it may be used in place of survival time, as a surrogate, to evaluate treatments. This strategy is convenient because a shorter follow-up period is needed to evaluate Y rather than T. This also is a major rationale for conducting phase II trials, which conventionally are considered to be an efficient bridge between dose-finding trials and phase III trials.

A simple probability computation tells a very different story. By the *Law of Total Probability*, the unconditional survival probabilities of the two treatments are the weighted averages

$$Pr(T > 12 \mid X = 1) = \overline{F}_1(1)\pi_1 + \overline{F}_1(0)(1 - \pi_1) = .95 \times .15 + .90 \times .85 = .91$$

for E and

$$Pr(T > 12 \mid X = 0) = \overline{F}_0(1)\pi_0 + \overline{F}_C(0)(1 - \pi_0) = .90 \times .40 + .70 \times .60 = .78$$

for C. Thus, E has a much lower response rate but a much higher 12-month survival probability compared to C. If early response rates alone are used for comparing treatments, as done in many phase II clinical trials, one would be likely to conclude that E is greatly inferior to C, and thus not conduct a phase III trial. If, instead, survival time is used for treatment comparison, one would be very likely to reach the opposite conclusion, that E provides a substantial advance over C. The apparent paradox arises from the general fact that response is not a perfect surrogate for survival time. As a sensitivity analysis, note that, even if $\pi_1 = 0$, which says that patients treated with E never respond, this would imply that $Pr(T > 12 \mid X = 1) = \overline{F}_1(0) = .85$, so E still would have a higher 12-month survival probability than C. In this extreme case, response is completely irrelevant with regard to survival for treatment E.

From a causal perspective, this might occur because Y mediates the treatment effect on T but treatment also has a direct effect on T. In the extreme case, the only thing that matters is the direct effect of E on survival time since not only is Y not a surrogate for T, but Y has no causal effect at all. This may occur in settings where there is direct biological treatment effect on survival time not mediated by early response. For example, an immunological treatment may provide a long-term beneficial effect, but this is unlikely to be seen in terms of an indicator of an immune activation marker being above a particular fixed threshold one month after treatment, which is a typical way that early response may be defined. In such settings, using early response alone to screen and evaluate a new treatment, without also evaluating long-term benefit in terms of survival or PFS time, is a poor scientific strategy that leads to incorrect conclusions about long-term treatment effects. In such settings, the convention of defining an early binary variable to characterize efficacy is dysfunctional and misleading. This important issue will be addressed by several of the designs discussed in Chapter 3.

1.4 Elements of Practical Bayesian Statistics

Probability is the foundation on which all statistical inference rests, but the way in which probability and statistics are connected is controversial. For an outcome variable Y, Bayesian and frequentist statistical methods both rely on a probability model, $f(y \mid \boldsymbol{\theta})$, indexed by a parameter vector $\boldsymbol{\theta}$, to provide a formal basis for characterizing uncertainty about Y and statistical estimators and decision rules computed from a sample $\boldsymbol{Y} = (Y_1, \cdots, Y_n)$. While the general class of Bayesian nonparametric models, which will be discussed in Chapter 9, considers the probability function f itself to be the parameter, for most models, such as the binomial, normal, or exponential, $\boldsymbol{\theta}$ includes parameters like a distribution's mean, variance, or shape, and possibly covariate effects in a regression model.

The distinction between frequentist and Bayesian statistics begins with the fact that a frequentist considers parameters to be fixed and unknown, whereas a Bayesian considers parameters to be random. Consequently, a Bayesian model also requires a prior distribution $p(\boldsymbol{\theta} \mid \tilde{\boldsymbol{\theta}})$ to be specified, where $\tilde{\boldsymbol{\theta}}$ is a vector of fixed hyperparameters. The difference between the two approaches may be illustrated by juxtaposing the definitions of the two types of intervals, both denoted by $[L, U]$ for sample statistics L and U, that are used to quantify uncertainty about an estimator $\hat{\theta}$ of a parameter θ. In the binomial example, θ was the success probability π, but in general it may be any model parameter of interest, such as the mean of a normal distribution, the event rate under an exponential model, or the parameter multiplying a covariate Z in a regression model for $p(Y \mid Z, \boldsymbol{\theta})$. A frequentist 95% confidence interval (CI) $[L, U]$ for θ has the interpretation that, if the experiment that produced the data were repeated infinitely many times, and a 95% CI were computed from the data of each experiment, then 95% of the computed CI's would include θ between the statistics L and U. A Bayesian 95% CrI $[L, U]$ has the rather different interpretation that, given the data actually observed, the probability is .95 that the random θ is between L and U, formalized by the equation $\Pr(L < \pi < U \mid data) = .95$. While CrI's may be computed in various ways, for example to minimize $U - L$, the most common approach is to compute the bounds as L = the 2.5^{th} percentile and U = the 97.5^{th} percentile of the posterior of θ. A great advantage of Bayesian methods is that one can draw pictures of posterior distributions and use them to illustrate posterior probabilities of interest. For example, Figure 1.3 plots posteriors and 95% CrI's for each of four beta distributions, all having mean 2/3, with ESS values 6, 12, 24, and 48, illustrating how larger sample sizes provide greater reliability in terms of shorter CrI's.

Bayes' Law, also called *The Law of Reverse Probability*, can be presented in terms of events or probability distributions. For random events A and B, the *Law of Total Probability*, which was applied earlier, may be written generally as follows. Denote the respective complements by A^c = [not A] and B^c = [not B], and write the conditional probability of A given B as $\Pr(A \mid B)$, and of A given B^c as $\Pr(A \mid B^c)$. Since $\Pr(B^c) = 1 - \Pr(B)$, the *Law of Total Probability* in terms of A and B is

$$\Pr(A) = \Pr(A \mid B)\Pr(B) + \Pr(A \mid B^c)\{1 - \Pr(B)\}. \tag{1.1}$$

For example, when studying the effects of chemotherapy, if R = Response and T = Toxicity, and these two events are positively associated, as might be the case with chemotherapy, this is formalized by the inequality $\Pr(R \mid T) > \Pr(R \mid T^c)$. As a numerical example, suppose that $\Pr(R \mid T) = .40$, $\Pr(R \mid T^c) = .20$, and $\Pr(T) = .30$. Equation (1.1) implies that the unconditional probability of response is $\Pr(R) = .40 \times .30 + .20 \times .70 = .26$. This sort of computation may be elaborated by conditioning on the events of a more refined partition, such as B_1 = [No Toxicity], B_2 = [Mild Toxicity], B_3 = [Moderate Toxicity], and B_4 = [Severe Toxicity], so that $\sum_{j=1}^{4} \Pr(B_j)$

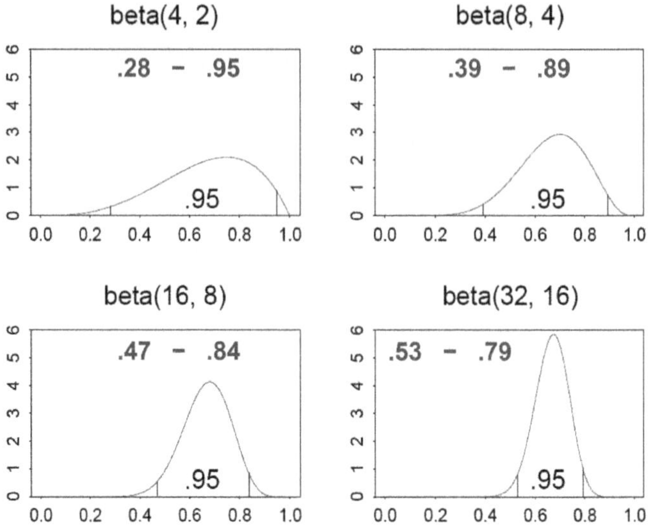

FIGURE 1.3
95% credible intervals for four beta distributions with mean 2/3.

= 1 and the more general form of equation (1.1) in this case is

$$\Pr(R) = \sum_{j=1}^{4} \Pr(R \mid B_j)\Pr(B_j).$$

If the partition $\{B_1, B_2, B_3, B_4\}$ is refined to be real-valued toxicity severity Z on a scale of 0 to 100 with distribution $f_T(z)$ for $0 \le z \le 100$, then

$$\Pr(R) = \int \Pr(R \mid Z = z) f_T(z) dz.$$

Bayes' Law exploits the *Law of Total Probability* and the definition of conditional probability to provide a way to compute $\Pr(B \mid A)$ from $\Pr(A \mid B)$, $\Pr(A \mid B^c)$, and $\Pr(B^c)$. That is, it shows how to reverse the order of conditioning from $\Pr(A \mid B)$ to $\Pr(B \mid A)$. It may be derived by using the definition $\Pr(A \mid B) = \Pr(A \text{ and } B)/\Pr(B)$, and the fact that $\Pr(A \text{ and } B) = \Pr(B \text{ and } A)$. This implies that $\Pr(B \mid A) = \Pr(B \text{ and } A)/\Pr(A)$. These facts may be used to derive *Bayes' Law*, which is given by the equation

$$\Pr(B \mid A) = \frac{\Pr(A \mid B)\Pr(B)}{\Pr(A)} = \frac{\Pr(A \mid B)\Pr(B)}{\Pr(A \mid B)\Pr(B) + \Pr(A \mid B^c)\Pr(B^c)}. \quad (1.2)$$

The French mathematician Pierre-Simon Laplace independently discovered this probability equality. Despite applications to solve problems in medicine, cryptography, warfare, and numerous other settings, Bayes' Law has been the center of controversy for centuries. A very interesting historical account is given by Mc Grayne (2011).

As a simple illustration, in the chemotherapy example given above,

$$\begin{aligned} \Pr(T \mid R) &= \frac{\Pr(R \mid T)\Pr(T)}{\Pr(R \mid T)\Pr(T) + \Pr(R \mid T^c)\Pr(T^c)} \\ &= \frac{.40 \times .30}{.40 \times .30 + .20 \times .70} = .46. \end{aligned}$$

Since the unconditional probability is $\Pr(T) = .30$, Bayes' Law says that, if it is known that a response occurred, then the probability of toxicity is increased from .30 to .46. Bayes' Law thus provides a way to update one's uncertainty about an event using the knowledge that another, related event has occurred.

Applications of Bayes' Law often give nonintuitive answers. A well-known example is interpreting the result of a test for a disease. Denote D = [have the disease], T^{pos} = [the test is positive], and its complement by T^{neg} = [the test is negative]. The *sensitivity* of the test is $\Pr(T^{pos} \mid D)$, the probability of correctly giving a positive result if a person has the disease. The *specificity* of the test is $\Pr(T^{neg} \mid D^c)$, the probability of correctly giving a negative result if a person does not have the disease. This becomes quite interesting if one asks the question, "If someone tests positive, what is the probability that they actually have the disease?" If the prevalence of the disease, $\Pr(D)$, is known, the question can be answered by applying Bayes' Law. Suppose that the test's sensitivity is $\Pr(T^{pos} \mid D) = .99$ and the specificity is $\Pr(T^{neg} \mid D^c) = .95$ for the disease HIV, which has a global prevalence in people aged 15 to 49 years of $\Pr(D) = .007$, or 7 in 1000. If someone in that age range tests positive, then the probability that they actually have the disease is

$$\Pr(D \mid T^{pos}) = \frac{\Pr(T^{pos} \mid D)\Pr(D)}{\Pr(T^{pos} \mid D)\Pr(D) + \Pr(T^{pos} \mid D^c)\Pr(D^c)}.$$

Since the probability of a false positive test is $\Pr(T^{pos} \mid D^c) = 1 - \Pr(T^{neg} \mid D^c) = 1 - .95 = .05$, the above equation gives

$$\Pr(D \mid T^{pos}) = \frac{.99 \times .007}{.99 \times .007 + .05 \times (1 - .007)} = .122.$$

So a positive test does not imply with certainty that the person has HIV, but rather that they have a 12.2% chance of having HIV, about 1 in 8. What the positive test has done is update the unconditional probability .007 to the posterior probability .122, which is a 17-fold increase. Of course, 12.2% is large enough to be worrying, and in practice, a second, possibly more reliable test may be warranted. The prevalence is very important in this computation. For example, if $\Pr(D) = .001$, then the answer becomes the much lower value $\Pr(D \mid T^{pos}) = .019$, or about 1 in 50.

For parametric Bayesian models, denote the likelihood function of observable Y and parameter vector $\boldsymbol{\theta}$ by $\mathscr{L}(Y \mid \boldsymbol{\theta})$. In practice, the likelihood is a parametric distribution $f(y \mid \boldsymbol{\theta})$ such as a binomial, normal, Poisson, or exponential. The prior may be denoted by $p(\boldsymbol{\theta} \mid \tilde{\boldsymbol{\theta}})$ where $\tilde{\boldsymbol{\theta}}$ is a vector of fixed hyperparameters. Bayes' Theorem then can be written as the equation

$$p(\boldsymbol{\theta} \mid Y, \tilde{\boldsymbol{\theta}}) = \frac{1}{c} \mathscr{L}(Y \mid \boldsymbol{\theta}) \times p\boldsymbol{\theta} \mid \tilde{\boldsymbol{\theta}}), \tag{1.3}$$

where dividing by the norming constant

$$c = c(Y, \tilde{\boldsymbol{\theta}}) = \int_{\boldsymbol{\theta}} \mathscr{L}(Y \mid \boldsymbol{\theta}) \times p\boldsymbol{\theta} \mid \tilde{\boldsymbol{\theta}}) d\boldsymbol{\theta}$$

ensures that the posterior $p(\boldsymbol{\theta} \mid y, \tilde{\boldsymbol{\theta}})$ is a probability distribution. For a sample $\boldsymbol{Y}_n = (Y_1, \cdots, Y_n)$ following a distribution $f(y \mid \boldsymbol{\theta})$, the likelihood in equation (1.3) is $\mathscr{L}(\boldsymbol{Y}_n \mid \boldsymbol{\theta}) = \prod_{i=1}^n f(Y_i \mid \boldsymbol{\theta})$, and this is the usual form used in practice. For all but a few tractable cases, c cannot be computed easily and the posterior is computed numerically as a large sample $\boldsymbol{\theta}^{(1)}, \cdots, \boldsymbol{\theta}^{(B)}$ using Markov chain Monte Carlo (MCMC) methods (Robert and Cassella, 1999; Gammerman and Lopes, 2006; Brooks et al., 2011). Since Bayesian inference is based on the posterior, in practice posterior statistics such as means or percentiles, or probabilities such as the 12-month survival probability $\Pr(\mu > 12 \mid data)$ for μ = mean survival time, can be computed as sample statistics from the MCMC posterior sample.

For the probability π characterizing a binomial distribution, $X_n \sim binom(n, \pi)$, the most commonly assumed prior for π under a Bayesian model is a beta distribution, which has two hyperparameters $a > 0$ and $b > 0$, denoted by $\pi \sim beta(a, b)$. The beta has mean $\mu = a/(a+b)$ and variance $\sigma^2 = \mu(1-\mu)/(\alpha+b+1)$. A very useful property of the binomial-beta model is that the posterior distribution also is beta, $\pi \mid X_n, n \sim beta(a+X_n, b+n-X_n)$. This is an example of a *conjugate* prior, since the posterior belongs to the same distributional family. This gives the posterior mean of the beta as $E(\pi \mid X_n, n) = (a+X_n)/(a+b+n)$, which is very easy to compute, and also shows that a sample of size n has increased the prior sum $a+b$ to the posterior sum $\alpha+b+n$. These values may be called the effective sample size, ESS. For example, if $a = b = .5$ then the prior $ESS = a+b = 1$ is increased to the posterior $ESS = (a+X_n)+(b+n-X_n) = n+1$. The posterior mean can be written in the equivalent form

$$E(\pi \mid X_n, n) = \left(\frac{a+b}{a+b+n}\right) \times \left(\frac{a}{a+b}\right) + \left(\frac{n}{a+b+n}\right) \times \left(\frac{X_n}{n}\right),$$

which is a weighted average of the prior mean $\mu = a/(a+b)$ and sample mean X_n/n, the usual frequentist estimator. For large n, the weight $(a+b)/(a+b+n)$ of μ becomes small, and the weight $n/(a+b+n)$ of the sample mean becomes close to 1. The prior thus has the effect of shrinking the sample mean X_n/n toward the prior mean μ, with this shrinkage becoming smaller for larger sample sizes. This illustrates the fact that all Bayesian estimators shrink a frequentist estimator toward a prior value. Essentially, shrinkage introduces some bias into the estimator while reducing its variance, which improves overall estimation quality.

A very useful property of Bayes' Law is that it can be applied repeatedly as data accumulate during an experiment. This is exploited by many of the clinical trial designs that will be described in Chapters 3 - 8. Using the binomial-beta model to illustrate this, suppose that the posterior is computed after successive cohorts of size 20. Starting with a beta(1, 1) prior for π = Pr(Response), and denoting the number of responders in the first cohort by $X_{20}^{(1)}$, the first posterior is

$$p(\pi \mid X_{20}^{(1)}) = beta(X_{20}^{(1)} + 1,\ \ 20 - X_{20}^{(1)} + 1).$$

If the trial is continued, this posterior may be used as the prior for the next stage. Denoting the number of responders out of the next 20 patients by $X_{20}^{(2)}$, due to conjugacy the second posterior is

$$p(\pi \mid X_{20}^{(1)}, X_{20}^{(2)}) = beta(X_{20}^{(2)} + X_{20}^{(1)} + 1,\ \ 20 - X_{20}^{(2)} + 20 - X_{20}^{(1)} + 1).$$

But since $X_{40} = X_{20}^{(1)} + X_{20}^{(2)}$ = the number of responders in the first 40 patients, this equals

$$p(\pi \mid X_{40}) = beta(X_{40} + 1, 40 - X_{40} + 1),$$

which is what one would obtain by computing the posterior once from a 40-patient trial. This sort of computation may be repeated for subsequent cohorts. In fact, it is easy to show that this computation is true quite generally for any distribution and prior, provided that the observations are independent. Computing such a sequence of posteriors as data are accumulated may be called *Bayesian Learning*. The four posteriors in Figure 1.3 illustrate this for successive sample sizes of n = 6, 12, 24, and 48.

For small to moderate sample sizes, the effect of shrinkage can be very important. For example, if $n = 3$ cancer patients are treated with a chemotherapy and π = Pr(severe toxicity), and it is assumed that $\pi \sim beta(.5, .5)$, then the respective posterior means for X_3 = 0, 1, 2, 3 toxicities are $\hat{\pi}^B = E(\pi \mid X_3, 3) = (.5 + X_3)/(1+3)$ = .125, .375, .625, .875. If one uses the posterior mean $\hat{\pi}^B$ as a Bayesian point estimator of π, then these four numerical estimates replace the corresponding frequentist estimates $\hat{\pi}^F = X_3/3$ = 0, .33, .67, 1.00. These numerical values show precisely how the posterior mean shrinks the empirical mean toward .50, but they also illustrate an advantage of the shrinkage provided by a Bayesian model. If no toxicities are observed, $X_3 = 0$, then the empirical point estimator is $\hat{\pi}^F = 0$, which says that toxicity is impossible. Similarly, $X_3 = 3$ gives the estimate $\hat{\pi}^F = 1$, which says that toxicity is certain to occur. Of course, these are ridiculous conclusions, based on only 3 observations. In contrast, the corresponding Bayesian estimates are .125 if $X_n = 0$, which says that there is a 12.5% estimated chance of toxicity, and .875 if $X_n = 3$, which says that there is an 87.5% estimated chance of toxicity. This illustrates how shrinkage toward a prior mean under a Bayesian model can help one to avoid making inferences that do not make sense. Of course, reporting a point estimator alone does not account for variability, which is very bad statistical practice. It is much more useful to report a Bayesian posterior 90% or 95% CrI along with the posterior mean. For example, the 95% CrI's are [0, .536] if $X_3 = 0$, and [.039, .823] if $X_3 = 1$, which reflect the fact that the prior ESS = 1 gives a posterior ESS = 4, and shows that the prior shrinks the respective empirical estimators 0 and .33 toward the prior mean .50.

There are a variety of methods for computing a frequentist CI for a probability from a small binomial sample, but none are entirely satisfactory and all involve some sort of *ad hoc* smoothing. Each method looks suspiciously like someone trying to be a Bayesian but not quite succeeding. In contrast, the equation $\Pr(L < \theta < U \mid data)$ = .95 is coherent, easy to understand, and the limits L and U are easy to compute. For example, in the R programming language, the formulas L = *qbeta(.025,a, b)* and U = *qbeta(.975,a, b)* give the 2.5^{th} and 97.5^{th} percentiles of a $beta(a,b)$ distribution, which are lower and upper 95% credible interval limits. In the beta-binomial computation, plugging in the posterior parameter values $a+R$ and $b+n-R$ gives L and U for a 95% posterior credible interval. Similar percentile computations can be done very easily using R or other languages for numerous other standard distributions, including the normal, Poisson, gamma, and exponential. For more complex models with posteriors computed using MCMC, L and U are obtained as the 2.5^{th} and 97.5^{th} percentiles of the posterior sample.

A practical advantage of Bayesian inferences is that they are exact, and can be applied for any sample size. In contrast, the frequentist distribution theory underlying most confidence intervals, sample statistics, and p-values relies on large sample theory, such as asymptotic properties of maximum likelihood estimators. For example, for treatment indicator X =1 for E and 0 for C and covariate Z = 1 for severe disease and 0 otherwise, a logistic regression model for the probability of response may be

$$\pi(X,Z \mid \boldsymbol{\beta}) = \frac{e^{\beta_0+\beta_1 X+\beta_2 Z}}{1+e^{\beta_0+\beta_1 X+\beta_2 Z}}.$$

Thus, for a randomized trial of E versus C, β_1 is the E-versus-C effect and β_2 is the effect of severe disease on the probability of response. For a small to moderate sample size such as n =30, the asymptotic theory that is used routinely to determine the distributions of the maximum likelihood estimators $(\hat{\beta}_0, \hat{\beta}_1, \hat{\beta}_2)$ is invalid because n is not large, so any computed p-values are incorrect. This is a pervasive problem in medical statistics when analyzing, for example, phase II clinical trial results.

Under a Bayesian model, one may establish a prior by eliciting values of the probabilities $p(1,1 \mid \boldsymbol{\beta})$, $p(1,0 \mid \boldsymbol{\beta})$, $p(0,1 \mid \boldsymbol{\beta})$, $p(0,0 \mid \boldsymbol{\beta})$, and then use nonlinear least squares or some other numerical method to solve for the prior means of $\beta_0, \beta_1, \beta_2$, and assume that $\beta_0, \beta_1, \beta_3$ follow independent normal distributions with these means and common variance σ_β^2 = 1, 5, or 10. To obtain an operational prior, σ_β^2 may be chosen by simulating a large sample of $\boldsymbol{\beta}$ vectors from the prior for each value of σ_β^2, and examining the histogram of $\pi(X,Z \mid \boldsymbol{\beta})$ for each (X,Z) pair to make sure that the prior is reasonable. A value of σ_β^2 that is too small may give an informative prior on each $p(X,Z \mid \boldsymbol{\beta})$ that is not appropriate. But if σ_β^2 is too large, then too much prior probability mass will be concentrated near 0 or 1 for each (X,Z) pair, which would give a prior saying that response is nearly impossible or nearly certain.

TABLE 1.4
Fitted Bayesian piecewise exponential regression model for PFS time in the randomized trial of FCB versus FB. Comorb = comorbidity score, CR = in complete remission at transplant, sd = standard deviation, PBE = probability of a beneficial effect = $\Pr(\beta_j < 0 \mid data)$.

	Posterior Quantities			
Variable	Mean	sd	95% CrI	PBE
FCB	-0.67	0.42	[-1.50, 0.16]	0.94
Age ≤ 60	-0.38	0.27	[-0.90, 0.14]	0.92
Unrelated Donor	0.09	0.22	[-0.34, 0.52]	0.33
AML	0.30	0.22	[-0.13, 0.74]	0.09
Int/Good Risk	-0.48	0.18	[-0.83, -0.12]	1.00
CR	-1.12	0.28	[-1.68, -0.58]	1.00
Cell Type Apheresis	-0.21	0.22	[-0.63, 0.22]	0.83
Comorb ≥ 3	0.29	0.26	[-0.21, 0.82]	0.14
FCB$\times$[Age ≤ 60]	-0.02	0.39	[-0.77, 0.76]	0.52
FCB$\times$ CR	0.48	0.39	[-0.28, 1.26]	0.11
FCB$\times$[Comorb ≥ 3]	0.61	0.39	[-0.15, 1.38]	0.06

Once an MCMC posterior sample is generated, it may be used to compute quantities such as $\Pr(\beta_1 > 0 \mid data)$, which is the posterior probability that π increases for larger values of the E-versus-C effect β_1, and it may be used to quantify how likely it is that E is superior to C. This approach, which can be used in a wide variety of regression settings, frees one from relying on invalid large sample distribution theory, and it also avoids the use of p-values to quantify the strength of the effects of X and Z on $\pi(X, Z \mid \boldsymbol{\beta})$. An application of this Bayesian approach to regression was given by Andersson et al. (2017). They reported results of a clinical trial to compare the effects of pretransplant conditioning regimens Fludarabine + Clofarabine + Busulfan (FCB) to Fludarabine + Busulfan (FB) on progression-free survival (PFS) time in patients with high-risk acute myelogenous leukemia (AML) or myelodysplastic syndrome (MDS) receiving an allogeneic stem cell transplant (allosct). Patients were randomized to receive FCB or FB, with 249 evaluable and 130 (52%) suffering failure (progression or death). A Bayesian piecewise exponential model was fit to the data, including the linear term $\eta(\mathbf{Z}, \boldsymbol{\beta})$ with the hazard function multiplied by $\exp\{\eta(\mathbf{Z}, \boldsymbol{\beta})\}$. The linear term included the variables given in Table 1.4, which also includes three treatment-covariate interaction terms. Since $\beta_j < 0$ corresponds to the variable Z_j being associated with a lower risk of failure, one may define the probability of a beneficial effect for Z_j to be PBE = $\Pr(\beta_j < 0 \mid data)$.

The fitted model in Table 1.4 suggests that, while FCB was effective, its advantage over FB may have been greatest in patients with active disease and NCR = [no CR] at the time of allosct, but less advantageous in patients in CR at transplant.

TABLE 1.5
Estimated probability of a beneficial effect of FCB-versus-FB within prognostic subgroups determined by the number of comorbidities, disease status at transplant (CR = complete remission versus NCR = active disease), and age.

Subgroup			PBE
Comorb ≤ 2	CR	Age$\leq$ 60	.73
Comorb ≤ 2	CR	Age> 60	.69
Comorb ≤ 2	NCR	Age$\leq$ 60	.98
Comorb ≤ 2	NCR	Age> 60	.92
Comorb ≥ 3	CR	Age$\leq$ 60	.13
Comorb ≥ 3	CR	Age> 60	.07
Comorb ≥ 3	NCR	Age$\leq$ 60	.38
Comorb ≥ 3	NCR	Age> 60	.20

To explore this, estimates of FCB-versus-FB effects within subgroups determined by number of comorbidities, disease status, and age, are summarized in Table 1.5. These results suggest that the survival benefit of adding clofarabine to the FB preparative regimen was greatest for patients with active disease, provided that they had ≤ 2 comorbidities. This was the case regardless of age, which had an additive effect.

2

Statistics and Causality

2.1 Causal Relationships

Statistical models and causal models are closely related, but they are not the same thing. This chapter will present some basic concepts and methods for causal analysis. Some useful books on the rationale and methods for causal inference are by Imbens and Rubin (2015), Rosenbaum (2017), and Brumback (2021), among many others. To establish some basic elements, consider a dataset containing two variables for each patient, a treatment indicator $X = 1$ for an experimental targeted agent, E, and $X = 0$ for a standard control therapy, C, and a clinical response indicator Y_R. One may formulate a regression model, $p(Y_R \mid X)$, which is a probability distribution for Y indexed by X that, if fit to the data, may be used to estimate the coefficient characterizing the E-versus-C treatment effect on Y_R. If the estimated coefficient is large, this may be interpreted to imply that the causal effect of the targeted agent on Y is large, that is, E is likely to be a better treatment than C.

But things often are not so simple. Because patients are heterogeneous, the model may be elaborated to include patient covariates, such as an indicator Z_1 of good versus poor prognosis and indicator Z_2 for the biomarker that E is designed to attack. Denoting $\mathbf{Z} = (Z_1, Z_2)$, the regression model becomes $p(Y_R \mid X, \mathbf{Z})$, and the fitted model can be used to make statistical inferences about associations between $(X, \mathbf{Z})$ and Y. Since many relationships among the variables are possible, one may consider several possible models for $p(Y_R \mid X, \mathbf{Z})$, fit each to the available data, and infer which model is most appropriate. Determining a model may be complicated, however, by the possibility that X and $\mathbf{Z}$ have synergistic or interactive effects on Y_R, and elements of (X, Z_1, Z_2) also may affect each other. Causal analysis provides methods for sorting this out, to the extent that it is possible based on what one knows or suspects may be the case for relationships among Y_R, X, Z_1, and Z_2.

In clinical practice, physicians consider $\mathbf{Z}$ when choosing whether to treat a patient with E or C. Since a response may be considered more likely if the biomarker targeted by E is present, a physician may be more likely to treat a patient with E if $Z_2 = 1$. Similarly, denoting an indicator of severe toxicity by Y_T, if E carries a higher risk of toxicity than C, then a physician may be less likely to treat the patient with E if they have a poor prognosis, that is, if $Z_1 = 0$. Consequently, the estimated coefficient of X does not reliably summarize the treatment effect on Y, since it does not account for the fact that Z_1 influenced the physician's choice of X. These

DOI: 10.1201/9781003474258-2

considerations also may imply that the outcome should be elaborated to include Y_T, so now it is the pair $\boldsymbol{Y} = (Y_R, Y_T)$, and the regression model is the bivariate probability distribution $p(\boldsymbol{Y} \mid X, \mathbf{Z})$ indexed by treatment X and covariates $\mathbf{Z}$. This illustrates the general fact that medical practice involves using $\mathbf{Z}$ to choose X, based on the fundamental knowledge that treatments may have both desirable and adverse effects. These issues complicate evaluating the causal effects of the elements of $(X, \mathbf{Z})$ on the elements of $\boldsymbol{Y}$. They also lead to consideration of what regression models actually represent from a causal viewpoint.

A pervasive problem is that many regression analyses published in medical journals ignore the difference between data from a designed experiment in which values of $(X, \mathbf{Z}, \boldsymbol{Y})$ have been determined by randomizing subjects between $X = E$ and $X = C$, and an observational dataset, representing clinical practice, where physicians used the values of $\mathbf{Z}$ to choose X. While the two datasets are structurally identical, from a causal viewpoint they are very different. For randomized trial data, the usual estimators of the treatment effect parameters for the marginals of Y_R and Y_T are unbiased for the causal E-versus-C effects. In a randomized clinical trial, the basic idea is that, by taking the treatment choice out of the physician's hands, randomization does not allow $\mathbf{Z}$ to affect the choice of treatment X, which eliminates bias. That is, it removes the causal effect $\mathbf{Z} \rightarrow X$. In this sense, randomization may be considered antithetical to medical practice. For each outcome, say Y_R, the estimator of the parameter multiplying X in its marginal regression model is exactly or approximately unbiased, so it gives a fair comparison for the E-versus-C treatment effect on the response rate. If observational data are analyzed in the same way as randomized trial data, the resulting estimators of the treatment effects are wrong, and can be extremely misleading. This is because the medical practice of using patient prognostic variables to choose treatments confounds the effects of X on Y_R or Y_T with effects of the covariates $\mathbf{Z}$ on X. If the goal is to compare treatments fairly, routine medical practice produces data that are at odds with valid statistical inference.

For example, suppose that a clinical practice database represents a setting where attending physicians were more likely to give E to patients with poor prognosis, and more likely to give C to patients with good prognosis. A possible motivation for this practice may be that, because poor prognosis patients have fewer treatment choices and higher risks of toxicity and death, trying an experimental agent is warranted. This confounds treatment and covariate effects, and gives C an unfair advantage over E when comparing treatments. Unfortunately, many practitioners do not understand the need for randomization to provide a basis for making valid comparative statistical inferences, or the consequences of not randomizing when assessing causal treatment effects. Regression analysis often is taught from a purely statistical viewpoint without any discussion of causality, and the important distinction between observational and randomized experimental data often is ignored. While it may appear that the process of writing down a regression model, fitting it to a dataset, and interpreting the resulting numerical parameter estimates is doing exactly the same thing for data from randomized and non-randomized studies, this is not true.

Establishing a causal relationship requires additional considerations beyond statistical estimation. It is a truism that statistical association does not imply causation.

There are many examples of this. A general class of examples are those where it is determined statistically that two variables, X and Y, are strongly associated, but there is a third variable Z that affects both X and Y. This was illustrated by the examples of Simpson's Paradox given in Chapter 1. It is well known that shoe size is positively associated with intelligence, but since both shoe size and intelligence increase with age, it does not make sense to say that a larger shoe size causes higher intelligence. There are plenty of stupid people with big feet. Age is the external confounder, and can be identified with just a bit of thought. In a medical setting involving treatment of a particular type of cancer, suppose that $X = 0$ for chemotherapy, $X = 1$ for a targeted agent, and the outcome is Y = survival time. The relationship $X \rightarrow Y$ represents the idea that treatment causally affects survival time. By definition, Z = prognosis also affects the distribution of the clinical outcome Y, so in any case $Z \rightarrow Y$. The process whereby physicians use prognosis to choose treatment is represented causally as the third relationship $Z \rightarrow X$, and in this case, the effect of treatment X on Y is confounded by the effect of the covariate Z on X. Here, Z confounds the effect of X on Y, but this should be obvious.

An interesting example is the well-known positive statistical association between Z = income and Y = family size. A causal question is whether parents with more money are more likely to have a larger number of children because they can afford to support them or, instead, parents with more children are more likely to earn more money in order to take care of their children. Formally, the causal question is whether $Z \rightarrow Y$ or $Y \rightarrow Z$. This oversimplifies things, however, since the two variables may be observed repeatedly over time, and both of these causal effects may explain observed values of Z and Y. Consider successive times $t_0 < t_1 < t_2 < t_3$, with causal effects represented as $Z(t_0) \rightarrow Y(t_1) \rightarrow Z(t_2) \rightarrow Y(t_3)$. This says that, for example, once a couple with one child achieves an income of $Z(t_0)$ at t_0, they feel financially comfortable enough to have a second child, producing a family size of $Y(t_1) = 4$ (2 parents + 2 children) at t_1. This, in turn, may cause the couple to earn $Z(t_2)$ = more money to support their family, and this larger income later may cause them to feel that they can afford a third child, so $Y(t_3) = 5$ (2 parents + 3 children) at t_3. It also may be the case that there are direct effects $Y(t_0) \rightarrow Y(t_1)$ and $Y(t_1) \rightarrow Y(t_2)$, due to the fact that the couple's positive experiences with their children over time motivate them to have more children. The longitudinal structure in this example, where repeated values of Z and Y are observed over time, mimics a multistage treatment setting where successive treatments are chosen based on a patient's past history.

A useful paradigm for establishing causality between a treatment or exposure variable X and an outcome Y requires that three conditions must be satisfied. The first condition is the statistical requirement that reliable data must show a non-trivial *association* between X and Y. This accounts for the fact that X and Y exhibit random variation, which is why probability modeling is central to statistical thinking. For two numerical valued variables, this may be either a large positive or negative correlation. If X is a binary variable, association may take the form of reliable estimates showing that the two distributions $p(Y \mid X = 1)$ and $p(Y \mid X = 0)$ are substantively different, based on a sample large enough to establish that the difference is unlikely to be due to the play of chance. If X is deterministic, for example season of the year,

{Fall, Winter, Summer, Spring}, and Y is the mean daily temperature, which has substantial randomness, then there is no probability distribution on X, while Y has a probability distribution $p(Y \mid X)$. The second requirement is *temporality*, that X should be observed before Y. The idea is that, in order for a value of X to be likely to lead to, or cause, a value of the outcome Y, one must be able to observe X first, and then observe Y. The third requirement, sometimes called *nonspuriousness*, is much harder to establish. This requires that a putative causal relationship $X \to Y$, obtained by establishing association and temporality, is not due to other factors or variables that may affect X, Y, or both, and thus explain their relationship. This also is referred to as the assumption that there are *no external confounders*, or no alternative explanations. If X is observed before Y, a statistical estimate establishes that X and Y are strongly associated, and external confounders have been ruled out, then there must be some *mechanistic* explanation of the observed association and temporal relationship. Providing a mechanistic explanation may rely on basic knowledge from an area such as biology, physics, engineering, economics, or behavioral science, depending on the particular setting.

Establishing a plausible mechanistic explanation for a possible casual association may be difficult, however. For example, it may be argued that the positive association between poverty and criminal behavior is due to the need of poor people to have enough money to survive. A simple example is a poor person stealing food, or robbing a liquor store to obtain money to buy food. While association and temporality are obvious, there are other factors involved, such as age, race, intelligence, and educational level, but knowing these additional variables cannot do away with this mechanistic explanation. In the more extreme example where X = [shoot Bill in the head with a gun] and Y = [Bill dies after being shot], it may seem obvious that $X \to Y$. However, someone accused of murder for shooting Bill in the head may argue "I didn't kill Bill. The bullet killed him. I just loaded the gun, aimed it, and squeezed the trigger". While this may seem irrational, it may be argued that it actually makes sense. To see this, suppose that we change things and say that someone punched Bill in the nose, and Bill then had a heart attack and died. The argument then may become, "I didn't kill Bill. I just punched him in the nose. He died from a heart attack, probably because he had a weak heart". After all, most people who are punched in the nose do not die from it. But the argument is more plausible now that shooting Bill has been replaced by a punch in the nose. This may be taken a step further, by replacing the punch with the verbal insult "You are stupid, ugly, and you don't know how to accessorize", followed by Bill, upon hearing the insult, having a fatal heart attack. The argument that Bill's death actually was caused by a bad heart now becomes more plausible. It is not nice to insult people, but it is hard to convict someone of murder by insult. In these examples, each argument relies on a mechanistically plausible causal chain of two or more events occurring in sequence over time. Taken to an extreme, one may use the *reductio ad absurdum* argument that, if Bill had never been born, then he could not have died, regardless of what events may have occurred in between, so being born is what caused him to die.

People often devote a great deal of thought and effort to establishing mechanistic explanations of causal effects. Arguably, dogs do the same thing. Ivan Pavlov was a

Russian behavioral scientist who conducted a famous experiment that showed ringing a bell can cause a dog to salivate. This was done by ringing a bell just prior to the time the dog was fed, so the dog learned that the ringing bell would be followed by someone giving it food, and naturally the dog would salivate when it heard the bell. One can easily imagine the dog's mechanistic explanation for why the dinner bell caused humans to bring food. Based on the dog's observed evidence, the pooch concluded, quite sensibly, that the bell stimulated the "feed the hungry dog" portion of the human brain. This mechanistic explanation is similar to an immunologist believing that administering a monoclonal antibody that attacks a breast cancer tumor showing overexpression of the HER2/neu gene will be likely to kill cancer cells and thus cause the patient to survive longer. Using this reasoning, a physician may administer the monoclonal antibody to a HER2/neu positive patient based on the belief that the antibody is likely to cause the patient to live longer. Suppose that a disease remission is seen in a breast cancer patient given the monoclonal antibody, but the patient also prayed to God to cure their disease. While the physician may believe that the monoclonal antibody caused the remission, the patient may believe that the remission occurred because their prayers were answered by God. Additionally, an immunologist or physician also may have seen published studies showing that a person with a positive attitude is more likely to have a stronger immune system that is likely to do a better job of killing cancer cells. This may lead them to believe that the combination of prayer and the monoclonal antibody, together, are likely to lead to a positive attitude which in turn triggers this later causal chain of events. This mechanistic reasoning is supported by a study that showed, if someone tells a subject who has volunteered that they will be touched by a hot metal rod, but then touches the subject with a metal rod that instead has been refrigerated and is cold, the subject is very likely to develop a blister where they were touched by the cold metal rod. This may be explained mechanistically by the subject's belief that they have been burned, and that their conscious brain has told this to their unconscious but highly intelligent immune system, which forms a blister to treat the burn that the subject's neurological system has told them about. In terms of the logic of their respective mechanistic explanations, the dog, physician, religious patient, and immunologist all have reasoned similarly. Their conclusions are based on their beliefs and what they have observed or learned about association, temporality, and plausible mechanisms for causal chains.

Figure 2.1 gives a simple DAG illustrating causal relationships that may arise in practice between the three nodes treatment, baseline patient prognostic variables, and an outcome variable. In the figure, treatments and prognostic variables each have direct effects on Y = PFS time, they do not interact with (affect) each other, and there are no other known variables that either mediate (come between) the treatment or prognostic variables and PFS time, or that affect PFS time or any of these variables. Figure 2.1 illustrates causal relationships that result when treatments are chosen by randomization, which ensures that neither observed patient variables nor unknown external confounders can affect treatment choice. In terms of the DAG, randomization removes any arrows from known or unknown covariates to treatment. How this is done will be discussed below.

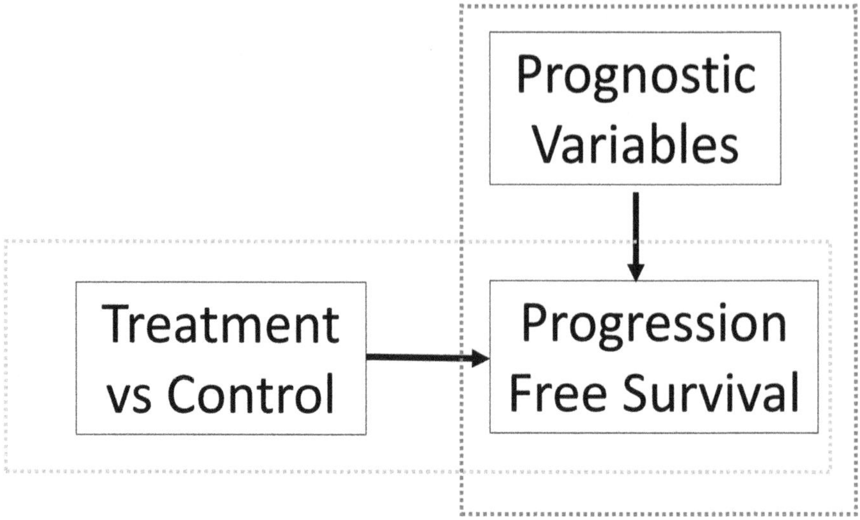

FIGURE 2.1
Causal diagram of direct treatment and prognostic variable effects on progression-free survival time, for data from a randomized clinical trial.

Since the use by physicians of patient covariates to guide their treatment choices would be reflected by an arrow from the prognostic covariates to treatments in the figure, one may think of randomization as taking the treatment choice away from the physician. This is ethical if all physicians participating in a randomized trial have equipoise between the treatments, and it provides data giving unbiased statistical estimators of the between-treatment effect. With randomization, an appropriate regression model has a linear term in the log hazard function for Y = PFS time taking the additive form $\eta = \alpha X + \boldsymbol{\beta Z}$, where $\boldsymbol{\beta Z} = \sum_{j=1}^{q} \beta_j Z_j$ is a linear combination of q baseline patient covariates on PFS time. This additive form is the most common relationship assumed for the linear term of a regression model when analyzing results of a randomized trial. The model may be elaborated by including treatment-covariate interactions, but these should be specified prior to conducting the trial and motivated by fundamental mechanistic considerations describing biological effects of the treatments.

Figure 2.2 illustrates a setting where a mediating variable is affected by both treatment and patient covariates. An example of this type of causal relationship is given by Msaouel et al. (2022), who described evaluation of the comparative treatment effect of the immune checkpoint inhibitor pembrolizumab ($X = 1$) versus placebo ($X = 0$) on PFS time in patients with renal cell carcinoma (RCC). Targeted agents like pembrolizumab are engineered to inhibit the ability of cancer cells to protect themselves from the patient's natural cancer-killing immune cells. In this setting, the pembrolizumab effect on PFS time is mediated by the immune

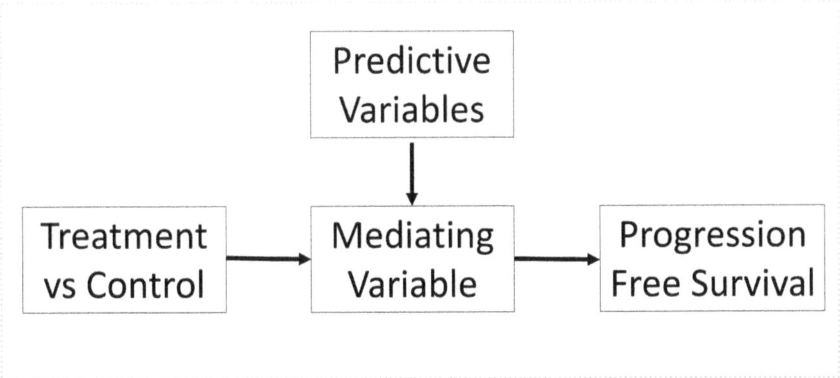

FIGURE 2.2
Causal diagram of treatment and predictive variable effects on progression-free survival time. Treatment and covariate effects are mediated by a biological variable.

microenvironment, which varies with RCC histologic subtype. Consequently, the effect of a given immunotherapy on the immune microenvironment and thus on PFS time can be expected to vary with disease histology. Thus, RCC histologic subtype is a predictive variable, since the treatment effect on the hazard of progression or death varies interactively with histologic subtype. The three most prevalent RCC histologies may be identified by the covariate $Z = 1$ for Clear cell, $Z = 2$ for Papillary, $Z = 3$ for Chromophobe, with $Z = 0$ for other histologies as a baseline category. To reflect the mediation of treatment effect by the histologic subtypes of the immune microenvironments, an appropriate regression model has a linear term taking the interactive form

$$\eta = \alpha X + \sum_{j=1}^{3} (\beta_j + \gamma_j X) I(Z = j).$$

In this model, α is the main pembrolizumab effect when $Z = 0$, the β_j's are the histological subtype main effects, the γ_j's are the pembrolizumab-histologic subtype interactions, and the total pembrolizumab-versus-placebo effect on PFS time for a patient with histology j is $\alpha + \gamma_j$.

Clinical trials in which comparative treatment effect estimates differ between subgroups are not uncommon, especially in settings where the subgroups are defined by biomarkers targeted by one or more of the treatments being studied. In some settings, there is a qualitative interaction wherein a treatment has effects that go in opposite directions for different subgroups. Douillard et al. (2010) reported such effects in reporting a phase III randomized trial of panitumumab, an anti-EGFR monoclonal antibody, combined with the standard chemotherapy FOLFOX4, versus FOLFOX4 alone in patients with *de novo* metastatic colorectal cancer. In the trial's randomization, patients were stratified by whether they had the KRAS mutation (+) or not (–), with treatment comparisons planned in each of these two complementary subgroups. In KRAS – patients, panitumumab + FOLFOX4 improved average PFS time, with estimated hazard ratio $\widehat{HR} = 0.80$ (95% CI, 0.66 to 0.97) while in KRAS +

patients average PFS was shorter in the panitumumab + FOLFOX4 arm compared to the FOLFOX4 arm, with an estimate $\widehat{HR}$ = 1.29 (95% CI, 1.04 to 1.62). Thus, the treatment-KRAS interactive effect on PFS time was qualitative since panitumumab + FOLFOX4 was preferable for KRAS – patients, while FOLFOX4 was preferable for KRAS + patients.

A similar but much larger qualitative treatment-subgroup interaction was seen by Mok et al. (2009), who presented the results of a phase III randomized trial of gefitinib, an oral epidermal growth factor receptor (EGFR) tyrosine kinase inhibitor, versus the chemotherapy combination carboplatin + paclitaxel for the treatment of advanced non-small cell lung cancer in East Asian patients. In the subgroup of 261 patients with the EGFR gene mutation (EGFR+), PFS time was much longer for patients who received gefitinib compared to carboplatin + paclitaxel, with an estimated hazard ratio $\widehat{HR} = 0.48$ (95% CI, 0.36 to 0.64). In 176 patients without the EGFR mutation (EGFR−), PFS was much worse with gefitinib, with an estimated $\widehat{HR}$ = 2.85 (95% CI 2.05 to 3.98). These results strongly suggested that single-agent gefitinib was a better treatment than chemotherapy in EGFR+ patients, but the chemotherapy was better than gefitinib in EGFR– patients. It is important to bear in mind that, in order to make these inferences, the trial design had to have two important properties. The first was that patients were randomized, and the second was that both EGFR+ and EGFR– patients were enrolled.

Gefitinib has one target, HER1/EGFR. It is engineered to interrupt the signaling pathway to this target and thereby reduce NSCLC tumor cell proliferation. This mechanistic biological explanation implies that gefitinib should not be effective in EGFR− patients, so from a causal viewpoint the observed qualitative interaction is not surprising. Based on this biological knowledge, it may be argued that the design of the gefitinib trial should have restricted enrollment to EGFR+ patients only, since the structure of gefitinib and the molecular biology predict that there is no reason that EGFR– patients should benefit. This is based on the assumption that gefitinib should behave exactly as it was designed and thus cannot provide anti-disease activity for EGFR– patients. However, as noted earlier, many designed molecules turn out to have multiple biological effects, beyond hitting their nominal target, and thus they may attack a disease through an unplanned mechanism. This implies generally that, in practice, there may be unknown biomarkers for anti-disease activity, and a subgroup of putatively biomarker-negative patients may benefit from a targeted agent. Presumably, these considerations motivated also enrolling EGFR– patients in the randomized trial reported by Mok et al. (2009).

These ideas also apply in a trial of a "dirty" molecule, E, that has been engineered to attack two targets, represented by the biomarker indicators $\mathbf{Z} = (Z_1, Z_2)$. These give a total of four subgroups, denoted by $(+,+)$ if (Z_1, Z_2) = (1,1), by $(+,-)$ if (Z_1, Z_2) = (1,0), and so on. For each treatment $X = 1$ for E and $X = 0$ for C, there are four parameters $\theta_{X,\mathbf{Z}}$, and consequently four comparative treatment effects $\Delta_{\mathbf{Z}} = \theta_{1,\mathbf{Z}} - \theta_{0,\mathbf{Z}}$, one for each biomarker vector $\mathbf{Z}$. This suggests that E-versus-C estimates should be computed and comparisons carried out in each of the four subgroups defined by $\mathbf{Z}$. Such subgroup-specific inferences must be planned before a trial is conducted, however, to avoid *post hoc* analyses where subgroups are chosen based on the observed data. This practice is difficult to defend because it increases the risk of making false positive conclusions in subsets where is no true E-versus-C effect, and sometimes is referred to as *data dredging*. This structure can be elaborated for more than two targets, but the number of subgroups within which E and C are compared may become unmanageable due to small subgroup sample sizes. For example, a randomized trial with 500 patients and 16 subgroups determined by four binary biomarkers may have about 31 patients per biomarker subgroup.

In Chapter 8, a Bayesian group sequential adaptive enrichment design, proposed by Park et al. (2022), will be discussed that uses adaptive variable selection applied to a vector of biomarkers to identify a subpopulation of E-sensitive patients, and restricts enrollment to that subpopulation. The design repeatedly does biomarker selection and restricts enrollment, so that the adaptively defined E-sensitive subgroup may change throughout the trial as data accumulate.

Figure 2.3 illustrates a setting that may arise when analyzing observational data. Westreich and Greenland (2013) discussed how a regression model fit to data arising from an epidemiological study, also known as a "population-based" study, may misrepresent the effects of what they refer to as "secondary covariates". Such observational studies commonly are reported in epidemiological or medical journals with a first table, "Table 1", summarizing the distributions of variables within exposure or treatment subgroups, and a second table, "Table 2", summarizing a fitted regression model. For the sake of illustration, in its simplest form, the regression model includes a so-called "exposure variable", X, a covariate, Z, and an outcome variable Y. An example is an observational study of the effect of human immunodeficiency virus (HIV) on the risk of stroke, where X indicates being HIV positive, Y indicates stroke, and Z may include individual patient variables such as age, sex, smoking, body mass index, etc. that may be related to stroke. For simplicity in this illustration, consider a single Z variable. Westreich and Greenland (2013) described what they called the "Table 2 Fallacy" where the exposure X is a risk factor of primary interest and Z is considered to be a secondary variable. They argued that, from a causal viewpoint, a fitted regression model aiming to account for the effect of X on Y and also the effect of Z on Y may misrepresent the effect of Z if Z also affects exposure X. In its simplest form, where X and Z are single indicator variables, the case discussed by Westreich and Greenland (2013) refers to the top causal diagram in Figure 2.3, where X affects Y while Z affects both X and Y. Thus, X partially mediates the effect of Z on Y. This structure is seen very commonly with the observational data that is the focus of epidemiological studies. If one fits a regression model with linear term taking the commonly assumed additive form

$$\eta(X,Z) = \alpha_0 + \alpha_1 X + \alpha_2 Z$$

to the observational data, then α_2 misrepresents the total effect of Z, since Z has both a direct effect on Y and also an effect on Y that is mediated through X. This is the so-called "Table 2 Fallacy". In a medical setting, if X is a treatment indicator with $X = 1$ for treatment A and 0 for treatment B, and physicians used Z to choose treatment, then the primary focus is on how X affects Y, rather than the effect of the secondary variable Z. In any case, if only observational, non-randomized data are

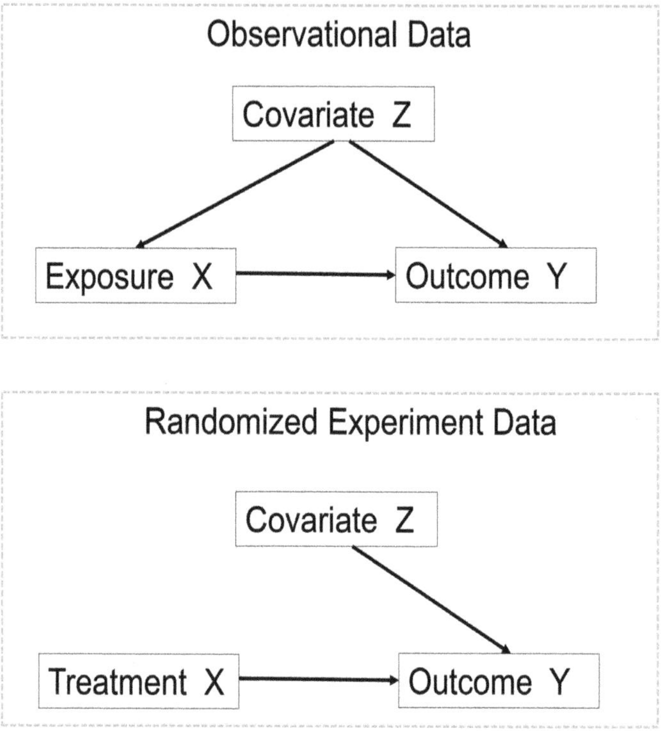

FIGURE 2.3
Causal diagrams of the effects of a covariate Z and an exposure or treatment variable X on outcome Y. The top DAG describes an observational study, where Z affects both X and Y, so X partially mediates the effect of Z on Y. The bottom DAG describes a randomized clinical trial where the effects of Z and a treatment variable X on Y both are direct, and the covariate Z has no effect on treatment X.

available, then in order to estimate a between-treatment effect or an exposure versus non exposure effect, one must account for the confounding effects of covariates such as Z by applying some form of bias correction, such as IPTW, pair matching, or G estimation.

2.2 Inverse Probability of Treatment Weighting

A common misunderstanding among some members of the medical research community is that the so-called "Table 2 Fallacy" also applies to regression analysis of data from a randomized clinical trial. This is not true. This misunderstanding is especially problematic among individuals who are accustomed to seeing analyses of data from observational studies, and who do not understand why randomization is done, or how it affects causal relationships. As explained above, randomization ensures that no covariate Z can have any effect on the choice of treatment X. In terms of the causal diagram, randomization removes the arrow from Z to X in the top diagram in Figure 2.3, turning it into the bottom diagram, where X does not mediate the effect of Z on Y. Consequently, in a fitted regression model, reported in Table 2 or anywhere else, there is no ambiguity about the effect of Z on Y, and there is no "Table 2 Fallacy". With randomization, a conventional additive regression model is appropriate, and α_2 is the unambiguous causal effect of Z on Y. The main reason for including Z in the regression model is to improve the precision of the estimator of the treatment effect, although the effect of Z on Y also may be of interest. Moreover, in a randomized medical study, there cannot be an effect of X on Z, simply because the covariate Z is a baseline variable measured before the treatment X is chosen by the randomization. In contrast, for an epidemiological study, this temporal relationship may not exist. When patients are randomized between treatments in a medical study, these arguments hold quite generally for multiple covariates, $\mathbf{Z} = (Z_1, \cdots, Z_p)$. A regression model with linear term taking the additive form

$$\eta(X, \mathbf{Z}) = \alpha_0 + \alpha_1 X + \sum_{j=1}^{p} \alpha_{2,j} Z_j$$

gives an unbiased estimate of the treatment effect α_1 and correct estimates of the baseline covariate effects on Y, due to the facts that randomization does not allow any Z_j to affect X, and X cannot go back in time and affect any Z_j.

Because conventional statistical methods are inadequate to obtain an unbiased estimator of a treatment or exposure variable effect when analyzing non-randomized, observational data, a statistical method to correct for bias must be applied. The following framework for linking statistical inference and causal analysis was established by Rubin (1974, 1978, 2005), and can be traced back to ideas discussed by Neyman (1923, 1934). Index two treatments for a disease by $X = 1$ and $X = 0$, and consider the following thought experiment for comparing the treatments fairly. Imagine making two identical physical copies of a patient, treating one copy with $X = 1$, and treating the other copy with $X = 0$. Denote the *potential outcomes* that would be observed by $Y(1)$ for $X = 1$ and $Y(0)$ for $X = 0$. For each treatment $x = 0$ or 1, the potential outcome $Y(x)$ is defined only if $X = x$. The *causal treatment effect* on the patient is defined to be $Y(1) - Y(0)$. While one cannot make two identical copies of a patient, if one actually wants to compare treatments then this thought experiment leads to some very useful statistical methods. Since only one of the two treatments can be given to a patient, only one of $Y(0)$ or $Y(1)$ can be observed, and the causal effect $Y(1) - Y(0)$ never can be observed. This is known as the *Fundamental Problem of Causal Inference* (Holland, 1986), and is an example of the difference between fantasy and reality. Continuing the thought experiment, to keep track of real and imaginary observations, if a patient's actual treatment is $X = 1$ then $Y(1)$ is observed, it is called the *factual outcome*, and $Y(0)$, which is not observed, is called the *counterfactual outcome*. Similarly, if the patient's actual treatment is $X = 0$ then the observed $Y(0)$ is the factual outcome and unobserved $Y(1)$ is the counterfactual outcome. It will be shown below that this imaginary construction actually is extremely useful for making statistical inferences about $\Delta = \mu_1 - \mu_0$ using observational data. This is because, under very reasonable assumptions, the potential outcomes, $Y(1)$ and $Y(0)$, can be linked to the observed outcome, Y.

This thought experiment mimics what a physician does when choosing between two treatments for an individual patient. Still, things are more complicated in practice. As discussed above, a physician typically uses the vector $\mathbf{Z} = (Z_1, \cdots, Z_q)$ of the patient's covariates to help choose a treatment. To account for heterogeneity, for a patient with covariates $\mathbf{Z}$, the *conditional average treatment effect* (CATE) is defined in terms of expected values of the potential outcomes, as

$$\Delta^{CATE}(\mathbf{Z}) = E\{Y(1) \mid \mathbf{Z}\} - E\{Y(0) \mid \mathbf{Z}\}.$$

The *average treatment effect (ATE)* is computed by averaging $\Delta^{CATE}(\mathbf{Z})$ over the covariate distribution, $p_{\mathbf{Z}}(\mathbf{z})$,

$$\Delta^{ATE} = E\{Y(1)\} - E\{Y(0)\} = \int_{\mathbf{z}} \Delta^{CATE}(\mathbf{z}) p_{\mathbf{Z}}(\mathbf{z}) d\mathbf{z}.$$

This covariate-weighted average is defined using a Lebesgue integral, which accommodates any combination of discrete and continuous covariates in $\mathbf{Z}$. In particular, for fully discrete $\mathbf{Z}$, the above integral can be expressed as a sum. In the following, for brevity, dependence of the various quantities on parameters is suppressed in the notation. However, it should be kept in mind that Δ^{CATE}, Δ^{ATE}, Δ, and other means typically are formulated as parametric functions.

In some settings, interest may focus on the *average treatment effect among the treated* (ATT), or the *average treatment effect among the untreated* (ATU), defined as the conditional expectations

$$ATT = E(\Delta^{ATE} \mid X = 1) \quad \text{and} \quad ATU = E(\Delta^{ATE} \mid X = 0).$$

For example, consider a study having the primary goal to evaluate the effect of a low carbohydrate diet (LCD, $X = 1$), compared to a regular diet (RD, $X = 0$) on a subject's weight change, Y, over a one year time period, compared to their baseline weight, in a particular population. Interest may focus on the impact of the LCD on subjects who actually followed the LCD, which is the ATT. The ATU is the effect that the LCD would have had on subjects who followed an RD, sometimes called the "I told you so" effect. The ATT and ATU are special cases of evaluating Δ^{ATE} in a subpopulation of interest defined by $\mathbf{Z}$. This might be useful in settings where, following the study, a goal is to make dietary recommendations to particular subpopulations of people. For example, a subpopulation might be children, defined as having age ≤ 18.

Conventional statistical definitions of treatment effects are formulated in terms of the observable outcome, covariates, and treatment indicators as

$$\Delta(\mathbf{Z}) = E\{Y \mid \mathbf{Z}, X = 1\} - E\{Y \mid \mathbf{Z}, X = 0\} = \mu_1(\mathbf{Z}) - \mu_0(\mathbf{Z}),$$

for a subject with covariates $\mathbf{Z}$. Averaging over the covariate distribution gives the population mean treatment effect

$$\Delta = E\{\Delta(\mathbf{Z})\} = \int_z \Delta(z) p_{\mathbf{Z}}(z) dz = \mu_1 - \mu_0.$$

The following three causal assumptions are used to connect the definitions of treatment effects defined in terms of potential and observed outcomes, and these relationships are used to show that $\Delta^{ATE} = \Delta$. This equality says that the causal effect defined in terms of the potential outcomes equals the usual difference between population means defined in terms of observed outcomes. To derive this equality, first denote probabilistic independence of variables or events A and B by $A \perp\!\!\!\perp B$, conditional independence of A and B given C by $A \perp\!\!\!\perp B \mid C$, and let $I(A)$ denote the indicator of the event A. For each patient covariate vector $\mathbf{Z}$ and treatment $X \in \{0,1\}$, the assumptions connecting the potential and actual outcomes, used to make causal inferences, are as follows.

The Three Assumptions for Making Causal Inferences

1. Consistency: $Y = XY(1) + (1-X)Y(0)$
2. Strong Ignorability: $X \perp\!\!\!\perp \{Y(1), Y(0)\} \mid \mathbf{Z}$
3. Positivity: $0 < P(X \mid \mathbf{Z}) < 1.$

Consistency says that, if a patient actually receives treatment X, then the observed outcome Y must equal the potential outcome $Y(X)$, otherwise the construction does not make sense. Thus, if $X = 1$ then $Y = Y(1)$ is observed, so $Y(1)$ is the factual outcome and $Y(0)$ is the unobserved counterfactual outcome. Similarly, if $X = 0$ then $Y = Y(0)$ is observed and $Y(1)$ is the unobserved counterfactual. The strong ignorability assumption says that, given the patient's covariates, the treatment assignment is independent of the potential outcomes. That is, one cannot know the possible future outcomes when choosing a treatment. In epidemiological settings, where X may refer to exposure rather than treatment, strong ignorability may be called *exchangeability*, which says that the future values $Y(1)$ and $Y(0)$ are conditionally independent of the exposure variable. Positivity says that both treatments are possible, since if this were not true for some value of X then it would not make sense to consider it.

Showing that $\Delta^{ATE} = \Delta$ connects expected values of causal treatment effects defined in terms of the potential outcomes $Y(0)$ and $Y(1)$ to treatment effects defined conditionally in terms of the observable variable Y. To establish this, it is useful to think of Δ^{ATE} and Δ as averages over $p_{\mathbf{Z}}(z)$ of their covariate-specific conditional means, $\Delta^{CATE}(\mathbf{Z})$ and $\Delta(\mathbf{Z})$. The idea is to compare means between subjects treated with either $X = 1$ or $X = 0$ within strata defined by $\mathbf{Z}$, and then average over the distribution of $\mathbf{Z}$. By consistency,

$$\begin{aligned} \mu_1(\mathbf{Z}) = E\{Y \mid X = 1, \mathbf{Z}\} &= E\{XY(1) + (1-X)Y(0) \mid X = 1, \mathbf{Z}\} \\ &= E\{XY(1) \mid X = 1, \mathbf{Z}\}, \end{aligned}$$

which by strong ignorability equals $E\{Y(1) \mid \mathbf{Z}\}$, so $\mu_1(\mathbf{Z}) = E\{Y(1) \mid \mathbf{Z}\}$. Similarly, $\mu_0(\mathbf{Z}) = E\{Y(0) \mid \mathbf{Z}\}$. Taking the differences and averaging over the marginal distribution of **Z** gives

$$\Delta^{ATE} = \int \{E[Y(1) \mid z] - E[Y(0) \mid z]\} p_{\mathbf{Z}}(z)dz = \int \{\mu_1(\mathbf{Z}) - \mu_0(\mathbf{Z})\} p_{\mathbf{Z}}(z)dz = \Delta.$$

This provides a basis for defining an unbiased estimator of Δ^{ATE}, based on a sample $\{(Y_i, \mathbf{Z}_i, X_i),\ i = 1, \cdots, n\}$ of observed outcomes, covariates, and treatments.

To define an estimator for Δ^{ATE} that corrects for bias, denote the ith subject's *propensity score* by $\pi_i = \pi(\mathbf{Z}_i) = Pr(X_i = 1 \mid \mathbf{Z}_i)$. A requirement to construct the estimator is that the π_i's are known, or can be estimated from the available data. In a medical setting, $\pi(\mathbf{Z}_i)$ reflects the use of covariates by physicians to choose treatments, so the data may reflect selection bias. Under positivity, the *inverse probability of treatment weighted* (IPTW) estimator of Δ is defined as

$$\hat{\Delta}^{IPTW} = \frac{1}{n}\sum_{i=1}^{n} Y_i \left\{ \frac{X_i}{\pi(\mathbf{Z}_i)} - \frac{1 - X_i}{1 - \pi(\mathbf{Z}_i)} \right\}. \tag{2.1}$$

This may be considered a two-sample extension of the estimator proposed by Horvitz and Thompson (1952). To show that $\hat{\Delta}^{IPTW}$ is unbiased, suppressing the index i, by consistency

$$E(YX) = E\{Y(1)X + Y(0)(1-X)X\} = E(Y(1)X).$$

By iterated expectation conditioning on **Z**, this equals $E_{\mathbf{Z}}\{E[Y(1)X \mid \mathbf{Z}]\}$, which by strong ignorability equals

$$E_{\mathbf{Z}}\{E[Y(1) \mid \mathbf{Z}]E[X \mid \mathbf{Z}]\} = E_{\mathbf{Z}}\{E[Y(1) \mid \mathbf{Z}]\}\pi(\mathbf{Z}).$$

Again by iterated expectation, this equals $E\{Y(1)\}\pi(\mathbf{Z})$. Similarly, $E\{Y(1-X)\} = E\{Y(0)\}(1-\pi(\mathbf{Z}))$.

Denote the number of subjects in the sample treated with $x = 0$ or 1 by n_0 and n_1, so $n = n_0 + n_1$. For data from a fairly randomized study, all $\pi_i = 1/2$ and n_0 and n_1 both follow a binomial distribution with parameters n and 1/2, denoted $n_x \sim Binom(n, 1/2)$ for $x = 0, 1$. In this case, showing unbiasedness is simple since the treatment assignment probabilities do not involve the $\mathbf{Z}_i$'s. The first sum in expression (2.1) equals $2n_1\overline{Y}_1/n$. Since E(n_1) = $(1/2)n$ and n_1, n_0 are independent of the Y_i's, it follows that E$(2n_1\overline{Y}_1/n)$ = E$(\overline{Y}_1)$ = μ_1. Similarly, the second sum has an expected value μ_0, so randomization ensures that E$(\hat{\Delta}^{IPTW})$ = $\mu_1 - \mu_0 = \Delta$. This shows that, if the goal is to obtain an unbiased estimator of Δ, then, rather than performing the impossible experiment of making two physical copies of each patient, one may simply flip a fair coin to choose each patient's treatment. This derivation also is valid for unbalanced randomization, for example, if randomization is done with probabilities 2/3 for an experimental treatment $X = 1$ and 1/3 for a control $X = 0$. However, someone who performs such a study really does not have equipoise, so arguably should not be randomizing patients between treatments. If the final data show that $\bar{Y}_1$ is much smaller than $\bar{Y}_0$, the fact that 2/3 of the patients in the sample intentionally were treated with $X = 1$ should be worse than embarrassing. Ethical problems with unbalanced randomization are discussed by Hey and Kimmelman (2014), among others.

When patients are not randomized and the available data are observational, the π_i's are not known, but can be estimated by fitting a parametric model for the propensity scores, $\{\pi_i, i = 1, \cdots, n\}$, to the treatment assignment and covariate data $\{(X_i, \mathbf{Z}_i), i = 1, \cdots, n\}$. Commonly used propensity score models take the form

$$link\{\pi_i(\mathbf{Z}_i, \boldsymbol{\alpha})\} = \alpha_0 + \sum_{j=1}^{q} \alpha_j Z_j,$$

where *link* may be the probit, logit, or complementary log-log. The fitted propensity score model yields an estimator $\hat{\boldsymbol{\alpha}}$ and thus $\hat{\pi}_i = \pi_i(\mathbf{Z}_i, \hat{\boldsymbol{\alpha}})$. These may be used as plug-in estimators to compute an approximate version of (2.1) which, under reasonable assumptions, gives an asymptotically unbiased estimator (Lunceford and Davidian, 2004). IPTW provides a practical tool for doing approximately unbiased estimation of a comparative treatment effect using observational data, by exploiting the idea that treatments were assigned using patient covariates. Since, in practice, the numerical values of some propensity estimates $\hat{\pi}_i$ may be near 0 or 1, to avoid the numerical instability that this may cause, one may use the *normalized IPTW estimator*

$$\hat{\Delta}^{NIPTW} = \frac{\sum_{i=1}^{n} Y_i X_i / \hat{\pi}_i}{\sum_{i=1}^{n} X_i / \hat{\pi}_i} - \frac{\sum_{i=1}^{n} Y_i (1 - X_i)/(1 - \hat{\pi}_i)}{\sum_{i=1}^{n} (1 - X_i)/(1 - \hat{\pi}_i)}.$$

Many other variants of IPTW estimators have been proposed. Reviews are given by Tsiatis (2006) and Austin and Stuart (2015).

2.3 Generalized Computation Methods

Generalized (G) computation methods were introduced by Robins (1986a) to correct for bias due to time-dependent confounding in observational studies with longitudinal data. G computation methods are a family including the G formula, which will be given below in the one-stage and two-stage cases, as well as marginal structural models and structural nested models. G methods accommodate settings where treatment may be given and patient covariates observed repeatedly over time, and the covariates may affect both treatment choice and the outcome of interest. Because each treatment is chosen adaptively using time-varying covariates and possibly previous treatment choices, the sequence of treatment decision rules may be considered a dynamic treatment regime. In an epidemiological study, the role of treatment in a medical study is played instead by an exposure variable, such as exposure to air pollution in a study of respiratory disease.

First consider a simple version of G computation for single-stage observational data $\mathscr{D}_n = \{(Y_i, \mathbf{Z}_i, X_i), i = 1, \cdots, n\}$. Recall that, under the assumptions of consistency, strong ignorability, and positivity, the conditional means of the potential and observed outcomes are connected by the equation $E\{Y(x) \mid \mathbf{Z}\} = E(Y \mid \mathbf{Z}, X = x)$, for all $\mathbf{Z}$ and $x = 0, 1$. Applying *The Law of Total Probability*, this implies the following equation:

G Computation Equation for Single-Stage Data

$$E\{Y(x)\} = \int_{\mathbf{z}} E(Y \mid \mathbf{z}, X = x) p_{\mathbf{Z}}(\mathbf{z}) d\mathbf{z}, \qquad \text{for } x = 0, 1. \tag{2.2}$$

In (2.2), the unconditional mean of the potential outcome $Y(x)$ for each treatment $X = x$ is computed by averaging the conditional mean $E[Y \mid \mathbf{z}, X = x]$ of the observed outcome over the covariate distribution $p_{\mathbf{Z}}(\mathbf{z})$.

The G computation formula requires one to assume a robust regression model for $p(Y \mid \mathbf{z}, X = x)$, called the *Q-model*. The Q model is assumed to be correct, as in any regression analysis, and it should be a robust model that reflects the structure of the DAG for the data. The purpose of the Q model is to provide a good fit to the $(Y, \mathbf{Z}, X)$ data that can be used to compute reliable predictions of potential outcomes for each treatment or exposure variable. The Q model may take any parametric, semiparametric, or nonparametric form. As a parametric model, it may include both main effects and treatment-covariate interactions of any order. A wide array of Q models are possible, including a neural net, a regression spline (Eubank, 1999), a Bayesian adaptive regression tree (BART) (Chipman et al., 2010a), or a Bayesian dependent Dirichlet process (MacEachern and Müller, 1998; MacEachern, 1999; Müller and Rodriguez, 2013; Müller and Mitra, 2013).

The G computation formula requires a sequence of statistical computations that, intuitively, mimic the thought experiment wherein one makes two identical copies of each subject, treats one copy with $X = 0$ and the other with $X = 1$, observes the outcomes $Y_i(0)$ and $Y_i(1)$, and computes the sample mean $n^{-1}\sum_{i=1}^{n}\{Y_i(1) - Y_i(0)\}$ as the causal estimator. The statistical process exploits the regression model to compute estimates of $Y_i(1)$ and $Y_i(0)$ as predicted values. It begins by fitting the Q model to $\mathscr{D}_n$ to account for all possible effects of confounding variables in $\mathbf{Z}$ that may modify the treatment effect on Y. The fitted Q model is used to predict each subject's potential outcomes, $\hat{Y}_i(0)$ and $\hat{Y}_i(1)$ for $i = 1, \cdots, n$. The means of the set of $2n$ predicted potential outcomes $\{(\hat{Y}_i(0), \hat{Y}_i(1)) : i = 1, \cdots, n\}$ then are used to estimate the causal effect Δ, by treating the predicted values like observed potential outcomes. An algorithm using the G formula for single-stage data is as follows:

G Formula Computational Algorithm for Single-Stage Data

Step 1. Fit a Q model $p(Y \mid \mathbf{Z}, X)$ to $\mathscr{D}_n$, and denote the fitted model by $\hat{p}(Y \mid \mathbf{Z}, X)$. For a parametric model $p(Y \mid \mathbf{Z}, X, \boldsymbol{\theta})$, frequentist estimation may be done by plugging in an estimator $\hat{\boldsymbol{\theta}}$ to obtain $\hat{p}(Y \mid \mathbf{Z}, X) = p(Y \mid X, \mathbf{Z}, \hat{\boldsymbol{\theta}})$.

Step 2. For each subject $i = 1, \cdots, n$ and each treatment $x = 0$ and $x = 1$, use the fitted Q model $\hat{p}(Y \mid \mathbf{Z}_i, X = x)$ to simulate two predicted values, $\hat{Y}_i(0)$ and $\hat{Y}_i(1)$, which are used as estimates of the two counterfactual outcomes for the ith subject. This gives a synthetic dataset of size $2n$, with the two pairs $(X_i = 1, \hat{Y}_i(1))$ and $(X_i = 0, \hat{Y}_i(0))$ for each i.

Step 3. Compute the estimator $\hat{\Delta} = n^{-1}\sum_{i=1}^{n}\{\hat{Y}_i(1) - \hat{Y}_i(0)\}$.

Step 2 provides a statistical solution to The Fundamental Problem of Causal Inference by using a fitted robust regression model to simulate each subject's potential outcomes. Step 3 averages each $E(Y \mid \mathbf{Z}, X = x)$ over the distribution of $\mathbf{Z}$ in (2.2) by using the empirical covariate distribution, weighting the sample covariates $\mathbf{Z}_1, \cdots, \mathbf{Z}_n$ equally, each with weight $1/n$. To estimate $sd(\hat{\Delta})$ in order to make inferences, such as computing a confidence interval for Δ or a p-value for a test of the hypothesis $\Delta = 0$ of no causal effect, a usual estimator is not correct, because it fails to account for fitting the Q model and using it to estimate the two potential outcomes for each $i = 1, \cdots, n$. To estimate $sd(\hat{\Delta})$, a sampling-based method, such as a bootstrap, may be used, based on the entire statistical process in steps 1 - 3 of the G computation algorithm. An alternative way to estimate Δ from the synthetic dataset is to use the covariate-weighted average of the n imputed causal effects,

$$\frac{\sum_{i=1}^{n} \{\hat{Y}_i(1) - \hat{Y}_i(0)\} \hat{p}_{\mathbf{Z}}(\mathbf{Z}_i)}{\sum_{i=1}^{n} \hat{p}_{\mathbf{Z}}(\mathbf{Z}_i)},$$

where $\hat{p}_{\mathbf{Z}}(\cdot)$ is an estimate of the covariate distribution.

Next, consider settings with longitudinal data, where a treatment X may be given repeatedly over time, or an epidemiological study where X is a time-varying exposure variable. Since treatment may be given more than once, a treatment sequence that may include adaptive decision rules, or dynamic treatment regime (DTR), may be the focus. The objective is to estimate the average or marginal effect of a treatment or treatment sequence on Y. In such settings, possibly time-varying covariates $\mathbf{Z} = (Z_1, \cdots, Z_k)$ may affect both the sequence of X values Y, and thus confound the effect of X on Y. To obtain an unbiased estimator of the causal effect of a treatment or a sequence of treatments on a final outcome Y, it is necessary to correct for possible confounding effects of covariates over time.

In a landmark paper, Hernan et al. (2000) described an observational study of HIV-positive men with the aim to evaluate the effect of the treatment zidovudine (AZT) on Y = survival time. Over time, physicians monitored each patient's Z = CD4 lymphocyte counts and used these values to decide whether to administer AZT. Because a higher CD4 count was associated with a higher risk of death, physicians were more likely to give AZT to patients with higher CD4 counts. Consequently, over time, the effects of AZT and CD34 count on survival time were confounded. The paper described methods to correct for the confounding effects of using CD34 counts to choose patients' treatments, when the statistical goal is to estimate the causal effect of AZT on survival. Hernan et al. (2000) showed that a conventional regression analysis using a Cox model incorrectly concluded that giving AZT reduced expected survival time, because this analysis ignored confounding due to the fact that physicians used a higher CD4 count as a trigger to give a patient AZT. Essentially, the conventional survival analysis misinterpreted the biological effect of a higher CD4 count on survival time as the effect of giving AZT. A bias-corrected analysis led to the opposite conclusion, that AZT increased survival time.

To describe the G computation algorithm and establish the main ideas for analyzing longitudinal data, consider a setting with two stages, s = 0 and 1. Let $\mathbf{Z}_0$ and $\mathbf{Z}_1$ denote covariate vectors observed at the two stages. In practice, $\mathbf{Z}_0$ includes baseline covariates, and $\mathbf{Z}_1$ may include updated elements of $\mathbf{Z}_0$ as well as intermediate outcomes. Let X_0 denote the initial treatment and X_1 the stage 1 treatment, where each $X_s = 0$ or 1, and let Y denote the primary outcome. The inferential goals are to compare marginal effects between the four possible two-stage strategies, $(X_0,X_1) =$ (0,0), (0,1), (1,0), or (1,1), on Y. This requires removing confounding effects of $\mathbf{Z}_0$ on X_0 and of $(\mathbf{Z}_0,\mathbf{Z}_1)$ on X_1. Denote a realization of (X_0,X_1) by (x_0,x_1), and the potential outcome of the two-stage treatment strategy (x_0,x_1) by $Y(x_0,x_1)$. Consistency now says that $Y = Y(x_0,x_1)$ if $(X_0,X_1) = (x_0,x_1)$. The strong ignorability assumption is made at each time point, conditional on past covariates and treatments. For two stages, the sequential strong ignorability assumptions are

$$Y(x_0,x_1) \perp\!\!\!\perp X_1 \mid X_0,\mathbf{Z}_1,\mathbf{Z}_0 \tag{2.3}$$

and

$$Y(x_0,x_1) \perp\!\!\!\perp X_0 \mid \mathbf{Z}_0. \tag{2.4}$$

Equation (2.3) says that, given the covariates and stage 0 treatment, the future potential outcome does not depend on the stage 1 treatment. Equation (2.4) says that, given the baseline covariates, the future potential outcome does not depend on the stage 0 treatment. These assumptions imply, respectively, that

$$E\{Y(x_0,x_1) \mid X_1,X_0,\mathbf{Z}_1,\mathbf{Z}_0\} = E\{Y(x_0,x_1) \mid X_0,\mathbf{Z}_1,\mathbf{Z}_0\}$$

and

$$E\{Y(x_0,x_1) \mid X_0,\mathbf{Z}_0\} = E\{Y(x_0,x_1) \mid \mathbf{Z}_0\}.$$

Positivity says that each treatment is possible at each stage,

$$0 < P(X_1 = 1 \mid X_0 = x_0,\mathbf{Z}_1,\mathbf{Z}_0) < 1, \quad \text{for} \quad x_0 = 0,1,$$

and

$$0 < P(X_0 = 1 \mid \mathbf{Z}_0) < 1.$$

To account for the sequences of treatments, covariates, and outcomes, the joint likelihood is factored into the product of conditional distributions,

$$p(y,x_1,\mathbf{z}_1,x_0,\mathbf{z}_0) = p(y \mid x_1,\mathbf{z}_1,x_0,\mathbf{z}_0)p(x_1 \mid \mathbf{z}_1,x_0,\mathbf{z}_0)p(\mathbf{z}_1 \mid x_0,\mathbf{z}_0)p(x_0 \mid \mathbf{z}_0)p(\mathbf{z}_0).$$

This factorization accounts for the distributions of the covariates at each stage as well as the outcomes.

In the two-stage case, the *G computation formula* computes the mean of each potential outcome $Y(x_0,x_1)$ as the average over the covariate distributions of the conditional mean of the observed outcome given $(X_0,X_1) = (x_0,x_1)$. This is the

G Computation Formula for Two-Stage Data

$$E\{Y(x_0,x_1)\} = \int_{\mathbf{z}_0,\mathbf{z}_1} E(Y \mid x_1,\mathbf{z}_1,x_0,\mathbf{z}_0)p(\mathbf{z}_1 \mid x_0,\mathbf{z}_0)p(\mathbf{z}_0)d\mathbf{z}_0 d\mathbf{z}_1. \tag{2.5}$$

To derive equation (2.5), for any (x_0, x_1),

$$E\{Y(x_0,x_1)\} = E[E\{Y(x_0,x_1) \mid \mathbf{Z}_0\}] = E[E\{Y(x_0,x_1) \mid \mathbf{Z}_0, X_0 = x_0\}].$$

where the first equality is by iterated expectation the second equality is by strong ignorability at stage 0. Again applying iterated expectation, this equals

$$= E[E\{E(Y(x_0,x_1) \mid \mathbf{Z}_0, X_0 = x_0, \mathbf{Z}_1) \mid \mathbf{Z}_0, X_0 = x_0\}],$$

which by strong ignorability at stage 1 equals

$$= E[E\{E(Y(x_0,x_1) \mid \mathbf{Z}_0, X_0 = x_0, \mathbf{Z}_1, X_1 = x_1) \mid \mathbf{Z}_0, X_0 = x_0\}].$$

By consistency, this equals

$$= E[E\{E(Y \mid \mathbf{Z}_0, X_0 = x_0, \mathbf{Z}_1, X_1 = x_1) \mid \mathbf{Z}_0, X_0 = x_0\}],$$

and by the definition of expectation, this equals

$$\int_{\mathbf{z}_0, \mathbf{z}_1} E(Y \mid \mathbf{z}_0, x_0, \mathbf{z}_1, x_1) p(\mathbf{z}_1 \mid x_0, \mathbf{z}_0) p(\mathbf{z}_0) d\mathbf{z}_0 d\mathbf{z}_1,$$

which establishes the G computation formula (2.5). To compute this requires estimates of the regression model $p(Y \mid x_0, \mathbf{z}_1, x_1, \mathbf{z}_0)$, the distribution $p(\mathbf{z}_1 \mid x_0, \mathbf{z}_0)$ of the updated covariates, where $\mathbf{Z}_1$ may include an intermediate outcome, and the baseline covariate distribution $p(\mathbf{z}_0)$.

Figure 2.4 illustrates the DAG in the two-stage setting, with one-dimensional Z_0 and Z_1. The causal effects in the figure may be described as follows.

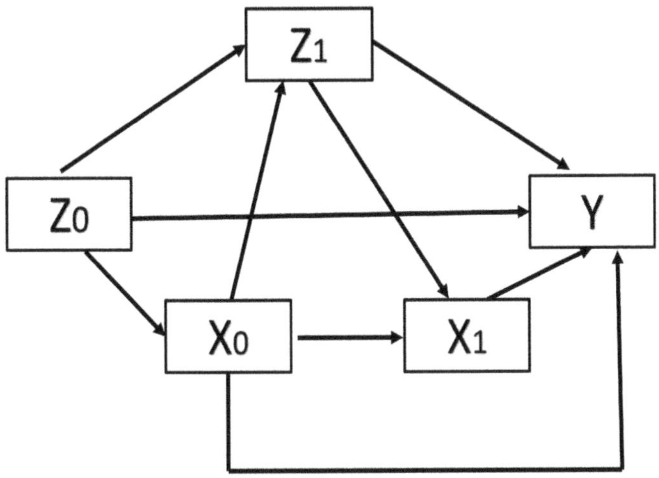

FIGURE 2.4
Causal diagram of the possible effects of treatments X_0 given at stage 0 and X_1 given at stage 1, and stage 0 baseline covariates $\mathbf{Z}_0$ and updated stage 1 covariates $\mathbf{Z}_1$. Y is the final outcome.

Description of the Causal Effects in Figure 2.4

1. The baseline covariates, $\mathbf{Z}_0$ may be used by the physician to choose the frontline treatment X_0, and $\mathbf{Z}_0$ has direct effects on both Y and X_0.
2. $\mathbf{Z}_0$ is updated to $\mathbf{Z}_1$, and possibly may include an intermediate outcome.
3. $\mathbf{Z}_1$ has direct effects on both X_1 and Y.
4. The frontline treatment X_0 and second treatment X_1 both have direct effects on Y, and X_0 has an effect on X_1 that may reflect a physician accounting for the stage 0 treatment when choosing the stage 1 treatment.

The causal effects $\mathbf{Z}_0 \rightarrow X_0$ and $\mathbf{Z}_1 \rightarrow X_1$ represent a physician's use of available covariate data to guide treatment decisions, where $\mathbf{Z}_1$ may include an intermediate outcome. For example, if $Z_{0,R}$ indicates early response, then $Z_{0,R} = 0$ (no response) may motivate the use of a salvage therapy X_1, whereas if $Z_{0,R} = 1$ (response) then X_1 may be a consolidation therapy given to maintain response. In addition to direct effects, there also may be interactive effects among combinations of two or more variables on Y. Assuming that there are no unobserved external confounders, to estimate the marginal effects of (X_0, X_1) on Y, a statistical method must correct for the confounding effects due to the observed variables $(\mathbf{Z}_0, \mathbf{Z}_1)$.

Before presenting the structure to correct for confounding in the multistage setting, it is worthwhile to mention how this problem may be avoided when comparing multistage treatment strategies. A simple and effective solution is to randomize each patient between prespecified treatments at each stage of the therapeutic process. Each patient is randomized to an initial treatment, and later re-randomized between treatments at a later stage, with this process possibly continued, depending on the medical setting. For example, in a study designed to compare multistage therapies for a particular cancer, patients might be randomized at stage 0 between $X_0 = 0$ (standard therapy) and $X_0 = 1$ (targeted therapy). In stage 1, a patient who has responded, $Z_{0,R} = 1$, may be re-randomized between $X_1(Z_{0,R}) = X_1(1) = 0$ (standard consolidation therapy) and $X_1(Z_{0,R}) = X_1(1) = 1$ (targeted consolidation therapy). A patient who has not responded, $Z_{0,R} = 0$, may be may be re-randomized between $X_1(Z_{0,R}) = X_1(0) = 0$ (standard salvage therapy) and $X_1(Z_{0,R}) = X_1(0) = 1$ (targeted salvage therapy). That is, the stage 1 treatments are chosen adaptively based on the early response indicator $Z_{0,R}$, with each adaptive choice a randomization between two appropriate therapies. Thus, (X_0, X_1) is a pair of adaptive, random treatment assignment rules, and is a two-stage DTR (Tsiatis et al., 2021). This design is an example of a sequential multiply randomized adaptive randomized trial (SMART) (Murphy and Bingham, 2009; Almirall et al., 2014). Due to the randomization, there are no causal arrows from $\mathbf{Z}_0$ or $\mathbf{Z}_1$ to X_0 or X_1 in a DAG, and conventional statistical estimators of the effects of each possible realization of the treatment pair (X_0, X_1) on Y are appropriate. These effects may be parameters obtained from a regression model that also accounts for effects of $(\mathbf{Z}_0, \mathbf{Z}_1)$ additively.

In the two-stage setting, causal effects are defined in terms of mean potential outcomes, $E\{Y(x_0, x_1)\}$, for the set of possible treatment pairs, (x_0, x_1). If each of x_0 and x_1 can take two possble values, then there are four possible pairs and several different contrasts. For three or more stages, the number of potential outcomes and contrasts can become unmanageable, however. In the two-stage treatment setting, the possible contrasts between pairs of different two-stage strategies may be represented as follows.

Possible Contrasts Between Pairs of Two-Stage Strategies

1. Compare (1,1) to (0,1), where only the stage 0 treatment is varied with fixed $X_1 = 1$.
2. Compare (1,1) to (1,0), where only the stage 1 treatment is varied with fixed $X_0 = 1$.
3. Compare (1,0) to (0,1), the two strategies where the treatments given in each of the two stages are different.
4. Compare (1,1) to (0,0), giving treatment 1 in both stages versus giving treatment 0 in both stages.

Using the same arguments as in the one-stage setting, with the four possible values of the potential outcome $Y(x_0,x_1)$ now playing the roles played by $Y(0)$ and $Y(1)$ in the one-stage setting, consistency ensures that the means of the observed and potential outcomes are the same, formally

$$E\{Y(x_0,x_1)\} = E\{Y \mid X_0 = x_0, X_1 = x_1\}.$$

Recall that strong ignorability says that, since a future potential outcome is not known, it cannot be used to choose a treatment. Simply put, at any point in time one cannot know the future when choosing a treatment. For multiple stages, this is extended to the assumption of *sequentially ignorable treatment assignment*. In the two-stage setting, this says that strong ignorability is assumed at each stage. Working backwards from stage 1 to stage 0, the strong ignorability assumption for stage 1 is that, given the stage 0 treatment and stage 1 covariates, the potential outcome and stage 1 treatment are conditionally independent,

$$X_1 \perp\!\!\!\perp Y(x_0,x_1) \mid \mathbf{Z}_1, X_0,$$

for all two-stage treatment regimes (x_0,x_1). That is, one cannot know a future outcome when choosing a treatment. This implies that

$$E\{Y(x_0,x_1) \mid X_0, X_1, \mathbf{Z}_1\} = E\{Y(x_0,x_1) \mid X_0, \mathbf{Z}_1\}.$$

Strong ignorability for the stage 0 treatment takes the same form as the assumption in the single-stage case, and is given by

$$X_0 \perp\!\!\!\perp Y(x_0,x_1) \mid \mathbf{Z}_0,$$

for all (x_0,x_1), which implies

$$E\{Y(x_0,x_1) \mid X_0, \mathbf{Z}_0\} = E\{Y(x_0,x_1) \mid \mathbf{Z}_0\}.$$

Positivity now is assumed for the treatment probabilities at each of the two stages, with $0 < P(X_1 = 1 \mid \mathbf{Z}_1, X_0) < 1$ for stage 1 and $0 < P(X_0 = 1) < 1$ for stage 0.

In the longitudinal setting, the joint distribution of Y and the treatments and covariates is factored using a standard conditional probability calculation, under the assumption that the treatments and covariates in each stage are random quantities. This factorization is

$$\begin{aligned} p(y, x_1, z_1, x_0, z_0) &= p(y \mid x_1, z_1, x_0, z_0) p(X_1 = x_1 \mid z_1, X_0 = x_0, z_0) \\ &\times \; p(z_1 \mid X_0 = x_0, z_0) p(X_0 = x_0 \mid z_0) p(z_0). \end{aligned}$$

In this setting, $\mathbf{Z}_0$ is a vector of baseline covariates, but $\mathbf{Z}_1$ may include interim outcomes, such as an early response indicator, so $p(z_1 \mid X_0 = x_0, z_0)$ is a conditional interim outcome distribution. The unconditional expected value of the observed final outcome Y is computed by averaging over the treatment and covariate distributions,

$$\begin{aligned} E(Y) &= \sum_{x_0=0}^{1} \sum_{x_1=0}^{1} \int_{z_0, z_1} y p(y \mid x_1, z_1, x_0, z_0) p(X_1 = x_1 \mid z_1, X_0 = x_0, z_0) \\ &\times \; p(z_1 \mid X_0 = x_0, z_0) p(X_0 = x_0 \mid z_0) p(z_0) dz_1 dz_0. \end{aligned}$$

To evaluate the expected value of the potential outcome $Y(x_0, x_1)$, due to conditioning on $(X_0, X_1) = (x_0, x_1)$, in the above equation the terms for the probability distributions of X_1 and X_0 are dropped. Under the three causality assumptions, the conditional expectation can be computed using the following equation, which is called the *G formula*.

G Computation Formula for Two-Stage Data

$$E\{Y(x_0, x_1)\} = \int_{z_0} \int_{z_1} E(Y \mid x_1, z_1, x_0, z_0) p(z_1 \mid X_0 = x_0, z_0) p(z_0) dz_1 dz_0. \quad (2.6)$$

Because the right hand side of equation (2.6) is a Lebesgue integral, it takes the form of a sum for discrete $(\mathbf{Z}_0, \mathbf{Z}_1)$. The G equation connects the mean of the potential outcome $Y(x_0, x_1)$ to the mean of the observed outcome Y. It provides an explicit way to compute an estimator for the mean of $Y(x_0, x_1)$ for each treatment pair (x_0, x_1), by averaging over the covariate distribution. The G formula thus provides a way to compute an estimated mean for any contrast $E\{Y(x_0, x_1) - Y(x_0', x_1')\}$ of interest to compare two-stage treatment regimes.

The G equation for two stages can be derived similarly to the earlier derivation of the equality $E\{Y(x)\} = E\{Y \mid X = x\}$ for $x = 0, 1$ that connects the potential outcomes to the observed outcome in the single-stage setting. Repeatedly applying the strong ignorability assumption and the law of iterated expectation,

$$\begin{aligned} E\{Y(x_o, x_1)\} &= E[E\{Y(x_o, x_1) \mid X_0\}] \\ &= E[E\{Y(x_o, x_1) \mid X_0 = x_0\}] \\ &= E[E\{E[Y(x_o, x_1) \mid X_0 = x_0, X_1 = x_1, \mathbf{Z}_1, \mathbf{Z}_0] \mid X_0 = x_0\}]. \end{aligned}$$

By consistency, this equals

$$E[E\{E[Y \mid X_0 = x_0, X_1 = x_1, \mathbf{Z}_1, \mathbf{Z}_0] \mid X_0 = x_0\}].$$

This can be computed as the Lebesgue integral

$$\int_{\mathbf{z}_0}\int_{\mathbf{z}_1} E(Y \mid X_0 = x_0, X_1 = x_1, \mathbf{z}_1) p(\mathbf{z}_1 \mid X_0 = x_0, \mathbf{z}_0) p(\mathbf{z}_0) d\mathbf{z}_1 d\mathbf{z}_0,$$

which for discrete covariates is the sum

$$\sum_{\mathbf{z}_0}\sum_{\mathbf{z}_1} E(Y \mid X_0 = x_0, X_1 = x_1, \mathbf{z}_1) p(\mathbf{z}_1 \mid X_0 = x_0, \mathbf{z}_0) p(\mathbf{z}_0).$$

The computational steps to apply the G estimation algorithm for longitudinal data follow those given before in the one-stage case, with appropriate extensions. For the two-stage case, the following algorithm generalizes X to two consecutive treatments (X_0, X_1) and Z to $(\mathbf{Z}_0, \mathbf{Z}_1)$.

G Computation Steps for Two-Stage Data

Step 1. Fit a robust Q model, $p(Y \mid X_0, X_1, \mathbf{Z}_0, \mathbf{Z}_1)$, to the observed data, and denote the fitted model by $\hat{p}(Y \mid X_0, X_1, \mathbf{Z}_0, \mathbf{Z}_1)$. The Q model may be constructed to reflect a DAG for the variables in the dataset.

Step 2. For each treatment pair (x_0, x_1), and each subject $i = 1, \cdots, n$ in the dataset, compute a predicted value $\hat{Y}_i(x_0, x_1)$ of the subject's potential outcome by substituting $(X_0, X_1) = (x_0, x_1)$ into the fitted Q model, $\hat{p}(Y \mid X_0 = x_0, X_1 = x_1, \mathbf{Z}_0, \mathbf{Z}_1)$. Doing this prediction for each subject $i = 1, \cdots, n$ produces a synthetic dataset of size $4n$, consisting of the imputed values $\hat{Y}_i(x_0, x_1)$ for each treatment pair (x_0, x_1) = (1,1), (1,0), (0,1), and (0,0).

Step 3. For a given contrast $\psi = E\{Y_i(x_0, x_1) - Y_i(x_0', x_1')\}$, the estimator using the synthetic data is $n^{-1}\sum_{i=1}^{n}\{\hat{Y}_i(x_0, x_1) - \hat{Y}_i(x_0', x_1')\}$.

In practice, the two-stage structure may be considerably more complicated if several different types of outcomes can be observed at each stage, and there may be more than two stages, so Y must be extended appropriately. Analysis of dataset with a much more general longitudinal structure, arising from the multistage treatment of patients with acute leukemia, will be described below, in Chapter 9.

For $m+1$ stages, denote $\boldsymbol{X} = (X_0, \cdots, X_m)$, $\boldsymbol{x} = (x_0, \cdots, x_m)$, $\mathbf{Z} = (\mathbf{Z}_0, \cdots, \mathbf{Z}_m)$, and $\mathbf{z} = (\mathbf{z}_0, \cdots, \mathbf{z}_m)$, and denote the potential outcome for treatment sequence $\boldsymbol{x}$ by $Y(\boldsymbol{x})$. In this case, each $\mathbf{Z}_s$ for $s > 0$ may include interim outcomes, so the likelihood For each interim stage $r = 1, \cdots, m-1$, denote the current covariate vector by $\mathbf{Z}_r = (\mathbf{Z}_1, \cdots, Z_r)$, and similarly for the other vectors. For discrete $\mathbf{Z}$, the G computation formula is as follows.

G Equation for Longitudinal Data

$$E\{Y(\boldsymbol{x})\} = \sum_{\mathbf{z}}\left\{E(Y \mid \boldsymbol{X} = \boldsymbol{x}, \mathbf{Z} = \mathbf{z})\prod_{j=1}^{m} p(Z_j = z_j \mid \boldsymbol{X}_{j-1} = \boldsymbol{x}_{j-1}, \mathbf{Z}_{j-1} = \mathbf{z}_{j-1})\right\}.$$

This more general equation may be derived similarly to the derivation in the two-stage case. The formulation accommodates settings where $\mathbf{Z}$ may include intermediate outcomes, such as early response or disease progression when Y is survival time. An application of G computation for estimating the mean overall survival time for each of a set of DTRs for treating acute leukemia will be given in Chapter 10.

Bayesian G Computation

Bayesian G computation follows the same causal rationale as the frequentist formulation, with modifications to accommodate prior specification and basing inferences on a posterior distribution rather than plug-in estimators. Bayesian prediction provides a natural framework to compute the potential outcome for a future patient given each treatment and covariate vector. Shrinkage provided by the prior may improve estimation accuracy for limited sample sizes, and Bayesian estimators under robust Q functions, such as Bayesian nonparametric regression models or BARTs, often have desirable frequentist properties, such as reliable bias correction for observational data. Moreover, computations may be done using readily available Bayesian software. See, for example, Keil et al. (2018).

Consider the single-stage case, with a vector of stochastic covariates $\mathbf{Z}$ and data $\mathscr{D}_n = \{(Y_i, X_i, \mathbf{Z}_i), i = 1, \cdots, n\}$. Bayesian model fitting includes the usual specification of a prior on $\boldsymbol{\theta}$ and MCMC sampling to obtain a posterior. A Bayesian G estimation algorithm is based on the prediction of potential outcomes under the posterior, since the causal idea of estimating potential future outcomes for a given treatment $X = 0$ or 1 coincides with the Bayesian idea of predicting future outcomes for these treatments, given the observed data. As in the frequentist setting, the final step in the computation averages over the distribution of $\mathbf{Z}$, which may be the empirical baseline covariate distribution.

For a parametric regression sampling model, $p(Y \mid x, \mathbf{z}, \boldsymbol{\theta})$, the posterior predictive distribution of the future potential outcome $Y(x)$ for a subject treated with x may be obtained by averaging the sampling model over the posterior, and then carrying out the G estimation computation by averaging over the covariate distribution. This is used to define a *Bayesian G computation formula* for the potential outcome $Y(x)$ under treatment x, treating the observed outcome Y, covariates $\mathbf{Z}$, and parameters $\boldsymbol{\theta}$ as random. This may be computed by first averaging the parametric distribution of $(Y, \mathbf{Z})$ given $\boldsymbol{\theta}$ and x over the posterior of $\boldsymbol{\theta}$, and then averaging over the distribution of $\mathbf{Z}$, as

$$
\begin{aligned}
p(Y(x) \mid \mathscr{D}_n) &= \int_{\mathbf{z}} p(y \mid x, \mathbf{z}, \mathscr{D}_n) p(\mathbf{z}) d\mathbf{z} \\
&= \int_{\mathbf{z}} \left\{ \int_{\boldsymbol{\theta}} p(y \mid x, \mathbf{z}, \boldsymbol{\theta}) p(\mathbf{z}) p(\boldsymbol{\theta} \mid \mathscr{D}_n) d\boldsymbol{\theta} \right\} d\mathbf{z}. \qquad (2.7)
\end{aligned}
$$

The integrand $p(y \mid x, \mathbf{z}, \mathscr{D}_n)$ is the Bayesian posterior predictive distribution of the observed outcome given the treatment, covariates, and data. When $\mathbf{Z}$ is a vector of baseline covariates, $p(\mathbf{z} \mid \mathscr{D}_n) = p(\mathbf{z})$ and the outer integral over $\mathbf{z}$ in (2.7) may be computed empirically by taking the sample mean over the observed covariate vectors, giving each $\mathbf{Z}_i$ weight $1/n$. The posterior predictive mean of the potential outcome $Y(x)$ is then computed from the predictive distribution of Y as

$$E\{Y(x) \mid \mathscr{D}_n)\} = \frac{1}{n}\sum_{i=1}^{n}\int_y y\, p(y \mid x, \mathbf{Z}_i, \mathscr{D}_n)dy.$$

The causal effect $\Delta = E\{Y(1) - Y(0) \mid \mathscr{D}_n\}$ may be estimated using the difference between the means of the two posterior predictive distributions.

For the Bayesian computation, first a large sample $\boldsymbol{\theta}^{(1)}, \cdots, \boldsymbol{\theta}^{(B)}$ is generated from the posterior $p(\boldsymbol{\theta} \mid \mathscr{D}_n)$. For each $\boldsymbol{\theta}^{(b)}$, $b = 1, \cdots, B$, treatment $x = 0, 1$, and covariate vector $\mathbf{Z}_i$, $i = 1, \cdots, n$ a predicted value $Y_i^{(b)}(x)$ is simulated from the sampling model $p(Y \mid X = x, \mathbf{Z}_i, \boldsymbol{\theta}^{(b)})$. The simulated samples $\{Y_i^{(1)}(1), \cdots, Y_i^{(B)}(1)\}$ and $\{Y_i^{(1)}(0), \cdots, Y_i^{(B)}(0)\}$ of predicted values then can be used to estimate the potential outcome for each x as an average over the posterior sample of parameters and the sample of observed covariates,

$$\hat{Y}(x) = \frac{1}{n}\sum_{i=1}^{n}\frac{1}{B}\sum_{b=1}^{B} Y_i^{(b)}(x)$$

for $x = 0$ or 1, and the causal effect estimator is $\hat{Y}(1) - \hat{Y}(0)$. A useful aspect of this construction is that, because a large sample representing the posterior predictive distribution of $Y(x)$ is available for each x, a 95% or 99% credible interval may be computed readily for the causal effect estimator.

If a predicted outcome is desired for a given x and fixed $\mathbf{Z} = \mathbf{z}$, corresponding to a future patient with a particular covariate vector treated with x, then, rather than averaging over the $\mathbf{Z}_i$'s, one may compute the mean of the *posterior predictive distribution*,

$$E(Y(x) \mid \mathbf{z}, \mathscr{D}_n) = E(Y \mid x, \mathbf{z}, \mathscr{D}_n) = \int_{\boldsymbol{\theta}} y\, p(y \mid x, \mathbf{z}, \boldsymbol{\theta}) p(\boldsymbol{\theta} \mid \mathscr{D}_n) d\boldsymbol{\theta}.$$

To compute this using the posterior sample described above, one may generate a sample $\{Y^{(1)}(x, \mathbf{z}), \cdots, Y^{(B)}(x, \mathbf{z})\}$ of predicted values from the B densities $p(Y \mid X = x, \mathbf{z}, \boldsymbol{\theta}^{(b)})$, $b = 1, \cdots, B$. This gives the estimator as the posterior predictive mean

$$\hat{Y}(x, \mathbf{z}) = B^{-1}\sum_{b=1}^{B} Y^{(b)}(x, \mathbf{z}),$$

and the sample of B predicted values also may be used to compute a 95% confidence interval.

Aside from specifying a prior and a robust parametric regression model, steps for implementing a Bayesian G computation are as follows. While this is given in terms of one stage of therapy and a single treatment, the notation can be generalized for a multistage dynamic treatment regime as done earlier in the frequentist case. The steps assume that the covariate distribution is estimated empirically.

Step 1 Fix treatment $X = 0$ or 1.

Step 2 Sample parameters $\boldsymbol{\theta}^{(1)}, \cdots, \boldsymbol{\theta}^{(B)}$ from the posterior $p(\boldsymbol{\theta} \mid \mathscr{D}_n)$.

Step 3 For each $\boldsymbol{\theta}^{(b)}$, $b = 1, \cdots, B$, and $\mathbf{Z}_i, i = 1, \cdots, n$, and x, simulate a predicted value $\hat{Y}_i^{(b)}(x)$ from the likelihood $p\{Y \mid X = x, \mathbf{Z}_i, \boldsymbol{\theta}^{(b)}\}$.

Step 4 Average the simulated sample $\{\hat{Y}_i^{(1)}(x), \cdots, \hat{Y}_i^{(B)}(x)\}$ of predicted values to obtain $\hat{Y}_i(x) = B^{-1}\sum_{b=1}^{B} Y_i^{(b)}(x)$, and average this over the empirical covariate distribution, giving each $\mathbf{Z}_i$ weight $1/n$, to obtain $\hat{Y}(x) = n^{-1}\sum_{i=1}^{n} \hat{Y}_i(x) = (nB)^{-1}\sum_{i=1}^{n}\sum_{b=1}^{B} Y_i^{(b)}(x)$.

Step 5 Repeat Steps 2 – 4 for each x = 0 and 1, and compute $\hat{Y}(1) - \hat{Y}(0)$ as the estimated treatment effect.

3

Precision Dose Optimization

3.1 Dose Finding Based on Time to Toxicity

Bayesian phase I and phase I-II clinical trial designs use sequentially adaptive rules based on the data from previously treated patients to choose doses, dose pairs of two-agent combinations, or (dose, schedule) combinations for successive patient cohorts. These designs also include rules to stop accrual to doses, combinations, or schedules that have been found to be unsafe or ineffective. In phase I, the outcome typically is a vector of grades for each of several qualitatively different types of toxicity, observed within a prespecified follow up period. Most often, these are combined by defining one binary indicator of unacceptably severe toxicity, conventionally called Dose Limiting Toxicity (DLT), or simply "Toxicity". For phase I-II trials, Toxicity and Efficacy each may be characterized by a variety of binary, ordinal, or time-to-event (TTE) variables. Many examples are given by Yuan et al. (2016). Most phase I trials are based on a binary indicator, Y, that toxicity occurs during a follow-up interval $[0, t_F]$, for some fixed time t_F. The value of t_F may vary widely depending on the clinical setting. For example, in a trial to evaluate rapid treatment of acute ischemic stroke, t_F may be 24 or 48 hours, with toxicity defined in terms of cerebral bleeding and edema. In most oncology settings, evaluating Y takes a nontrivial amount of time. An oncology trial of a targeted agent or cellular immunotherapy may require t_F = 30 or 60 days to evaluate both toxicity and anti-disease effect. In radiation therapy trials, t_F may be six months because adverse regimen-related events may not be seen until several months from the start of therapy.

If t_F is large relative to the trial's accrual rate, then a severe logistical problem arises when attempting to apply adaptive rules that use the (dose, Y) data from previous patients to choose doses for new patients. For example, suppose that the follow-up interval for evaluating whether toxicity occurs is $t_F = 10$ weeks, the accrual rate is one patient per week, and the cohort size is three. Ideally, the dose of the second cohort, consisting of patients 4, 5, and 6 in order of enrollment, is chosen using the (dose, Y) data from the first cohort of patients 1, 2, and 3. A patient's outcome can be scored as "Toxicity" ($Y = 1$) at the time that it occurs, but it can be scored as "No Toxicity" ($Y = 0$) only if the patient has been followed for 10 weeks without toxicity occurring. Consequently, the Y values of the first cohort of three patients may not be known until weeks 11, 12, or 13 of the trial. If so, an adaptive rule cannot be used to choose a dose for the patients in the second cohort at the times that they are.

DOI: 10.1201/9781003474258-3

accrued, which can be expected to be during weeks 4, 5, and 6. Moreover, beyond the second cohort, five more patients may be expected to be enrolled by week 11, so the problem worsens over time. It is not obvious how one should choose any of the new patients' doses adaptively based on previously treated patients' (dose, toxicity) data. This sometimes is called the "late onset toxicity" problem.

Various approaches have been proposed for constructing adaptive rules that can deal with late-onset toxicities in a way that is practical and gives a design with good properties, but unfortunately, most are unsatisfactory. One solution is simply to make each new cohort of patients wait until all previous patients' toxicity outcomes have been evaluated fully before choosing the new cohort's dose adaptively. From a clinical perspective, delaying treatment in this way is very undesirable, and often is unacceptable. A different solution is to treat each new patient without delay at the current recommended dose based on the completely observed (dose, Y) data. This may be very unsafe, however, since if a design has escalated to a dose that later, when more data are observed, will be determined to be unacceptably toxic then, using this "do not wait" rule, a large number of patients may be treated at that toxic dose. Another solution is to not wait and treat each new patient one dose level below the current recommended dose, which may be called "one dose level down" rule. Unfortunately, this rule may substantially increase the risk that patients will be treated with an ineffective dose, and thus have a lower chance of achieving a response. Thall et al. (1999) proposed a very simple approach, called the "look ahead" method, wherein if all possible outcomes of patients treated but not yet fully evaluated will not alter the adaptive rule's decision, then that decision is made immediately in order to allow new patients to be treated more quickly. In practice, the look-ahead rule is of little use early in the trial, but may become more useful later when a sufficient amount of data have been accumulated. Bekele et al. (2008) proposed a Bayesian method to protect patient safety based on predictive probabilities of toxicity, but simulations showed that this method is impractical because it requires accrual to be suspended repeatedly. Finally, one may treat newly accrued patients off protocol, rather than making them wait so that an adaptive rule can be applied, but this approach may produce an unacceptably long trial duration. A discussion of these approaches is given by Jin et al. (2014).

Cheung and Chappell (2000) proposed the *time-to-event continual reassessment method* (TiTE-CRM) for phase I trials based on time to toxicity, rather than a binary outcome. The TiTE-CRM was the first practical solution to the late-onset toxicity problem in phase I trials. It is a useful generalization of the CRM (O'Quigley et al., 1990), which assumes that a binary toxicity outcome can be observed very quickly. The TiTE-CRM is much more realistic because it uses possibly right-censored time-to-toxicity data from patients who have been treated but whose toxicity outcomes may or may not have been fully evaluated at the time a new patient is accrued. Each patient's outcome data at a given evaluation consist of either the time at which toxicity occurred, or the patient's time without toxicity, up to a specified maximum follow-up time t_F. For example, if $t_F = 10$ weeks and it is known that a patient has gone 8 weeks without toxicity, this information is used by the TiTE-CRM. Intuitively, the patient's dose is given partial credit for the patient lasting 8 weeks without toxicity out of the 10-week follow-up period. In contrast, at 8 weeks the CRM would consider this patient's outcome to be not yet evaluated.

To account for both patient time and calendar (trial) time, let $u(t) \in [0,\ t_F]$ be the patient's follow-up time at any calendar time t during the trial when a dose must be chosen for a newly enrolled patient. Let $Y(u(t))$ be the indicator that the patient has toxicity by follow-up time $u(t)$. Thus, $Y(t_F) = 1$ if the patient experienced toxicity at any time during the follow-up interval $[0, t_F]$, and 0 if not. At trial time t, denote the accrual time of the ith patient by $a_i \leq t$, so the patient's follow-up time is $u_i(t) = \min\{t - a_i, t_F\}$. Since $Y_i(u_i(t))$ is the indicator that patient i has toxicity by trial time t, $Y_i(t_F) = Y_i = 1$ indicates that the ith patient has toxicity by follow-up time t_F. This is precisely the data structure used to construct the Kaplan and Meier (1958) estimator of a time-to-event survival function, based on data with maximum follow-up time t_F for each patient.

Denote the standardized dose by x, the model parameter vector by $\boldsymbol{\theta}$, and the toxicity probability for patients treated with dose x by $\pi(x, \boldsymbol{\theta}) = \Pr(Y(t_F) = 1 \mid x, \boldsymbol{\theta})$. For example, given raw doses $d_1 < d_2 < \cdots < d_K$ studied in the trial, one may define standardized doses as $x_j = (d_j - \bar{\boldsymbol{d}})/s(\boldsymbol{d})$, for $j = 1, \cdots, K$, denoting the mean dose by $\bar{\boldsymbol{d}}$ and the standard deviation by $s(\boldsymbol{d})$. There are various other methods for standardizing doses, such as $x_j = d_j/d_K$. Indexing patients in order of enrollment by $i = 1, \cdots, n$, denote the available data from the trial at trial time t by $\mathscr{D}_t$. Cheung and Chappell (2000) defined the *working likelihood function*

$$\mathscr{L}^w(\boldsymbol{\theta} \mid \mathscr{D}_t) = \prod_{i=1}^{n} \pi(x_{[i]}, \boldsymbol{\theta})^{Y_i(u_i(t))} \left\{1 - \frac{u_i(t)}{t_F}\pi(x_{[i]}, \boldsymbol{\theta})\right\}^{1 - Y_i(u_i(t))}$$

where $x_{[i]}$ denotes the ith patient's standardized dose. The term $u_i(t)/t_F$ in $\mathscr{L}^w$ down-weights $\pi(x_{[i]}, \boldsymbol{\theta})$ for the working likelihood contribution of a patient who has not yet experienced toxicity, that is, for whom $Y_i(u_i(t)) = 0$. At full follow-up $u_i(t) = t_F$, if $Y_i(t_F) = 0$, i.e. the patient did not experience toxicity by the end of follow-up, then the likelihood contribution is the usual term $1 - \pi(x_{[i]}, \boldsymbol{\theta})$. If, for example, a patient has been followed for time $t_F/2$ without toxicity, then the patient's working likelihood contribution is $1 - \frac{1}{2}\pi(x_{[i]}, \boldsymbol{\theta})$, which is larger than $(1 - \pi(x_{[i]}, \boldsymbol{\theta}))$, so in this sense the working likelihood gives an increasingly larger partial credit for a patient going through the follow-up period without toxicity.

Any reasonable parsimonious parametric function for π may be used. Defining the linear term $\eta(x, \boldsymbol{\theta}) = \alpha + e^{\beta}x$, for real-valued α and β, and $\boldsymbol{\theta} = (\alpha, \beta)$, two commonly used regression models are the logistic,

$$\pi(x, \boldsymbol{\theta}) = \frac{\exp\{\eta(x, \boldsymbol{\theta})\}}{1 + \exp\{\eta(x, \boldsymbol{\theta})\}},$$

and the probit,

$$\pi(x, \boldsymbol{\theta}) = \Phi(\eta(x, \boldsymbol{\theta})),$$

where $\Phi(\cdot)$ denotes the standard normal cdf.

The TiTE-CRM exploits the fact that $\Pr(Y_i(t_F) = 1 \mid u_i(t), Y_i(u_i(t)) = 0)$ decreases monotonically to 0 as $u_i(t)$ goes from 0 to t_F. The longer that a patient has gone without toxicity during the follow-up period, the larger the probability becomes that the patient will not experience toxicity by t_F. To see why this is true, for simplicity suppress i and t, and denote $\tilde{Y}$ = the time to toxicity, $\tilde{Y}^o = \min\{\tilde{Y}, u\}$ = the observed time to toxicity or last follow-up, and $1 - \pi_{t_F} = \Pr(\tilde{Y} > t_F)$ = the probability of no toxicity by the follow-up time t_F. For $0 \le u \le t_F$,

$$\begin{aligned} Pr(\tilde{Y} \le t_F \mid \tilde{Y}^o = u \le t_F) &= \frac{Pr(\tilde{Y} \le t_F \text{ and } \min\{\tilde{Y}, u\} = u \le C)}{Pr(\min\{\tilde{Y}, u\} = u \le t_F)} \\ &= \frac{Pr(u \le \tilde{Y} \le t_F)}{Pr(u \le \tilde{Y})} = 1 - \frac{1 - \pi_{t_F}}{Pr(u \le \tilde{Y})}. \end{aligned}$$

As u increases from 0 to t_F, $Pr(u \le \tilde{Y})$ decreases from 1 to $1 - \pi_{t_F}$, so the conditional probability must decrease to 0 at $\tilde{Y}^o = u = t_F$.

Because the TiTE-CRM uses right-censored time-to-toxicity, it has greatly superior properties compared to phase I designs that reduce toxicity to a binary variable. A discussion of the TiTE-CRM is given by Cheung (2011). A very useful extension of the TiTE-CRM is the late-onset Efficacy-Toxicity (LO-ET) design proposed by Jin et al. (2014). The LO-ET design does dose-finding based on both time-to-toxicity and time-to-efficacy. The LO-ET design extends the phase I-II EffTox design of Thall and Cook (2004) by replacing binary response and toxicity with time-to-event outcomes subject to administrative right censoring. It also deals with the difficult case where the efficacy outcome is not an event time, but rather is a binary response indicator observed at the end of a fixed follow-up period, such as 90 days.

A major limitation of the TiTE-CRM is that it does not account for settings where patients may be heterogeneous, with different dose-toxicity functions that depend on prognostic subgroups. To address this issue, Chapple and Thall (2018) proposed a phase I dose-finding design, called Sub-TiTE, that generalizes the TiTE-CRM to accommodate prespecified patient subgroups by allowing different dose-toxicity hazard functions within subgroups. This provides a basis for assigning possibly different "optimal" doses, at the same point in the trial, to patients who belong to different subgroups. Additionally, rather than fixing the pre-specified subgroups, the Bayesian model underlying Sub-TiTE includes latent subgroup membership variables that are used to adaptively combine subgroups found to have similar dose-toxicity relationships. The latent subgroup-based methodology also allows the possibility of later re-splitting previously combined subgroups, if subsequent data show that they are different. This adaptive process will be described below. Computer simulations, also summarized below, that compare Sub-TiTE to the TiTE-CRM, show that, if the actual dose-toxicity relationships differ between subgroups, then ignoring the heterogeneity and applying the TiTE-CRM to select one nominally "optimal" dose for all patients often is likely to choose a greatly suboptimal dose in one or more subgroups. A different, very simple alternative approach is to use the TiTE-CRM to conduct a separate trial within each subgroup. This is very inefficient, however, due to small subgroup sample sizes, and it is a particularly bad approach if two or more subgroups are truly homogeneous with the same dose-toxicity curves, since it does not borrow strength between similar subgroups. Sub-TiTE addresses all of these problems.

The following Bayesian dose-toxicity model provides a basis for adaptively accounting for subgroups during a trial, and assigning subgroup-specific doses. It includes a latent variable structure that is a prototype for any setting where one wishes to account for possible subgroup-specific treatment effects on one or more outcome variables, and do adaptive decision-making on that basis. Index G prespecified subgroups by $\mathscr{G} = \{0, 1, ..., G-1\}$, where subgroup $g = 0$ will be used as a baseline. Denote the ith patient's subgroup by $g_i \in \mathscr{G}$. Denote the probability that the ith patient has toxicity by follow-up time t_F by

$$\pi_i(x_{[i]}, g_i, \boldsymbol{\theta}) = \text{logit}^{-1}\{\eta_i(x_[i], g_i, \boldsymbol{\theta})\},$$

where the linear term is assumed to take the form

$$\eta_i(x_{[i]}, g_i, \boldsymbol{\theta}) = \alpha + \sum_{g=1}^{G-1} \alpha_g I(g_i = g) + \exp\left\{\beta + \sum_{g=1}^{G-1} \beta_g I(g_i = g)\right\} x_{[i]}.$$

The parameter vector is

$$\boldsymbol{\theta} = (\alpha, \alpha_1, \cdots, \alpha_{G-1}, \beta, \beta_1, \cdots, \beta_{G-1}),$$

and all $2G$ entries of $\boldsymbol{\theta}$ are real-valued. Under this definition, $\pi_i(x_{[i]}, 0, \boldsymbol{\theta}) = \text{logit}^{-1}(\alpha + e^{\beta} x_{[i]})$ for the baseline subgroup. The subgroup-specific intercept parameters $\{\alpha_g,\ g = 1, \cdots, G-1\}$ are deviations from the baseline subgroup intercept α_0, and the subgroup-specific dose effects $\{\beta_g,\ g = 1, \cdots, G-1\}$ are deviations from the baseline dose effect β. Denote the dataset at trial time t by

$$\mathscr{D}_t = \left\{(x_{[i]}, g_i, u_i(t), Y_i(u_i(t))) : i = 1, \cdots, n(t)\right\}.$$

Extending Cheung and Chappell (2000), a *generalized working likelihood* based on data from $n(t)$ patients at trial time t is given by

$$\mathscr{L}^w(\boldsymbol{\theta} \mid \mathscr{D}_t) = \prod_{i=1}^{n(t)} \pi(x_{[i]}, g_i, \boldsymbol{\theta})^{Y_i(u_i(t))} \left\{1 - \frac{u_i(t)}{t_F} \pi(x_{[i]}, g_i, \boldsymbol{\theta})\right\}^{1-Y_i(u_i(t))}.$$

This provides a basis for subgroup-specific dose-finding based on possibly right-censored time-to-toxicity data.

Subgroup Clustering Model and Algorithm

While it is useful to account for patient heterogeneity in terms of prespecified subgroups, it may be the case that the dose-toxicity functions are the same for two or more subgroups. To address this possibility, adaptive clustering of similar subgroups is done using a two-level hierarchical prior on $\boldsymbol{\theta}$. For each $g = 1, \cdots, G-1$, a *latent subgroup membership variable* ζ_g is defined that takes on a value in $\mathscr{G}$. For α and β, corresponding to the baseline subgroup $g = 0$, normal priors are assumed,

$\alpha \sim N(\tilde{\alpha}_0, \tilde{\sigma}_\alpha)$ and $\beta \sim N(\tilde{\beta}_0, \tilde{\sigma}_\beta)$, where $\tilde{\sigma}_\alpha$ and $\tilde{\sigma}_\beta$ are suitably large so that the priors are non-informative. For each $g = 1, \cdots, G-1$, defining the priors of the α_g's and β_g's requires the point mass probability function $\delta_\xi(\omega) = 1$ if $\omega = \xi$ and 0 if $\omega \neq \xi$. The level 1 priors are defined conditionally, given the ζ_g's, as the spike-and-slab mixtures

$$\alpha_g \mid \zeta_g \sim I(\zeta_g = g) N(\tilde{\alpha}_g, \tilde{\sigma}_\alpha) + I(\zeta_g \neq g) \sum_{m=0}^{G-1} \delta_{\alpha_m}(\alpha_g) I(\zeta_g = m)$$

and

$$\beta_g \mid \zeta_g \sim I(\zeta_g = g) N(\tilde{\beta}_g, \tilde{\sigma}_\beta) + I(\zeta_g \neq g) \sum_{m=0}^{G-1} \delta_{\beta_m}(\beta_g) I(\zeta_g = m).$$

The "spikes" of these priors are the probability point masses of 1 on the α_g's and β_g's, and the "slabs" are the normal distributions.

As a conceptual device, the clustering algorithm may be thought of as being directional, in the sense that one subgroup may be clustered to another subgroup. By thinking of the subgroups as houses, it may be called the *House Party Algorithm* for clustering. If subgroup g is clustered to subgroup $m \neq g$, one may think of g as going to a party at the house of m, so both g and m are together at a party in house m, and house g becomes empty. Additionally, no subgroup may go to (be clustered to) an empty house, since every party must have a host. Since the algorithm allows the possibility of re-splitting subgroups that previously were combined, this means that someone may attend a party but later leave. To see how the latent variables are used to implement the House Party algorithm, for each $g = 1, \cdots, G-1$, if the latent subgroup membership variable $\zeta_g = m$ for some subgroup $m \neq g$, then (1) subgroup g is clustered to subgroup m, and (2) $\alpha_g = \alpha_m$ and $\beta_g = \beta_m$ with probability 1. As an illustration, for $G = 3$, in a step of the MCMC suppose that $\zeta_1 = \zeta_2 = 1$. Then $E(\beta_1) = \tilde{\beta}_1$, the original prior mean, but $\beta_2 = \beta_1$. If $\zeta_1 = \zeta_2 = 2$, then the roles of the two parameters are reversed with $E(\beta_2) = \tilde{\beta}_2$, and $\beta_1 = \beta_2$. If $\zeta_g = 0$, then $\alpha_g = \beta_g = 0$, and subgroup g is clustered to the baseline subgroup 0. For convenience, the baseline subgroup 0 is not clustered with any other subgroup. That is, a party may be held at the baseline subgroup's house, but subgroup 0 cannot attend a party at any other house. Define the random subgroup index set $S = \{m : \zeta_m = m\} \cup \{0\}$. If a latent subgroup membership variable $\zeta_g \neq g$, implying that subgroup g has been clustered to another subgroup, then its index g is not in S. To account for the fact that some subgroups may be clustered to other subgroups, denoting the cardinality of S by $|S|$, this may vary from $|S| = G$ if no subgroups have been clustered (complete heterogeneity) to $|S| = 1$ if all subgroups have been clustered into one subgroup (complete homogeneity), which for convenience is indexed by 0. The level 2 prior is given by $\Pr(\zeta_g = g) = .9$, with the remaining probability mass .1 distributed uniformly over the other indices in S, formally $\Pr(\zeta_g = m) = .1/|S|$ for each $m \in S$ with $m \neq g$. For example, if $G = 3$ and $\zeta_1 = 1$ to start, then $\Pr(\zeta_2 = 2) = .9$, and $\Pr(\zeta_2 = 1) = \Pr(\zeta_2 = 0) = .1/2 = .05$. Similarly, if $\zeta_2 = 2$ then $\Pr(\zeta_1 = 1) = .9$ while $\Pr(\zeta_1 = 2) = \Pr(\zeta_1 = 0) = .05$. If $\zeta_1 = 0$ then $\Pr(\zeta_2 = 1) = 0$, $\Pr(\zeta_2 = 2) = .9$, and $\Pr(\zeta_2 = 0) = .1$.

During the MCMC posterior computations, to implement the House Party Algorithm, several different types of clustering move proposals may be made using the latent variables. The following MCMC proposal steps underscore the facts that the cluster configuration is random, and the subgroup clusters may change adaptively during the MCMC.

MCMC Proposal Steps

1. Cluster All Subgroups Together: Propose setting all $\zeta_g = m$ for some randomly chosen m, with all $\boldsymbol{\theta}_g = \boldsymbol{\theta}_m$, so that the G subgroups are homogeneous. Thus, all subgroups attend the party at the house of subgroup m. This is the "One Big Party For Everybody" proposal.

2. Uncluster All Subgroups: Propose setting all $\zeta_g = g$ with all $\boldsymbol{\theta}_g$ distinct. This creates a completely heterogeneous trial with G subgroups. In this case, the model still borrows strength across subgroups through α and β. In this case, no subgroup attends a party at any other subgroup's house. This is the "No Parties At All" proposal.

3. Randomly Draw One g From $\{0,..,G-1\}$: If its latent subgroup equals it's subgroup, $\zeta_g = g$, then propose clustering this to another random subgroup $m \neq g$ such that $\zeta_m = m$, by setting $\boldsymbol{\theta}_g = \boldsymbol{\theta}_m$ and $\zeta_g = m$. Otherwise, propose creating its own subgroup, by setting $\zeta_g = g$ and randomly choosing $\boldsymbol{\theta}_g$ around its previous value in the MCMC chain. This is the "Attend Any Party" proposal

4. Switch From One Cluster To Another: For g such that $\zeta_g \neq g$, randomly choose a subgroup k such that $\zeta_k = k$ and propose setting $\boldsymbol{\theta}_g = \boldsymbol{\theta}_k$. This is like g leaving one party to attend a different party at the house of subgroup k. This is the "Party Switching" proposal.

To see how subset clustering and re-splitting using the House Party algorithm may work in practice, Figure 3.1 illustrates a case in which $G = 4$ subgroups are specified initially, and the model in subgroup g is parameterized by θ_g. For simplicity, in this illustration, the subgroup indices are $\{1, 2, 3, 4\}$ and there is no subgroup 0. The successive MCMC iterations are labeled 1, 2, 3, and 4, but the illustrative subgroup configurations could occur at any consecutive points in the MCMC computation process. At iteration 2, since $\zeta_1 = \zeta_2 = 2$, subgroup 1 is clustered into subgroup 2, and there are three induced subgroups $\{1,2\}$, $\{3\}$, and $\{4\}$, with $S = \{2, 3, 4\}$. At iteration 3, since $\zeta_1 = \zeta_2 = \zeta_3 = 2$, subgroups 1 and 3 are clustered into subgroup 2, and there are two induced subgroups $\{1,2,3\}$ and $\{4\}$ with $S = \{2, 4\}$. At the fourth iteration, subgroup 1 has been removed from the cluster in subgroup 2 since $\zeta_1 = 1$, so the cluster $\{1, 2, 3\}$ is split into the two clusters $\{1\}$ and $\{2, 3\}$, the three induced subgroups are $\{1\}$, $\{2, 3\}$, and $\{4\}$, and $S = \{1, 2, 4\}$.

To implement SubTiTE for trial design and conduct, one must specify the doses, the subgroups, the follow up time t_F for observing toxicity, a fixed target toxicity probability π_g^* in each subgroup g, although these need not be different values, and a maximum overall sample size. A methodology for determining numerical hyperparameters for the prior from elicited toxicity probabilities as functions of dose and

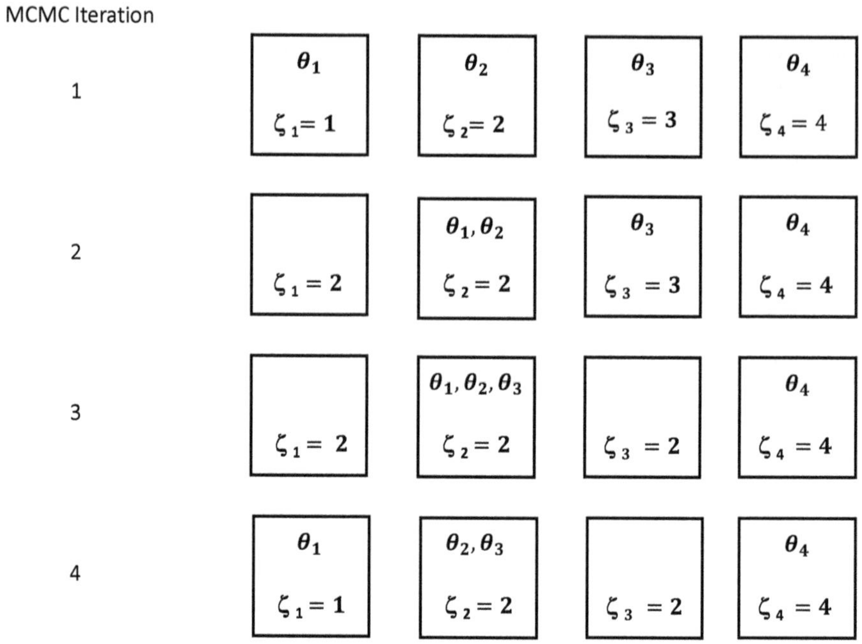

FIGURE 3.1
Possible sequential clusterings of $G = 4$ subgroups by the latent variables $\{\zeta_1, \zeta_2, \zeta_3, \zeta_4\}$, in four consecutive MCMC iterations. The model parameter vector for subgroup g is denoted by θ_g.

subgroup is given in Section 3 of Chapple and Thall (2018). For trial conduct, in each subgroup a starting dose is specified and an untried dose may not be skipped when escalating. In subgroup g, for fixed target toxicity probability π_g^* and current data $\mathscr{D}_t$ from the trial, the optimal dose is defined as that having posterior mean toxicity probability closest to π_g^*, formally

$$x_g^{opt} = \text{argmin}_{x_j} \left| \hat{\pi}(x_j, g, \mathscr{D}_t) - \pi_g^* \right|$$

where $\hat{\pi}(x_j, g, \mathscr{D}_t) = \text{E}\{\pi(x_j, g, \boldsymbol{\theta}) \mid \mathscr{D}_t\}$. To protect patient safety, this dose optimality criterion is superseded by a *subgroup-specific dose acceptability rule* that declares a dose x_j to be *unacceptably toxic in subgroup* g if

$$Pr\{\pi(x_j, g, \boldsymbol{\theta}) > \pi_g^* \mid \mathscr{D}_t\} > p_U \tag{3.1}$$

for some fixed upper probability cut-off p_U, such as .90 or .95. The rule (3.1) is similar to the subgroup-specific stopping rules given by Wathen et al. (2008) for futility and Lee et al. (2023) for toxicity. In subgroup g, if at least three patients have been treated at the lowest dose x_1 and their outcomes fully evaluated, then if (3.1) says that x_1 is unacceptable accrual to subgroup g is terminated. In each subgroup still open at the end of the trial, the optimal dose for that subgroup is selected based on the posterior computed from the final data.

Chapple and Thall (2018) provided a computer simulation study in which Sub-TiTE was compared to (1) using the TiTE-CRM while ignoring subgroups, and (2) using the TITE-CRM separately within each subgroup, called Sep-TITE. The simulation design was a stylized version of a phase I radiation therapy (RT) trial for advanced non-small cell lung cancer, with $G = 2$ subgroups corresponding to the radiation modalities proton beam (g = 0) and conventional non–proton beam (g = 1). Five RT doses 10, 20, 30, 50, and 70 Gy were studied, with an evaluation period t_F = 6 months, and the same target π^* = .30 was used for both subgroups. The three dose-finding methods were compared using two subgroup-specific criteria, computed for each g. The first was $Psel_g$ = proportion of simulated trials in which the truly optimal dose in subgroup g was selected. The second was the subgroup-specific average over the B simulated trials of the absolute deviations between the true toxicity probability of the selected optimal dose x_g^b in subgroup g, denoted by $\pi^{true}(x_g^b, g)$ for simulated trial $b = 1, \cdots, B$, and $\pi_g(x_g^{opt})$ = the best possible dose in subgroup g, i.e the dose in that subgroup with true toxicity probability closest to π^*. This criterion is given by

$$\Delta_g = \frac{1}{B}\sum_{b=1}^{B} \left|\pi^{true}(x_g^b, g) - \pi_g(x_g^{opt})\right|. \tag{3.2}$$

For each g, a larger value of $Psel_g$ and a smaller value of Δ_g are more desirable, with $\Delta_g = 0$ corresponding to the best possible dose x_g^{opt} in subgroup g always being chosen.

To illustrate the behavior of Sub-TiTE and compare it to Sep-TiTE and the TiTE-CRM, an abbreviated version of Table 3.2 of Chapple and Thall (2018) is given here in Table 3.1. In these simulations, there were $G = 2$ subgroups and five doses to mimic the RT trial. The simulation scenarios represented a wide variety of possible subgroup-specific dose-toxicity probability functions that may be seen in practice. For example, in scenario 5 the lowest dose is unacceptably toxic for subgroup 0, with a true toxicity probability 0.50, while the lowest dose is optimal in subgroup 1.

Table 3.1 shows that, in scenario 1, Sep-TITE and TiTE-CRM pick the optimal dose for subgroup 0 most often, but Sub-TITE picks the optimal dose for subgroup 1 with a much higher probability. In scenario 2, the optimal dose for subgroup 0 is picked by Sep-TITE most often, but Sub-TITE performs best for subgroup 1. The TiTE-CRM design does a very poor job of picking the optimal dose for both subgroups in scenario 2, because it ignores the patient heterogeneity. In scenario 3, TiTE-CRM picks the optimal dose for subgroup 0 with the highest probability, slightly higher than Sub-TITE, but for subgroup 1, TiTE-CRM performs very poorly. When considering both subgroups, the total probability of selecting the optimal dose for the two subgroups, $Psel_0 + Psel_1$, is highest for Sub-TITE. While the TITE-CRM is best in scenario 4, where the subgroups are truly homogeneous, when there is heterogeneity, it performs poorly in one or both subgroups. In several cases, Sep-TiTE performs as well as or better than Sub-TiTE in one subgroup, but has greatly inferior performance in the other subgroup.

TABLE 3.1
Simulation results for a trial with $G = 2$ subgroups. In Scenario 4 the two subgroups are homogeneous, and in Scenario 5 no dose is acceptable for subgroup 0. In subgroup g, $Psel_g$ = probability of selecting the truly optimal dose, and Δ_g = mean absolute difference between the π_g's of the selected dose and the optimal dose.

	Sub-TiTE		Sep-TiTE		TiTE-CRM	
g	$Psel_g$	Δ_g	$Psel_g$	Δ_g	$Psel_g$	Δ_g
			Scenario 1			
0	.53	.07	.59	.06	.59	.08
1	.64	.06	.47	.09	.39	.09
			Scenario 2			
0	.86	.04	.88	.04	.42	.19
1	.46	.06	.39	.07	.02	.14
			Scenario 3			
0	.75	.05	.60	.09	.79	.05
1	.74	.05	.74	.05	.19	.14
			Scenario 4			
0	.75	.04	.61	.07	.77	.04
1	.76	.04	.54	.08	.77	.04
			Scenario 5			
0	–	–	–	–	–	–
1	.84	.04	.85	.04	.43	.17

To examine how the adaptive subgroup combination and re-splitting of Sub-TiTE performs in a more complex setting, results from a second simulation study given by Chapple and Thall (2018) are summarized here in Table 3.2. In scenario 1, all subgroups are heterogeneous. The other extreme is given by scenario 4, where the four subgroups are homogeneous. General conclusions are that, when two, three, or four of $G = 4$ subgroups are homogeneous, Sub-TITE accurately combines these subgroups, and Sub-TiTE greatly outperforms both Sep-TITE and the TiTE-CRM in nearly all cases where there is between-subgroup heterogeneity. In an additional robustness study, Chapple and Thall (2018) simulated Sub-TiTE for time-to toxicity following a Weibull, exponential, uniform, or lognormal distribution, and found that its OCs change very little for the different distributions.

Computer Software A SAS computer package for design and implementation of the TiTE-CRM design is *Tite-CRM Version 8.1*, available online from the University of Michigan Center for Cancer Biostatistics. A computer package, *SubTite*, that implements the Sub-Tite design is available on CRAN.

TABLE 3.2
Simulation results for a trial with $G = 4$ subgroups. In Scenario 4 all subgroups are homogeneous. In subgroup g, $Psel_g$ = probability of selecting the truly optimal dose, and Δ_g = mean absolute difference between the π_g's of the selected dose and the optimal dose.

	Sub-TiTE		Sep-TiTE		TiTE-CRM	
g	$Psel_g$	Δ_g	$Psel_g$	Δ_g	$Psel_g$	Δ_g
			Scenario 1			
0	.46	.09	.53	.07	.23	.08
1	.60	.07	.43	.10	.77	.04
2	.55	.06	.45	.08	.18	.09
3	.53	.05	.61	.06	.00	.09
			Scenario 2			
0	.80	.07	.84	.05	.01	.38
1	.48	.06	.34	.08	.06	.41
2	.45	.06	.37	.08	.41	.06
3	.58	.07	.55	.08	.18	.09
			Scenario 3			
0	.68	.07	.57	.13	.94	.01
1	.67	.06	.73	.05	.17	.14
2	.72	.06	.55	.10	.94	.01
3	.75	.05	.56	.09	.94	.01
			Scenario 4			
0	.67	.05	.53	.08	.90	.02
1	.68	.06	.52	.09	.90	.02
2	.69	.05	.53	.08	.90	.02
3	.69	.05	.54	.08	.90	.02

3.2 Choosing Estimands

Dose finding relies on between-dose comparisons using estimated values of an estimand that has been chosen as the decision criterion for evaluating and comparing doses. When accounting for patient subgroups, an important aspect of this process is that different estimands may give different answers to the question of how much better one dose is compared to another within each subgroup. This is illustrated by the following examples.

First, consider the simple problem of comparing two doses, d=1 and d=2, in each of two prognostic subgroups, g = favorable (fav) or poor prognosis. To provide a medical context, suppose that a phase I-II dose-finding trial of a new investigational drug as salvage therapy for metastatic renal cancer has been completed, and the goal is to choose between the two doses in each subgroup based on both Y_D = time to death or Y_T = time to toxicity. For each outcome $j = D, T$ and each of the four the (d,g) pairs, assume that $[Y_j \mid d,g]$ follows an exponential distribution with hazard $h_j(d,g)$, cdf $F_j(t \mid d,g) = \Pr(Y_j \leq t \mid d,g)$, survivor function $\overline{F}_j(t \mid d,g) = 1 - F_j(t \mid d,g)$, and mean $\mu_j(d,g) = 1/h_j(d,g)$. Denote the $d = 2$ to $d = 1$ hazard ratio to compare doses in subgroup g by $HR_j(g) = h_j(2,g)/h_j(1,g)$.

Suppose that $d = 2$ yields greater benefit than $d = 1$ for both subgroups in terms of overall survival time, with estimates $\widehat{HR}_D(fav) = \widehat{HR}_D(poor) = .5$. Thus, subgroup has no effect on the HR for death, with $d = 2$ uniformly better than $d = 1$. Next, suppose that $d = 2$ is more likely than $d = 1$ to produce toxicities, with HR estimates $\widehat{HR}_T(fav) = \widehat{HR}_T(poor) = 1.33$, and this is stable across the subgroups. This implies that, for both survival time and time-to-toxicity, subgroup is a prognostic variable and is not predictive.

Figure 3.2 shows the survival probability and cdf curves for Y_D and Y_T, stratified by prognostic subgroups. Since the exponential distribution has median $\tilde{\mu} = \ln(2)\ \mu = .693\ \mu$, if patients with favorable prognosis have estimated median survival $\tilde{\mu}_D(1, fav) = 17.3$ months when treated with $d = 1$, then $\widehat{HR}_D(fav) = .5$ implies that the estimated median survival with $d = 2$ is approximately 34.7 months. This gives the difference between medians $\tilde{\mu}_D(2, fav)$ - $\tilde{\mu}_D(1, fav)$ = 34.7 - 17.3 = 17.4 months, and difference between means $\mu_D(2, fav)$ - $\mu_D(1, fav)$ = 50 - 25 = 25 months for favorable risk patients. Patients with poor prognosis have median survival time $\tilde{\mu}_D(2, poor) = 4$ months with $d = 2$ and $\tilde{\mu}_D(1, poor) = 2$ months with $d = 1$, for a difference in median survival times of 4 - 2 = 2 months, and difference in mean survival 5.8 - 2.9 = 2.9 months. This shows that, in terms of either mean or median survival time, the benefit of dose 2 over dose 1 is far greater for patients with favorable risk than poor risk. This is illustrated in the top row of Figure 3.2.

If comparisons are based on survival probabilities beyond a given milestone time point, however, the opposite conclusion is reached. The survival probability difference between $d = 2$ and $d = 1$ in subgroup g at 3 months is

$$\overline{F}_D(3 \mid 2, g) - \overline{F}_D(3 \mid 1, g).$$

While survival probabilities and hazards may seem to convey the same information under an exponential distribution, since

$$\overline{F}_D(t \mid d, g) = \exp\{-t\ h_D(d,g)\},$$

the following numerical example shows that this is not true. The 3-month survival probability difference for favorable-risk patients is

$$\overline{F}_D(3 \mid 2, fav) - \overline{F}_D(3 \mid 1, fav) = .942 - .887 = .055,$$

while this difference for poor-risk patients is

$$\overline{F}_D(3 \mid 2, poor) - \overline{F}_D(3 \mid 1, poor) = .592 - .350 = .242.$$

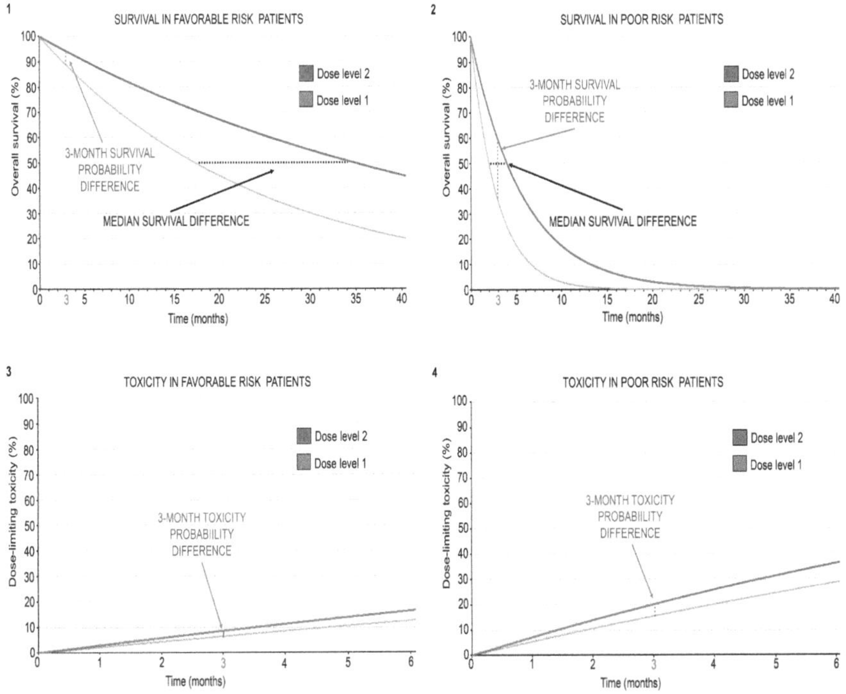

FIGURE 3.2
For dose levels $d = 2$ (red) and $d = 1$ (blue), survival curves of favorable risk (1) and poor risk (2) patients, and cumulative distribution functions for time to toxicity of favorable risk (3) and poor risk (4) patients,

Thus, in terms of 3-month survival probabilities, the benefit of d = 2 over d = 1 is much larger for poor-risk patients than for favorable patients. This numerical example illustrates the fact that, if one wishes to account for each patient's subgroup when comparing doses in order to choose a best personalized dose, how "best" is defined depends on the estimand used to evaluate performance, since different estimands can give different answers for the same endpoint, such as survival time.

These ideas provide a conceptual framework for constructing a utility function. The functional form of U should reflect the scale of interest, such as differences between means or survival probabilities at a milestone time point. The examples also show that, ideally, U should be tailored to account for key covariates or subgroups, since numerical between-dose effects may differ dramatically for different patient subgroups. This requires a family of utility functions indexed by patent subgroups, $\mathscr{G} = \{1, \cdots, G\}$. Denoting the utility of outcome $\boldsymbol{Y}$ in subgroup g by $U(\boldsymbol{Y}, g)$, the set $\{U(\boldsymbol{Y}, g), g \in \mathscr{G}\}$ is a *precision utility function family* (PUFF).

The phase I-II design in metastatic renal cancer described in the next section has outcomes Y_T = time to toxicity and ordinal categorical Y_E (Lee et al., 2021). To accommodate $\boldsymbol{Y} = (Y_E, Y_T)$, a subgroup-specific utility function was constructed. For subgroup $g = 1, \cdots, G$, denote the utility of each outcome $j = E, T$ in subgroup g by $U_j(Y_j, g)$. For each g, $U_{\rm T}(y_T, g)$ first was established, then $U_{\rm E}(y_E, g)$ was elicited for each level $y_E = 0, 1, \cdots, K-1$ of Y_E, and re-scaled to have values on the same numerical domain as $U_{\rm T}(y_T, g)$. The overall utility then was defined as the sum

$$U(\boldsymbol{Y}, g) = U_{\rm T}(Y_{\rm T}, g) + U_{\rm E}(Y_{\rm E}, g).$$

Utility construction was begun by eliciting the numerical utility, $U_{\rm T,max}$, of not observing toxicity during the follow up period $[0,\ t_F]$, and defining the parametric function

$$U_{\rm T}(Y_{\rm T}, g) \quad = \quad \begin{cases} U_{\rm T,max}\left(\frac{Y_{\rm T}}{t_F}\right)^{a_g} & \text{if } \ Y_{\rm T} < t_F, \\ U_{\rm T,max} & \text{if } \ Y_{\rm T} \geq t_F. \end{cases}$$

This function increases with Y_T over the follow-up period, from $U_{\rm T}(0, g) = 0$ at $Y_{\rm T} = 0$, i.e. immediate toxicity, to its maximum $U_{\rm T}(t_F, g) = U_{\rm T,max}$ at $Y_{\rm T} = t_F$, i.e. no toxicity by the end of follow up. The subgroup-specific shape parameter $a_g > 0$ may be determined during the utility elicitation process by showing several plots of $U_{\rm T}(y, g)$ as a function of both y and g on the interval $[0,\ t_F]$ to the physician(s) providing the utility, for each of a set of candidate a_g values. For the renal cancer trial, the elicited values were $U_{\rm T,max} = 140$ and $U_{\rm T}(70, 1) = U_{\rm T}(42, 2) = U_{\rm T}(28, 3) = 70$. Thus, toxicity at day 70 for Good prognosis, at day 42 for Intermediate prognosis, and at day 28 for Poor prognosis had identical utilities. The resulting shape parameters are $a_1 = 3.80$, $a_2 = 1.00$, and $a_3 = 0.63$. The functions $\{U_{\rm T}(y, g),\ g = 1, 2, 3\}$ are illustrated in Figure 3.3, which shows that, within each subgroup, an early occurrence of toxicity has a lower utility than a later occurrence. Toxicity-versus-no toxicity utility differences are largest for patients with a Favorable prognosis. A dose with a high toxicity probability is less likely to be optimal for Favorable prognosis patients than for Intermediate or Poor prognosis patients.

The efficacy utility $U_E(Y_E, g)$ was constructed by asking the following questions:

1. How much is the utility U_T penalized for the occurrence of PD?

2. Compared to the utility U_T when PD occurs, how much is U_T increased for SD, PR, and CR?

For the first question, the answer was to reduce U_T by half for PD, since this offsets the benefit obtained by the time without toxicity, but patients with PD who experienced toxicity later would obtain more benefit than those who quickly developed severe adverse events. For the second question, a CR would give $U_E(CR, g) = 140$ and SD would give $U_E(SD, g) = 20$ regardless of g. This is because CR is highly valuable to advanced renal cancer patients regardless of prognostic subgroup, whereas SD adds a small utility. For PR, $U_E(PR, g) = 60$, 90, and 120 for favorable, intermediate, and poor risk disease. This reflects the higher importance of achieving PR in aggressive poor risk renal cancer.

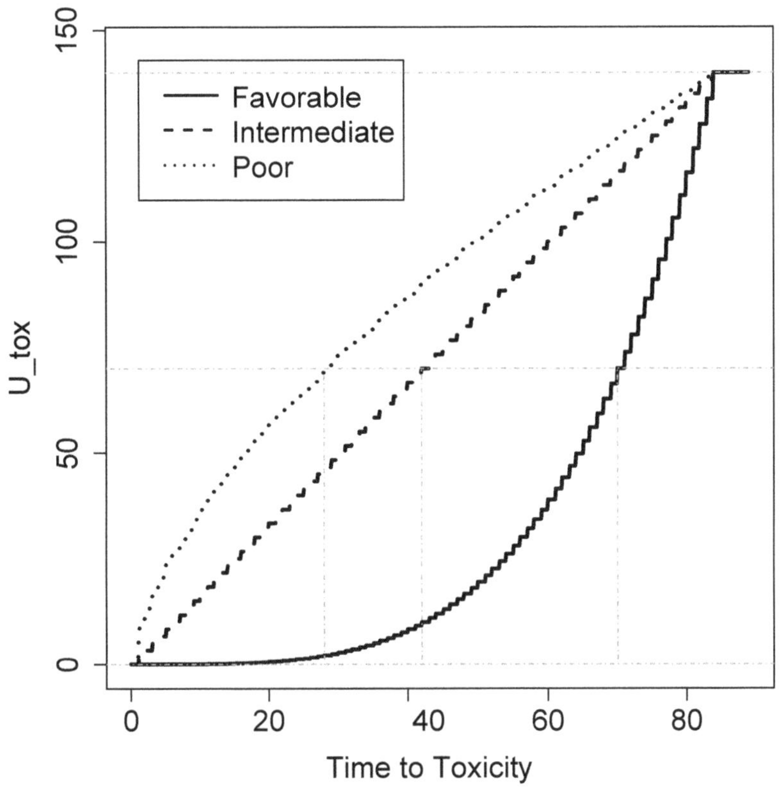

FIGURE 3.3
Prognostic subgroup-specific utilities of time to toxicity.

For each g, if $Y_E = 0$ (PD), then $U((Y_T, 0), g) = U_T(Y_{\mathrm{T}}, g)/2$. For efficacy outcomes other than PD, that is, Y_E = CR, PR or SD,

$$U(\boldsymbol{Y}, g) = U_T(Y_T, g) + U_E(Y_E, g),$$

with $U_E(1, g) = 20$ and $U_E(3, g) = 140$ for all g and $U_{\mathrm{E}}(2,1)$=60, $U_{\mathrm{E}}(2,2)$=90 and $U_{\mathrm{E}}(2,3)$=120. Thus, PD reduced the total utility by half of $U_{T,g}$, while the better efficacy outcomes, SD, PR, and CR, each increased the utility additively. The numerical value of $U_E(2, g)$ reflected the preference of PR relative to SD and CR in subgroup g, and a larger value of $U_E(2, g)$ made a higher dose more desirable. The maximum utility 280 was achieved by the best outcome, $(Y_T > C, Y_E = 3)$, where a toxicity event did not occur during the follow-up period and CR was observed. The worst outcome, $(Y_T < 1$ and $Y_E = 0)$, was assigned the minimum utility, 0. The PUFF was then defined as $\{U(\boldsymbol{Y}, g) : g = 1, 2, 3\}$.

3.3 Utility Functions for Dose Finding

The utility function described above was used by Lee et al. (2021) for a precision Bayesian phase I-II dose-finding design based on efficacy, toxicity, and patient prognosis. Indexing ordinal prognostic subgroups by $g = 1, \cdots, G$, with $g = 1$ indicating the lowest risk subgroup and $g = G$ the highest risk subgroup, the aim of the design is to choose a best dose for each g. Efficacy is an ordinal disease status variable, Y_E, evaluated at follow-up time t_F, with levels represented by the integer values $0, 1, \cdots, K-1$, where $Y_E = 0$ is the worst and $Y_E = K-1$ is the best disease status. The second outcome is time-to-toxicity, Y_T, monitored continuously during the follow-up period $(0, t_F]$. For solid tumors, commonly used values of Y_E are progressive disease (PD), stable disease (SD), partial response (PR), and complete response (CR), indexed respectively by $Y_E = 0$, 1, 2, and 3, with $K = 4$. This outcome structure often is used in oncology trials of treatments for solid tumors, and it generalizes "Case 2" of Jin et al. (2014), where Y_E is binary and it is observed at t_F, the end of the follow-up interval. Because in this case no information is available on Y_E for a patient until they have been followed for the full period $(0, t_F]$, adaptive decision-making based on $\boldsymbol{Y} = (Y_E, Y_T)$ is logistically difficult.

The dose-outcome model used by Lee et al. (2021) for this phase I-II design included latent subgroup membership variables used to cluster subgroups, and possibly later re-split them. This latent variable and adaptive decision structure is similar to that used by the Sub-TITE design. A key difference is that, because here the subgroups correspond to ordinal prognosis, only adjacent subgroups, g and $g+1$, may be combined adaptively. This is done if the interim data show, through the posteriors of the latent subgroup variables, that the distributions of $[\boldsymbol{Y} \mid d, g]$ and $[\boldsymbol{Y} \mid d, g+1]$ are close for all doses d. In this case, they are combined to form a new subgroup, $\{g, g+1\}$, and the model is simplified so that parameters corresponding to subgroups g and $g+1$ are identical. This process may be repeated, allowing more than two adjacent subgroups to be combined. Sets of combined subgroups are regarded as ordered prognostic clusters, and a subgroup later may be removed from a cluster if the subsequent data support this. By allowing adaptive subgroup clustering, the model borrows strength between subgroups, since if subgroups are combined the number of model parameters is reduced.

In contrast with Bayesian designs or methods for data analysis that base decisions on a utility function, $U(\boldsymbol{Y})$, the objective function specified by Lee et al. (2021) uses a precision utility, $U(\boldsymbol{Y}, g)$, that is a function of both the patient's outcomes and their prognostic subgroup. Thus, for example, $U(\boldsymbol{Y}, 1)$ and $U(\boldsymbol{Y}, 2)$ may differ for a given $\boldsymbol{Y}$, and the set $\{U(\boldsymbol{Y}, g), g \in \mathcal{G}\}$ is a PUFF. Construction of $U(\boldsymbol{Y}, g)$ is more detailed because it must be based on elicited numerical utilities for all combinations of $\boldsymbol{Y}$ obtained separately within each subgroup g. This refinement was motivated by the observation that, in some settings, physicians or patients may have different risk-benefit values for the same clinical outcome in different prognostic subgroups (Snapinn and Jiang, 2011). For example, in poor risk clear cell renal cell carcinoma patients, more aggressive therapies that carry a higher risk of toxicity may be preferable in order to achieve a greater anti-disease effect. This is because, for poor-risk patients, higher levels of toxicity are considered more acceptable if a CR or PR can be achieved. In contrast, higher toxicity is considered less acceptable for patients with better prognosis. The way in which risk-benefit trade-offs differ between prognostic subgroups depends on the particular clinical setting. For example, oncologists may prefer to treat poor prognosis metastatic triple-negative breast cancer patients with less intensive therapies in order to preserve better quality of life (Isakoff, 2010).

These examples require assigning numerical utilities to combinations of outcomes and prognostic subgroups, rather than constructing one utility function for all patients regardless of their prognosis. This refinement may provide a more realistic reflection of the risk-benefit trade-offs that actually are preferred in particular clinical settings. Results of a phase I-II dose-finding trial run using a precision utility function provide a more informed basis for planning a confirmatory randomized comparative trial that includes subgroup-specific decisions. The goal is to facilitate subgroup-specific decision-making by physicians in clinical practice.

The phase I-II design described here was motivated by a trial to optimize the dose of the targeted agent sitravatinib, a tyrosine kinase inhibitor (TKI), combined with fixed doses of the immune checkpoint inhibitors (ICIs) nivolumab and ipilimumab, in patients with metastatic clear cell renal cell carcinoma (ccRCC). While ICIs and TKIs may achieve durable responses in a subset of metastatic ccRCC patients, there is substantial heterogeneity between patients. The International Metastatic Renal-Cell Carcinoma Database Consortium (IMDC) prognostic risk score (Heng et al., 2013) is defined using both biomarkers and clinical variables, including anemia, thrombocytosis, neutrophilia, hypercalcemia, performance status, and time between diagnosis and treatment. IMDC score is used widely to classify ccRCC patients into three prognostic subgroups, Favorable (IMDC=0), Intermediate (IMDC=1 or 2), and Poor (IMDC $\geq$ 3). These subgroups may be indexed by g = 1, 2, and 3. For example, Heng et al. (2013) reported that metastatic ccRCC patients who received first-line VEGF-targeted treatment had survival times that differed substantially between the three IMDC risk subgroups. Favorable-risk ccRCC patients typically have indolent disease that can be managed successfully using less intensive treatments with low risks of toxicity. Patients with Poor risk IMDC scores have aggressive tumors that require more aggressive treatments that carry higher risks of toxicity (Zhang and George, 2020). The practice of using more aggressive therapies that carry higher risks of toxicity for patients with worse prognosis is very common in oncology, so in this regard the ccRCC trial may be considered a prototype for similar trials in other cancers. For the phase I-II design, a PUFF of subgroup-specific utilities was defined using the three IMDC risk subgroups. The ccRCC trial had follow–up period t_F = 84 days, with each patent's time-to-toxicity monitored continuously over this period and response evaluated at the end.

Let $n(t)$ denote the number of patients enrolled up to trial time t, and index patients by $i = 1, \cdots, n(t)$. Let $Y_{i,\mathrm{T}}^o$ denote the observed time from trial entry at time e_i to toxicity, $Y_{i,\mathrm{T}}$, or right-censoring, with $\delta_{i,\mathrm{T}} = 1$ if $Y_{i,\mathrm{T}}^o = Y_{i,\mathrm{T}}$ and $\delta_{i,\mathrm{T}} = 0$ if $Y_{i,\mathrm{T}}^o < Y_{i,\mathrm{T}}$. At trial time $t > e_i$, if $Y_{i,\mathrm{T}} \leq \min\{t - e_i, t_F\}$, then $Y_{i,\mathrm{T}}^o = Y_{i,\mathrm{T}}$ is the observed time of toxicity, and $\delta_{i,\mathrm{T}} = 1$; if $Y_{i,\mathrm{T}} > \min\{t - e_i, t_F\}$, then $Y^o_{i,\mathrm{T}}$ is the time of independent administrative right censoring and $\delta_{i,\mathrm{T}} = 0$. The event that toxicity has not occurred up to trial time t for the ith patient is $(Y^o_{i,\mathrm{T}} \le t_F, \delta_{i,\mathrm{T}} = 0)$. Because the ordinal disease status variable, $Y_{i,\mathrm{E}}$, is observed at trial time $e_i + t_F$, the model includes the indicator $\delta_{i,\mathrm{E}} = 1$ for any $t \ge e_i + F$, while for $t < e_i + t_F$, $Y_{i,\mathrm{E}}$ is not observed and $\delta_{i,\mathrm{E}} = 0$. Denote the doses standardized to have mean 0 and variance 1 by $\{d_1, \ldots, d_M\}$, the ith patient's dose by $d_{[i]}$, subgroup by g_i, and latent real-valued frailty vector by $\boldsymbol{\gamma}_i = (\gamma_{i,\mathrm{T}}, \gamma_{i,\mathrm{E}})$. It is assumed that $Y_{i,\mathrm{T}} \perp\!\!\!\perp Y_{i,\mathrm{E}} \mid \boldsymbol{\gamma}_i$ and a joint distribution of $(Y_{i,\mathrm{T}}, Y_{i,\mathrm{E}})$ is obtained by averaging over the distribution of $\boldsymbol{\gamma}_i$.

For the marginal toxicity event time distribution $p(Y_{i,\mathrm{T}} \mid g_i, d_{[i]}, \gamma_{i,\mathrm{T}})$, a proportional hazards (PH) model is assumed, with conditional hazard function

$$h_{\mathrm{T}}(t \mid g_i, d_{[i]}, \boldsymbol{\alpha}_T, \gamma_{i,\mathrm{T}}) = h_{0\mathrm{T}}(t) \exp\{\eta_{\mathrm{T}}(d_{[i]}) + \alpha_{\mathrm{T},g_i} + \gamma_{i,\mathrm{T}}\},$$

where $h_{0\mathrm{T}}(t)$ is a baseline hazard, and $\boldsymbol{\alpha}_T = (\alpha_{T,1}, \cdots, \alpha_{T,G})$ are prognostic subgroup effects on the risk of toxicity, with $\alpha_{\mathrm{T},1} = 0$ to ensure identifiability. A specific form of the dose-toxicity regression function $\eta_{\mathrm{T}}(d)$ will be given below. To ensure that h_{T} is non-decreasing in the ordinal risk subgroups for each dose, the constraint $\alpha_{\mathrm{T},1} \le \ldots \le \alpha_{\mathrm{T},G}$ is imposed. The survival function for $Y_{i,\mathrm{T}}$ is

$$S_{\mathrm{T}}(t \mid g_i, d_{[i]}, \boldsymbol{\alpha}_T, \gamma_{i,\mathrm{T}}) = \exp\left\{ - \int_0^t h_{\mathrm{T}}(v \mid g_i, d_{[i]}, \boldsymbol{\alpha}_T, \gamma_{i,\mathrm{T}}) dv \right\}.$$

For ordinal disease status or severity, such as $\{CR, PR, SD, PD\}$, the following multivariate probit regression model (Chib and Greenberg, 1998) is assumed. Denote subgroup-specific cutoffs $u_{g,0} < u_{g,1} < \cdots < u_{g,K}$ that are used by a probit model to define ordinal $Y_{i,\mathrm{E}}$ for each g, and let $\Phi(x \mid \mu, \sigma^2)$ denote the normal cdf with mean μ and variance σ^2. Given $\boldsymbol{\gamma}_i$ and $u_{g_i,k}$, for $k = 0, \ldots, K$, for a patent in subgroup g treated with dose d, the observed outcome is defined in terms of latent multivariate normal variables as

$$Y_{i,\mathrm{E}} = k \quad \text{if} \quad u_{g,k-1} < Z_{i,\mathrm{E}} \le u_{g,k},$$

where

$$Z_{i,\mathrm{E}} \sim N(\eta_E(d) + \alpha_{\mathrm{E},g_i} + \gamma_{i,\mathrm{E}}, \sigma^2),$$

$\boldsymbol{\alpha}_E = (\alpha_{E,1}, \cdots, \alpha_{E,G})$ is the vector of prognostic subgroup effects on $Y_{i,\mathrm{E}}$, and σ^2_π is a fixed hyperparameter. The conditional marginals are the multinomial probabilities induced by the distribution of $Z_{i,\mathrm{E}}$,

$$\begin{aligned}
\pi_k(g_i, d_{[i]}, \alpha_{\mathrm{E},g_i}, \gamma_{i,\mathrm{E}}, \boldsymbol{u}) &= \mathrm{P}(Y_{i,\mathrm{E}} = k \mid g_i, d_{[i]}, \boldsymbol{\alpha}_E, \gamma_{i,\mathrm{E}}, \boldsymbol{u}) \\
&= \Phi(u_{g_i,k} \mid \eta_{\mathrm{E}}(d_{[i]}) + \alpha_{\mathrm{E},g_i} + \gamma_{i,\mathrm{E}}, \sigma^2_\pi) \\
&- \Phi(u_{g_i,k-1} \mid \eta_{\mathrm{E}}(d_{[i]}) + \alpha_{\mathrm{E},g_i} + \gamma_{i,\mathrm{E}}, \sigma^2_\pi).
\end{aligned}$$

For added flexibility, the cut-offs $u_{g,k}$, $k = 2, \ldots, K-1$ are assumed to be random, fixing $u_{g,1} = 0$ and setting $u_{g,0} = -\infty$ and $u_{g,K} = \infty$ for all g, to ensure that

$$\sum_{k=0}^{K-1} \mathrm{P}(Y_{i,\mathrm{E}} = k \mid g_i, d_{[i]}, \gamma_{i,\mathrm{T}}) = 1.$$

The function η_{E} quantifies the dose effect in the probit function of Y_{E}, while subgroup effects are accounted for by $\{u_{g,k}\}$ and $\{\alpha_{\mathrm{E},g}\}$. Imposing the constraint $\alpha_{\mathrm{E},1} \geq \ldots \geq \alpha_{\mathrm{E},G}$ ensures that Pr(PD) is non-decreasing in the prognostic subgroups, since larger g corresponds to a worse prognosis. No restriction for g is imposed on $u_{g,k}$ for $k = 2, \ldots, K-1$, and $\mathrm{P}(Y_{i,\mathrm{E}} \leq k \mid g_i, d_{[i]}, \gamma_{i,\mathrm{T}})$ is not necessarily stochastically increasing in subgroups for $k > 0$. If desired, stochastic ordering of the distribution of Y_{E} in g can be imposed by requiring $u_{1,k} - \alpha_{\mathrm{E},1} \geq \ldots \geq u_{G,k} - \alpha_{\mathrm{E},G}$ for all k. An important property of the marginal models is that they do not assume any interaction between doses and subgroups, which allows the models to be more parsimonious and, in the ccRCC setting, reflects the fact that no studies have shown multiplicative interactions between IMDC score and treatment. If interactions are plausible, then a model including additional parameters to account for this may be used.

A flexible model for how the distribution of Y_j varies with d for each $j = E$ or T is given by the following function for the marginal model's linear term,

$$\eta_j(d \mid \boldsymbol{\beta}_j) = \frac{\beta_{j,3}}{1+\exp\{-\beta_{j,1}(10d - \beta_{j,2})\}},$$

denoting $\boldsymbol{\beta}_j = (\beta_{j,1}, \beta_{j,2}, \beta_{j,3})$, with dose multiplied by 10 to stabilize numerical computations. To ensure that η_j increases in d, $\beta_{j,1} \geq 0$, $\beta_{j,3} \geq 0$, and $\beta_{j,2}$ is real-valued. This implies monotonicity in dose, i.e. that $\mathrm{P}(Y_{\mathrm{T}} \leq F \mid g, d, \boldsymbol{\gamma})$, the probability of toxicity within the follow-up period $[0, t_F]$, and $\mathrm{P}(Y_{\mathrm{E}} \geq k \mid g, d, \boldsymbol{\gamma})$, the probability of level k or better response, both increase in d. This function may take a wide variety of different shapes, including an 'S'shape that allows the probabilities to reach a plateau at some dose. The parameter $\beta_{j,2}$ is the inflection point, and $\eta_j \to 0$ as $d \to -\infty$, while $\eta_j \to \beta_{j,3}$ as $d \to \infty$. Not treating a patient with the new drug, i.e. $d = 0$, corresponds to $\eta_j = 0$, and the $\eta_j(d)$ curve is bounded above by $\beta_{j,3}$.

Let $\boldsymbol{\theta}$ denote the vector of all model parameters and $\tilde{\boldsymbol{\theta}}$ the vector of all fixed hyperparameters. The joint likelihood of the outcomes of patient i, conditional on the latent frailty vector $\boldsymbol{\gamma}_i$, subgroup g_i, and dose $d_{[i]}$ is the product

$$p(y^o_{i,\mathrm{T}}, y_{i,\mathrm{E}}, \delta_{i,\mathrm{T}}, \delta_{i,\mathrm{E}} \mid g_i, d_{[i]}, \boldsymbol{\gamma}_i, \boldsymbol{\theta}, \tilde{\boldsymbol{\theta}}) =$$

$$h_{i,\mathrm{T}}(y^o_{i,\mathrm{T}} \mid g_i, d_{[i]}, \gamma_{i,\mathrm{T}})^{\delta_{i,\mathrm{T}}}\ S_{\mathrm{T}}(y^o_{i,\mathrm{T}} \mid g_i, d_{[i]}, \gamma_{i,\mathrm{T}})\ \pi_{y_{i,\mathrm{E}}}(g_i, d_{[i]}, \gamma_{i,\mathrm{E}})^{\delta_{i,\mathrm{E}}}.$$

The frailties $\{\boldsymbol{\gamma}_i\}$ account for heterogeneity between patients not explained by dose and subgroup, and correlations among the $\gamma_{i,j}$'s induce dependence between each patient's outcomes. The prior frailty distribution is $\boldsymbol{\gamma}_i \mid \Omega \sim iid\ \mathrm{N}_2(\mathbf{0}, \Omega)$ with random Ω, and the joint distribution of $\boldsymbol{Y}_i = (Y_{i,\mathrm{T}}, Y_{i,\mathrm{E}})$ is obtained by integrating over the frailty distribution,

$$p(\boldsymbol{y}_i \mid g_i, d_{[i]}, \boldsymbol{\theta}, \tilde{\boldsymbol{\theta}}) = \int_{\mathbb{R}^2} p(\boldsymbol{y}_i \mid g_i, d_{[i]}, \boldsymbol{\gamma}_i, \boldsymbol{\theta}, \tilde{\boldsymbol{\theta}}) \times p(\boldsymbol{\gamma}_i \mid \Omega) d\boldsymbol{\gamma}_i.$$

Subgroup Clustering Prior and Algorithm

Temporarily suppress the patient index, i. The following construction provides a framework for adaptively combining adjacent prognostic subgroups that have similar effects on $\boldsymbol{Y} = (Y_{\mathrm{T}}, Y_{\mathrm{E}})$ based on the observed data during the trial. This is done by introducing a second set of random latent variables, the *cluster membership indicators* $\boldsymbol{\zeta} = (\zeta_1, \cdots, \zeta_G)$, where each $\zeta_g \in \{1, \cdots, G\}$. Each set of one or more combined subgroups is a cluster, and the clusters are indexed consecutively by $r = 1, \cdots, R$, with $R \leq G$. For example, if $G = 3$, because only consecutive subgroups may be combined, there are four possible configurations of $\boldsymbol{\zeta} = (\zeta_1, \zeta_2, \zeta_3)$, given by $(1,1,1), (1,1,2), (1,2,2)$, and $(1,2,3)$. These define, respectively, R = 1, 2, 2, and 3 clusters. In the case $\boldsymbol{z} = (1,1,2)$, subgroups 1 and 2 are combined and subgroup 3 is distinct. The cluster $\{1,2\}$ is indexed by $r = 1$ and the singleton cluster $\{3\}$ is indexed by $r = 2$. To implement the clustering algorithm, $\zeta_1 = 1$ throughout, and it is assumed that

$$P(\zeta_g = \zeta_{g-1} \mid \zeta_{g-1}) = \xi_g \text{ and } P(\zeta_g = \zeta_{g-1} + 1 \mid \zeta_{g-1}) = 1 - \xi_g$$

for subgroups $g = 2, \cdots, G$. The prior of $\boldsymbol{\zeta}$ is

$$p(\boldsymbol{\zeta} \mid \boldsymbol{\xi}) = \prod_{g=2}^{G} \xi_g^{\mathrm{I}(\zeta_g = \zeta_{g-1})} (1 - \xi_g)^{1 - \mathrm{I}(\zeta_g = \zeta_{g-1})}. \tag{3.3}$$

Under this model, subgroup g joins the cluster to which subgroup $g-1$ belongs with probability ξ_g, in which case $\zeta_g = \zeta_{g-1}$. Subgroup g is in a separate cluster with probability $1 - \xi_g$, and in this case $\zeta_g = \zeta_{g-1} + 1$. This ensures a non-decreasing ordering of the cluster membership indicators, $\zeta_g \leq \zeta_{g+1}$, and subgroup g cannot belong to the cluster to which subgroup $g-2$ belongs unless subgroup $g-1$ also belongs to that cluster. The cluster membership indicator ζ_G of the highest risk subgroup is R, so both R and $\boldsymbol{z}$ are random. An illustration of how the sequential clustering algorithm might work for $G = 4$ ordinal prognostic subgroups is given in Figure 3.4. In a different setting where the subgroups are not ordinal, the ordering constraint on ζ_g is not imposed. In this case, any model-based clustering approach, such as the House Party Algorithm used in SubTITE or the random partitions in the SUBA design model Xu et al. (2016b), might be applied, although the model for $[Y_T, Y_E \mid d, g]$ would require appropriate modifications.

To ensure that subgroup effects are identical within each cluster, cluster-specific parameters $\alpha^\star_{\mathrm{T},r}$ and $\alpha^\star_{\mathrm{E},r}$ are defined for each cluster $r = 1, \cdots, R$, with vectors $\boldsymbol{\alpha}^\star_j = (\alpha^\star_{j,1}, \cdots, \alpha^\star_{j,R})$ for $j = T$ or E, and $\alpha_{j,g} = \alpha^\star_{j,\zeta_g}$. A cluster-specific cut-off vector $\boldsymbol{u}^\star$ is defined similarly, with each $u^\star_{r,k}$ the same for all subgroups in cluster r. This implies that $P(Y_{\mathrm{T}} \leq t \mid g, d, \boldsymbol{\alpha}^\star_T, \boldsymbol{\gamma})$ and $P(Y_{\mathrm{E}} = k \mid g, d, \boldsymbol{\alpha}^\star_E, \boldsymbol{\gamma})$ do not change for different g in the same cluster. Priors for $\boldsymbol{\alpha}^\star_j$ and $\boldsymbol{u}^\star$ must accommodate the facts that prognostic subgroup effects between adjacent subgroups may not be significantly different, while subgroups that correspond to very different prognoses are more likely to have distinct effects. Given the latent subgroup membership vector $\boldsymbol{z}$, normal distributions

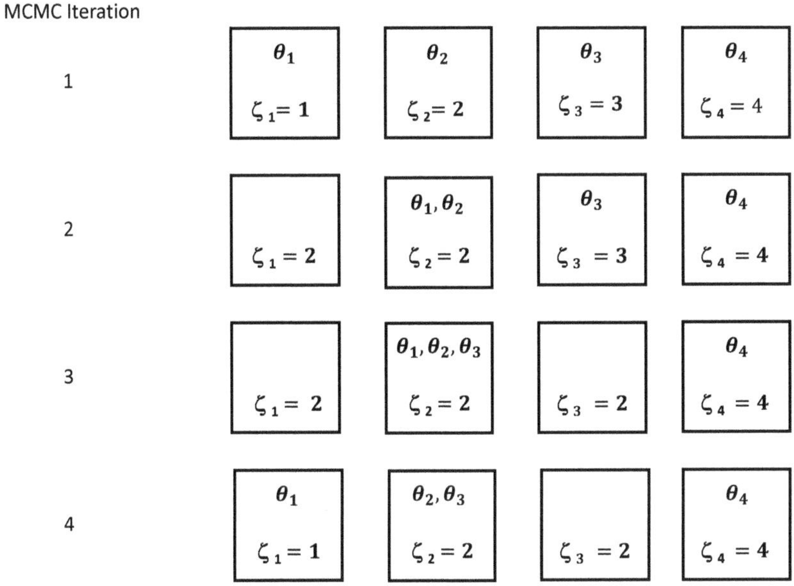

FIGURE 3.4
Illustration of possible sequential clusterings of $G = 4$ subgroups by the latent subgroup membership variables $\{\zeta_1, \zeta_2, \zeta_3, \zeta_4\}$, in four consecutive MCMC iterations. The model parameter vector for subgroup g is denoted by θ_g.

with order constraints on $\{\alpha^\star_{j,z}\}$ are assumed, given by

$$p(\boldsymbol{\alpha}^\star_T \mid \boldsymbol{z}, \bar{\boldsymbol{\alpha}}_{\mathrm{T}}, \boldsymbol{v}^2_{\mathrm{T}}) \propto \prod_{r=2}^{R} \exp\{-(\alpha^\star_{\mathrm{T},r} - \bar{\alpha}^\star_{\mathrm{T},r})^2/2v^2_{\mathrm{T}r}\} \mathrm{I}(\alpha^\star_{\mathrm{T},r} > \alpha^\star_{\mathrm{T},r-1}),$$

$$p(\boldsymbol{\alpha}^\star_{\mathrm{E}} \mid \boldsymbol{z}, \bar{\boldsymbol{\alpha}}_{\mathrm{E}}, \boldsymbol{v}^2_{\mathrm{E}}) \propto \prod_{r=1}^{R} \exp\{-(\alpha^\star_{\mathrm{E},r} - \bar{\alpha}^\star_{\mathrm{E},r})^2/2v^2_{\mathrm{E}r}\} \mathrm{I}(\alpha^\star_{\mathrm{E},r} < \alpha^\star_{\mathrm{E},r-1}),$$

where $\alpha^\star_{\mathrm{T},1} = 0$, $\alpha^\star_{\mathrm{E},0} = \infty$, $\bar{\boldsymbol{\alpha}}^\star_j = (\bar{\alpha}^\star_{j,1}, \ldots, \bar{\alpha}^\star_{j,R})$, and $\boldsymbol{v}^2_j = (v^2_{j,1}, \ldots, v^2_{j,R})$ are fixed hyperparameters. Due to the imposed ordering on $\{\alpha^\star_{j,r}\}$, higher-risk subgroups have larger probabilities of toxicity and PD. For $k = 2, \ldots, K-1$ and $r = 1, \ldots, R$, let $u^\star_{r,k} = u^\star_{r,k-1} + \rho_{r,k}$, and assume $\rho_{r,k} \sim \mathrm{Gamma}(\rho_{r-1,k}\kappa_{r,k}, \kappa_{r,k})$ independently with prior mean $\rho_{r-1,k}$ and variance $\rho_{r-1,k}/\kappa_{r,k}$. Fix $\rho_{0,k}$ and $\kappa_{r,k}$, for $k = 2, \ldots, K-1$ and $r = 1, \ldots, R$. The distributions of $\boldsymbol{\zeta}$, $\boldsymbol{\alpha}^\star_j$, and $\boldsymbol{u}^\star$ jointly define the distributions of $\boldsymbol{\alpha}_j$ and $\boldsymbol{u}$. In the posterior computations, $\boldsymbol{\zeta}$, $\boldsymbol{\alpha}^\star_j$ and $\boldsymbol{u}^\star$ are included as a parameter subvector of $\boldsymbol{\theta}$, replacing $\boldsymbol{\alpha}_j$ and $\boldsymbol{u}$.

Clustering is done to borrow strength between the subgroups within each cluster by combining subgroups having similar effect distributions, which reduces the number of model parameters. Doses are not chosen for clusters, but rather for each subgroup using its mean utility computed under the posterior distributions of $\alpha_{j,g}$ and $\boldsymbol{u}_g$. This formalizes common clinical practice, in which physicians may combine prognostic subgroups in an *ad hoc* manner to simplify their decision-making. In contrast, the adaptive clustering algorithm is a Bayesian statistical methodology that forms clusters empirically based on observed data. Because the final prognostic subgroup cluster distribution obtained from a trial may influence clinical decision-making and is based on observed data, it is an example of so-called "evidence-based medicine".

Priors for the remaining model parameters $h_{0T}(t)$, $\boldsymbol{\beta}_E$, $\boldsymbol{\beta}_T$, and Ω are specified as follows. For h_{0T}, a constant hazard over $(0,C)$ is assumed, with $h_{0T} \sim \text{Log-N}(\bar{h}_{0T}, v_h^2)$, since previous studies in ccRCC indicate that it is reasonable to assume a constant hazard during the follow-up period (Motzer et al., 2019a,b; Rini et al., 2019). If a time-varying hazard is more appropriate, then more complex models, such as a Weibull or piecewise exponential distribution, can be used. To ensure that the relationships of dose with the outcomes are monotonically increasing, normal or truncated normal distributions for $\beta_{j,\ell}$, $j = T,E$ and $\ell = 1,2,3$, are assumed, with $\beta_{j,1}$ and $\beta_{j,3}$ following normal priors truncated below at 0,

$$p(\beta_{j,\ell} \mid \bar{\beta}_{j,\ell}, \tau_{j,\ell}^2) \propto \exp\{-(\beta_{j,\ell} - \bar{\beta}_{j,\ell})^2/(2\tau_{j,\ell}^2)\}$$

for $\beta_{j,\ell} > 0$, $\ell = 1,3$, and $\beta_{j,2} \sim \text{N}(\bar{\beta}_{j,2}, \tau_{j,2}^2)$. To complete the prior specification of $\boldsymbol{\gamma}_i$, for fixed $\nu > 1$, it is assumed that

$$\Omega \mid \nu, \Omega^0 \sim \text{inv-Wishart}(\nu, \Omega^0)$$

where Ω^0 is a 2×2 positive definite hyperparameter matrix. Collecting terms,

$$\boldsymbol{\theta} = (h_{T0}, \boldsymbol{\beta}, \boldsymbol{\alpha}^\star, z, \boldsymbol{\rho}, \Omega)$$

denotes the vector of all model parameters, where

$$\boldsymbol{\beta} = \{\boldsymbol{\beta}_E, \boldsymbol{\beta}_T), \quad \boldsymbol{\alpha}^\star = \{\alpha_{j,r}^\star, j = T,E, r = 1,\ldots,R\},$$

and

$$\boldsymbol{\rho} = \{\rho_{r,k}, r = 1,\ldots,R, k = 2,\ldots,K-1\}.$$

For $j = E,T$, $\ell = 1,2,3$, $r = 1,\ldots,G$, and $k = 2,\ldots,K-1$, denote the hyperparameters $\bar{\beta} = \{\bar{\beta}_{j,\ell}\}$, $\boldsymbol{\tau}^2 = \{\tau_{j,\ell}^2\}$, $\boldsymbol{\xi} = (\xi_2,\ldots,\xi_G)$, $\bar{\boldsymbol{\alpha}}^\star = \{\bar{\alpha}_{j,r}^\star\}$, $\boldsymbol{v}^2 = \{v_{j,r}^2\}$, $\boldsymbol{\rho}_0 = \{\rho_{0,k}\}$, and $\boldsymbol{\kappa} = \{\kappa_{r,k}\}$. The vector of all fixed hyperparameters is

$$\tilde{\boldsymbol{\theta}} = (\bar{h}_{T0}, v_h^2, \bar{\beta}, \boldsymbol{\tau}^2, \boldsymbol{\xi}, \bar{\boldsymbol{\alpha}}^\star, \boldsymbol{v}^2, \boldsymbol{\rho}_0, \boldsymbol{\kappa}, \nu, \Omega^0).$$

Lee et al. (2021) established the prior hyperparameter vector $\tilde{\boldsymbol{\theta}}$ using data from the phase III CheckMate 214 trial described by Motzer et al. (2019b), and they also elicited prior probabilities from the clinical investigators. In the CheckMate 214 trial, patients were categorized into the Favorable ($g = 1$), Intermediate ($g = 2$), and Poor ($g = 3$) risk subgroups using their IMDC scores, as described above, and randomized between two treatments including combinations of *nivolumab* and *ipilimumab*. Table 1 of Motzer et al. (2019b) reports estimates of $\text{P}(Y_E = k)$ for $g = 1$ and $g > 1$ when no *sitravatinib* is given, i.e., $\eta_j = 0$ under the phase I-II model. The model was fit to pseudo data simulated from the elicited probabilities and the historical data, and posterior means were used to determine the prior's location hyper-parameters. Dispersion hyper-parameters were calibrated to reflect prior uncertainty, and effects of the calibrated $\tilde{\boldsymbol{\theta}}$ were examined with pseudo data simulated under various settings.

Given $\tilde{\boldsymbol{\theta}}$ and interim data $\mathscr{D}_{n(t)}$ at trial time t, the joint posterior of the parameters and patient random effects is

$$p(\boldsymbol{\theta}, \boldsymbol{\gamma} \mid \mathscr{D}_{n(t)}, \tilde{\boldsymbol{\theta}}) \quad \propto \quad p(\boldsymbol{\theta}, \boldsymbol{\gamma} \mid \tilde{\boldsymbol{\theta}}) \prod_{i=1}^{n(t)} p(y_{i,\mathrm{T}}^o, y_{i,\mathrm{E}}, \delta_{i,\mathrm{T}}, \delta_{i,\mathrm{E}} \mid d_{[i]}, \boldsymbol{\gamma}_i, \boldsymbol{\theta}, \tilde{\boldsymbol{\theta}}).$$

MCMC simulation is used to generate posterior samples of $(\boldsymbol{\theta}, \boldsymbol{\gamma})$ values, and since $\boldsymbol{z}$ is a subvector of $\boldsymbol{\theta}$, its posterior is obtained as a marginal of the posterior of $(\boldsymbol{\theta}, \boldsymbol{\gamma})$. This in turn provides a posterior on R, the clusters, and the cluster-specific subgroup effects $\boldsymbol{\alpha}^\star$.

In the design, posterior predictive (PP) mean utilities are used as optimality criteria for dose selection. Given data $\mathscr{D}_{n(t)}$ at trial time t, the PP mean utility of giving dose d_m to a future patient in subgroup g is

$$u_g(d_m \mid x, \mathscr{D}_{n(t)}) = \int_0^\infty \sum_{y_\mathrm{E}=0}^{K-1} U(\mathbf{y}, g) p(\mathbf{y} \mid x, d_m, \mathscr{D}_{n(t)}) dy_\mathrm{T}$$

$$= \int_{\boldsymbol{\theta}} \int_{\boldsymbol{\gamma}} \int_0^\infty \sum_{y_\mathrm{E}=0}^{K-1} U(\mathbf{y}, g) p(\mathbf{y} \mid x, d_m, \boldsymbol{\gamma}, \boldsymbol{\theta}) p(\boldsymbol{\gamma}, \boldsymbol{\theta} \mid \mathscr{D}_{n(t)}) dy_\mathrm{T} d\boldsymbol{\gamma} d\boldsymbol{\theta}.$$

Numerical computation of $u_g(d_m \mid x, \mathscr{D}_{n(t)})$ was based on the empirical posterior sample mean of $\boldsymbol{\theta}$ values simulated from $p(\boldsymbol{\theta} \mid \mathscr{D}_{n(t)}, \tilde{\boldsymbol{\theta}})$.

Because lower doses carry a higher risk of PD and higher doses carry a higher risk of toxicity, acceptable doses must satisfy the following safety requirements. In each subgroup, optimal dose selection and treatment assignment are restricted to the set of acceptable doses. The first safety constraint is that an untried dose may not be skipped when escalating within each subgroup. If $d_{m_g^{\max}(t)}$ is the highest dose that has been administered by trial time t in subgroup g, then the search for the optimal dose and the treatment assignment are restricted to the acceptable dose set

$$\mathscr{A}^{\text{Tried}}(g, t) = \{d_1, \ldots, d_{\min\{m_g^{\max}(t)+1, M\}}\}.$$

Toxicity and efficacy are monitored as follows. The probabilities of PD ($Y_\mathrm{E} = 0$) or severe toxicity during the $t_F = 84$-day follow-up in subgroup g are

$$\pi_\mathrm{E}(g, d_m, \boldsymbol{\theta}) = \Pr(Y_\mathrm{E} = 0 \mid g, d_m, \boldsymbol{\theta})$$

and

$$\pi_\mathrm{T}(g, d_m, \boldsymbol{\theta}) = \Pr(Y_\mathrm{T} \le t_F \mid g, d_m, \boldsymbol{\theta}).$$

For each outcome $j = E, T$, $\bar{\pi}_j(g)$ is an elicited fixed upper limit on $\pi_j(g, d_m, \boldsymbol{\theta})$, and $p^\star$ is a fixed cut-off probability. During the trial, if

$$\mathrm{P}\{\pi_j(g, d_m, \boldsymbol{\theta}) > \bar{\pi}_j(g) \text{ for } j = E \text{ or } T \mid \mathscr{D}_{n(t)}\} \; > \; p_U, \tag{3.4}$$

then d_m is *unacceptable for subgroup g*, due to an unacceptably high rate of PD or toxicity, and thus is not administered to any new patients enrolled in that subgroup. The set of acceptable doses at time t for subgroup g that do not satisfy (3.4) is denoted by $\mathscr{A}(g,t)$. The elicited values $\bar{\pi}_{\mathrm{T}}(g) = 0.40$ for all $g = 1,2,3$, and $\bar{\pi}_{\mathrm{E}}(1) = \bar{\pi}_{\mathrm{E}}(2) = 0.35$ for the Favorable and Intermediate subgroups and $\bar{\pi}_{\mathrm{E}}(3) = 0.20$ for the Poor subgroup were used for the ccRCC trial. The upper limit $\bar{\pi}_{\mathrm{T}}(g) = 0.40$ was set because dose-limiting toxicity probabilities $\geq .40$ generally are considered unacceptable in oncology dose-finding trials.

This was based on historical data on nivolumab and ipilimumab, so that adding sitravatinib could not be permitted to have worse safety. To obtain a design with high probabilities of stopping a truly unsafe or inefficacious dose, while still having high probabilities of selecting the best safe and efficacious dose, the cutoff $p_U = 0.85$ was obtained based on preliminary simulations.

The first patient enrolled in each subgroup in the ccRCC trial was treated at d_2, chosen by physicians, and the dose acceptability rules were applied thereafter. At trial time t, the safety and efficacy constraints together define a set of acceptable doses $\mathscr{A}(g,t) = \mathscr{A}^{\text{Tried}}(g,t) \cap \mathscr{A}^{\text{Accp}}(g,t)$ in $\{d_1,\ldots,d_M\}$ based on interim data $\mathscr{D}_{n(t)}$. Due to the late observation of efficacy at the final follow-up day t_F = 84, $\mathscr{A}^{\text{Accp}}(g,t)$ was based on toxicity only until ≥ 20 patients were fully followed up to 84 days. Patients were adaptively randomized among $d_m \in \mathscr{A}(g,t)$, with a patient in subgroup g assigned dose $d_m \in \mathscr{A}(g,t)$ with probability proportional to $1/\{n_{g,m}(t)+1\}$, where $n_{g,m}(t)$ = number of patients in subgroup g treated at dose d_m up to time t. If $\mathscr{A}(g,t) = \emptyset$, then no patients were enrolled in subgroup g. If $\mathscr{A}(g,t) = \emptyset$ for all g, then the trial would be terminated with no dose is selected, $d_{\text{sel}}(g) = None$, for all g. If the trial was not terminated early, subgroup-specific final optimal actions were taken at time $T_{\max} = e_{N_{\max}} + C$. Identifying $\mathscr{A}(g,T_{\max})$, let $d_{\text{sel}}(g) = None$ if $\mathscr{A}(g,T_{\max}) = \emptyset$. Otherwise, the optimal dose selected for subgroup g was

$$d_{\text{sel}}(g) = \underset{d_m \in \mathscr{A}(g,T_{\max})}{argmax} \; u(d_m \mid g, \mathscr{D}_{N_{\max}}).$$

Lee et al. (2021) performed a simulation study across 12 scenarios of the ccRCC design with $N_{\max} = 120$, assuming $G = 3$ subgroup probabilities $\boldsymbol{p}_g = (0.23, 0.60, 0.17)$ to reflect the historical data reported in Motzer et al. (2019b), $M = 5$ doses, $K = 4$ efficacy levels, and final follow up t_F = 84 days. Each scenario included true latent clustering variables $\boldsymbol{\zeta}^{\text{TR}} = (\zeta_1^{\text{TR}}, \zeta_2^{\text{TR}}, \zeta_3^{\text{TR}})$, with frailties $\boldsymbol{\gamma}_i^{\text{TR}} \sim iid\ \mathrm{N}_2(\mathbf{0}, \Omega^{\text{TR}})$. To simulate $Y_{i,\mathrm{T}}$, marginal probabilities $\mathrm{P}(Y_{i,\mathrm{T}} < t_F \mid g_i, d_m, \boldsymbol{\gamma}_i^{\text{TR}})$, for g_i, $d_{[i]} \in \{d_1,\ldots,d_5\}$ and $\boldsymbol{\gamma}_i^{\text{TR}}$ were specified, and $Y_{i,\mathrm{E}}$ from $\{0,\ldots,K-1\}$ was simulated using fixed probabilities $(\pi_{i,0}^{\text{TR}},\ldots,\pi_{i,(K-1)}^{\text{TR}})$ conditional on g_i, $d_{[i]}$ and $\boldsymbol{\gamma}_i^{\text{TR}}$. For $G = 3$, four configurations of $\boldsymbol{z}^{true}$ are possible due to subgroup ordinality. Each of $\boldsymbol{z}^{true} = (1,2,3)$, $(1,1,2)$, $(1,2,2)$, and $(1,1,1)$ was assumed in two scenarios.

The pattern of the true expected utilities varies with g in all scenarios. In Scenario 1, $d = 1$ is optimal for subgroups 1 and 2, but no dose is acceptable for subgroup 3 due to excessive probabilities of toxicity or PD. In Scenario 2, no dose is acceptable for any g. In Scenarios 3, 7, and 8, the optimal dose is the same for all g. In Scenario 3, true dose acceptability changes with g. For example, in Scenario 4, $d = 1$ is optimal for $g = 1$ and 2, whereas $d = 3$ is optimal for $g = 3$. In Scenarios 4 and 5, a higher risk subgroup has a higher optimal dose. In Scenario 6, $g = 1$ has optimal $d = 4$, but $d = 1$ is optimal for $g = 2$ and 3 due to large severe toxicity probabilities. A total of 500 trials with $N_{\max} = 120$ were simulated under each scenario.

The proposed design, called *D-Sub*, was compared to two designs, *D-Comb* which ignores patient subgroups, and *D-Sep* which runs a separate trial for each subgroup. Model parameters reflect the assumptions of D-Comb and D-Sep, but otherwise the models are similar to that of D-Sub. For D-Sep, all subgroup-specific factors are removed and the remaining parts of the model are unchanged. For D-Comb, $U_2(\boldsymbol{Y})$ is used as the common utility function since subgroup 2 has the highest prevalence. Under D-Comb, if a dose is identified as unacceptable, no patient is treated at that dose regardless of the patient's subgroup, and one optimal dose is recommended for all subgroups. The same stopping criterion upper limits $\bar{\pi}_j(g)$ and $U_g(\boldsymbol{Y})$ were used for D-Sub, with respective $N_{\max}$ = 28, 72, and 20 for the three subgroups. The designs were evaluated using the following criteria. In each subgroup g,

$p^{\text{unacc}}(g,d)$ = probability of identifying an unacceptable dose d with a truly excessive probability of severe toxicity or PD.

$p^{\text{sel}}(g,d)$ = probability of selecting the truly optimal dose d.

$n^{\text{trt}}(g,d)$ = mean number of patients treated at dose d.

Values of $p^{\text{unacc}}(g,d)$ and $p^{\text{sel}}(g,d)$ vary with g under D-Sub and D-Sep, but are the same for all g under D-Comb. Simulation results across 12 scenarios are summarized in Figure 3.5 to compare the performance metrics of D-Sub versus D-Comb, and in Figure 3.6, which compares D-Sub to D-Sep. Histograms are given for differences between $p^{\text{sel}}(g,d)$ for truly optimal doses, $p^{\text{unacc}}(g,d)$ for all doses, and $n^{\text{trt}}(g,d)$ for truly unacceptable doses. Positive differences in $p^{\text{sel}}(g,d)$ and $p^{\text{unacc}}(g,d)$, and negative differences in $n^{\text{trt}}(g,d)$, correspond to superior performance by D-Sub. The figures show that D-Sub is greatly superior to both D-Comb and D-Sep, in terms of both optimal dose selection and patient safety. For each subgroup, D-Sub is more likely to select subgroup-specific optimal acceptable doses, and more reliably identifies and drops doses that are either unsafe or have unacceptably low efficacy.

Computer Software A computer program, *Dose-finding-subgroup*, for implementing the design is available from the website *https://users.soe.ucsc.edu/ juheelee/*.

3.4 Dose Finding for Natural Killer Cells

Natural killer (NK) cells are lymphocytes that play a critical role in immune surveillance, and they are the body's first line of defense against viruses and newly transformed cancer cells (Rezvani and Rouce, 2015). In recent years, NK cells have been engineered to be used for cancer immunotherapy. A *chimeric antigen receptor*

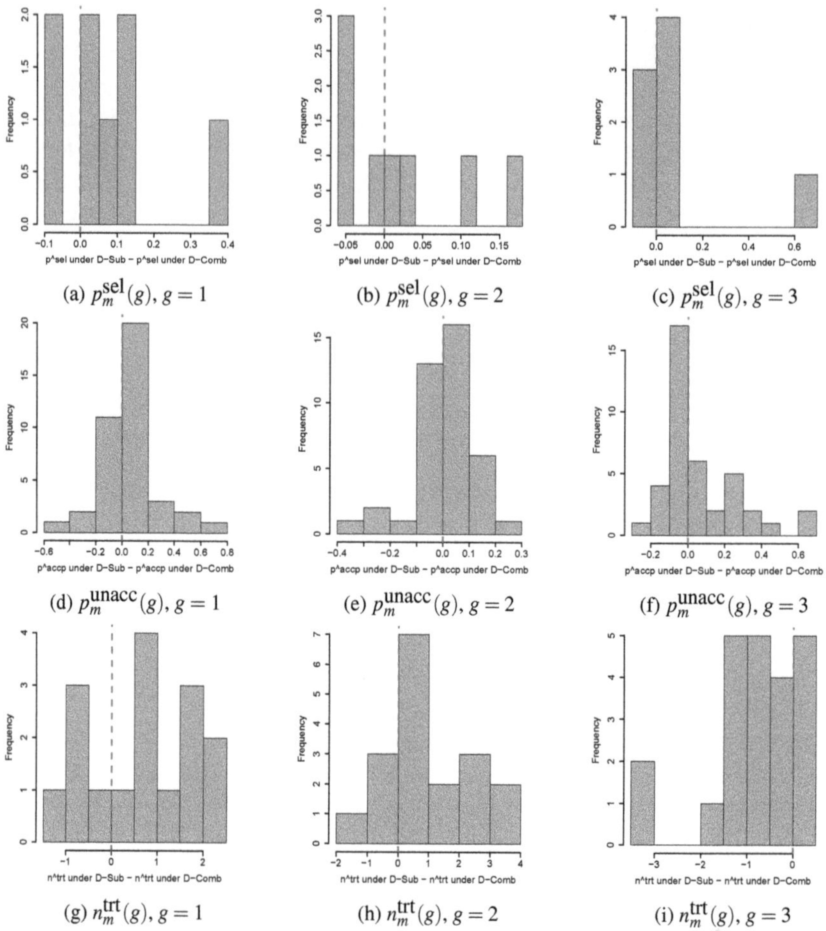

FIGURE 3.5
Histograms across 12 simulated scenarios comparing D-Sub to D-Comb in the ccRCC trial. Differences between best dose selection probabilities $p_m^{\text{sel}}(g)$ are in panels (a)-(c); between $p_m^{\text{unacc}}(g)$ for truly unacceptable doses in panels (d)-(f); and between sample sizes $n_m^{\text{trt}}(g)$ for the truly unacceptable doses in panels (g)-(i). Better performance of D-Sub than D-Comb corresponds to positive values for $p_m^{\text{sel}}(g)$ in panels (a)-(c), and for $p_m^{\text{unacc}}(g)$ in panels (d)-(f), and negative values for $n_m^{\text{trt}}(g)$ in panels (g)-(i). Left, middle and right columns are for subgroups $g = 1$(Favorable), $g = 2$ (Intermediate), and $g = 3$ (Poor).

(CAR) is designed to bind to specific proteins on cancer cells. NK cells are a natural choice for CAR engineering because they mediate cytotoxicity against tumor cells. Unlike T-cells, NK cells do not cause graft-versus-host disease (GVHD) in the allogeneic setting where the cells are not derived from the patient. CAR NK cells retain

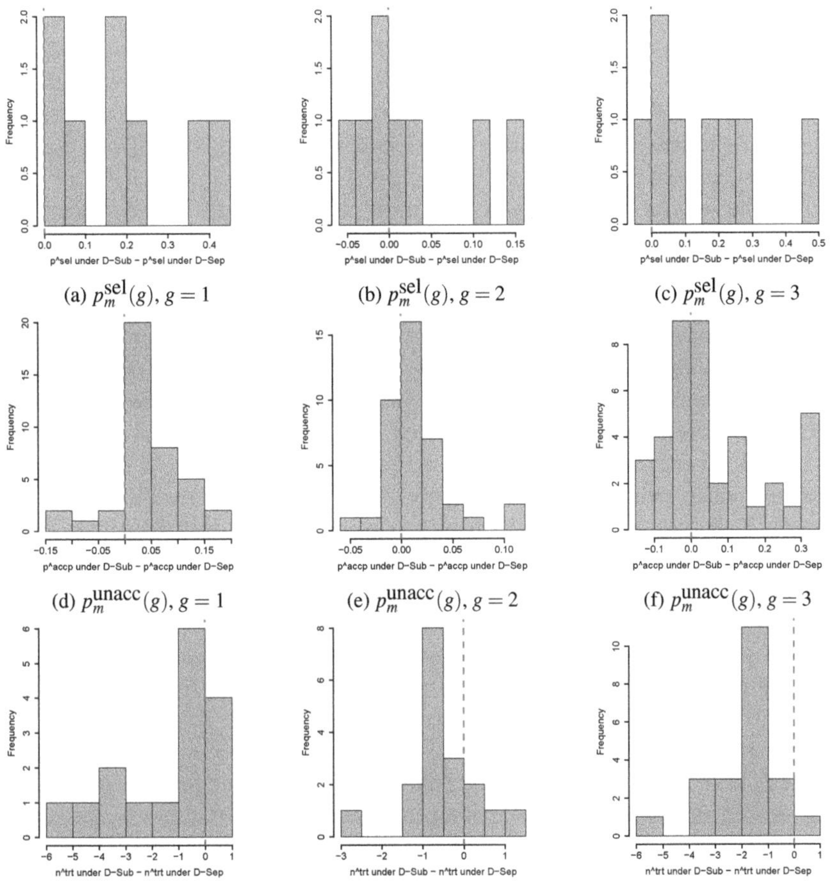

(a) $p_m^{\text{sel}}(g)$, $g=1$ (b) $p_m^{\text{sel}}(g)$, $g=2$ (c) $p_m^{\text{sel}}(g)$, $g=3$

(d) $p_m^{\text{unacc}}(g)$, $g=1$ (e) $p_m^{\text{unacc}}(g)$, $g=2$ (f) $p_m^{\text{unacc}}(g)$, $g=3$

(g) $n_m^{\text{trt}}(g)$, $g=1$ (h) $n_m^{\text{trt}}(g)$, $g=2$ (i) $n_m^{\text{trt}}(g)$, $g=3$

FIGURE 3.6
Histograms across 12 simulated scenarios comparing D-Sub to D-Sep. Differences between best dose selection probabilities $p_m^{\text{sel}}(g)$ are in panels (a)-(c); between $p_m^{\text{unacc}}(g)$ for truly unacceptable doses in panels (d)-(f); and between sample sizes $n_m^{\text{trt}}(g)$ for the truly unacceptable doses in panels (g)-(i). Better performance of D-Sub than D-Sep corresponds to positive values for $p_m^{\text{sel}}(g)$ in panels (a)-(c), and for $p_m^{\text{unacc}}(g)$ in panels (d)-(f), and negative values for $n_m^{\text{trt}}(g)$ in panels (g)-(i). Left, middle and right columns are for subgroups $g=1$(Favorable), $g=2$ (Intermediate), and $g=3$ (Poor).

their ability to recognize and target tumor cells through their native receptors, which may decrease the ability of cancer cells to downregulate the CAR target antigen.

The following design, proposed by Lee et al. (2019), aims to select optimal doses of umbilical cord blood-derived NK cells in an early phase dose-finding trial for patients with advanced hematologic diseases. The design accounts for the fact that patients in the trial were heterogeneous, with any of the three diseases chronic lymphocytic leukemia (CLL), acute lymphocytic leukemia (ALL), or non-Hodgkin's lymphoma (NHL). Patients also were classified as having either low or high bulk disease (LBD or HBD) so, together, disease type and bulk determined six prognostic subgroups. The trial's primary goal was to identify an optimal NK cell dose for each of the six subgroups.

The trial was far more complex than most phase I-II trials in that it accounted for five co-primary outcomes, each of which was a time-to-event variable subject to right-censoring. These were the times from cell infusion to death (D), disease progression (P), response (R), severe toxicity (T), and severe cytokine release syndrome (C). Since, if they occur, T and C are most likely to occur soon after cell infusion, they were monitored up to time $t_{F,T} = t_{F,C} = 100$ days. The remaining outcomes, P, R, and D were monitored for up to $t_{F,P} = t_{F,R} = t_{F,T} = 365$ days. The time Y_j to event j was independently censored at $t_{F,j}$ for each $j \in \{P,R,T,C,D\}$, death informatively censored all other events, and Y_P and Y_R were competing risks since at most one of P or R could occur. Thus, the Y_j's were highly interdependent, and the distribution of $\boldsymbol{Y}$ $= (Y_R, Y_C, Y_P, Y_T, Y_D)$ varied with both subgroup and NK cell dose. The four adverse event times (Y_C, Y_P, Y_T, Y_D) were assumed to be positively associated.

Denoting NK cell dose by d, and covariates that identify subgroup by $\mathbf{Z}$, the probability of death prior to day 100 is denoted by $\pi_D(d,\mathbf{Z}) = \text{P}(Y_D \leq 100 \mid d,\mathbf{Z})$ for all $(d,\mathbf{Z})$. In the design, if $\pi_D(d,\mathbf{Z})$ is unacceptably high compared to an elicited fixed upper limit $\bar{\pi}_D(\mathbf{Z})$ for subgroup $\mathbf{Z}$, then d is discontinued in $\mathbf{Z}$. A conventional "one size fits all" safety rule ignoring subgroups would discontinue d for all patients if an interim estimate of $\pi_D(d) = \text{P}(Y_D \leq 100 \mid d)$, defined under a simplified model that ignores $\mathbf{Z}$, is unacceptably high compared to a fixed upper limit. Such a rule produces a design with high probabilities of making incorrect decisions within subgroups if the outcome distributions vary with $\mathbf{Z}$, that is, if patients are heterogeneous. As a simple toy example, given historical death rates $\pi_D^{\mathcal{H}}(A) = 0.10$ and $\pi_D^{\mathcal{H}}(B) = 0.40$ with standard therapy for subgroups A (low risk) and B (high risk), if the interim estimate $\widehat{\pi}_D(d) = 0.25$ is obtained while ignoring subgroups, it may trigger the decision to stop all accrual to dose d, which is likely to be correct for subgroup A but incorrect for subgroup B.

A major logistical problem in conduct of the NK cell trial is that each patient's outcome is based on five event times monitored for up to 365 days. Each outcome is fully evaluated at its occurrence time, or censored by death, or administratively censored at the end of its 100 or 365 day follow-up period. To choose a dose adaptively for a new patient, it is very likely that some outcomes of previous patients have not been evaluated fully, and thus it is not feasible to suspend accrual to wait for full evaluation of all previous patients' data. This is similar to the "late onset toxicity" problem, but it is far more complex due to the fact that there are five outcomes that have complicated relationships.

In the NK cell trial design, it was assumed that Pr(CLL) = Pr(ALL) = Pr(NHL) = 1/3, Pr(LBD) = 1/3 and Pr(HBD)= 2/3. Given maximum sample size $N_{\max} = 60$, approximate expected subgroup sample sizes are (7, 7, 7, 13, 13, 13), and these subsamples are divided further among the NK cell doses. This severely limits the reliability of any method for subgroup-specific safety monitoring and optimal dose selection. A key fact underlying the model and design is that, based on knowledge about NK cell biology, the six hazard functions may not be monotone in NK cell dose. Finally, with five time-to-event outcomes, a major issue is defining what is meant by "optimal NK cell dose" in each subgroup, since this is not obvious. The following design addresses all of these issues.

For interim sample size $n(t) \leq N_{\max}$ at trial time t, index patients in order of enrollment by $i = 1, \cdots, n(t)$, with trial entry times $0 \leq e_1 \leq e_2 \leq \cdots \leq e_{n(t)}$. For patient i, the trial time of event j is $e_i + Y_{i,j}$, if it is observed. Let $Y^o_{i,j}$ denote the time of observing $Y_{i,j}$ or right-censoring, with $\delta_{i,j} = 1$ if $Y^o_{i,j} = Y_{i,j}$ and 0 otherwise. The following are the basic structures and assumptions underlying the design:

1. For $j = P, R$ or D, $Y_{i,j}$ is followed until $e_i + 365$, and $Y_{i,T}$ and $Y_{i,C}$ are followed until $e_i + 100$.
2. If $Y_{i,D} > t - e_i$ or $Y_{i,D} > 365$, then $Y^o_{i,D}$ is the time of independent right censoring ($\delta_{i,D} = 0$). If $Y_{i,D} < \min(t - e_i, 365)$, then $Y^o_{i,D} = Y_{i,D}$ is the observed time of death ($\delta_{i,D} = 1$).
3. For the nonfatal events $j = T$ and C, if $Y_{i,j} < \min\{t - e_i, t_{F,j}, Y_{i,D}\}$, then $Y^o_{i,j} = Y_{i,j}$ ($\delta_{i,j} = 1$) and otherwise $Y^o_{i,j}$ is the time of right-censoring ($\delta_{i,j} = 0$).
4. If $Y_{i,P} < \min\{t - e_i, t_{F,P}, Y_{i,R}, Y_{i,D}\}$, i.e. $Y_{i,P}$ is observed, then $Y^o_{i,P} = Y^o_{i,R} = Y_{i,P}$ and $(\delta_{i,P}, \delta_{i,R}) = (1, 0)$.
5. If $Y_{i,R} < \min\{t - e_i, t_{F,R}, Y_{i,P}, Y_{i,D}\}$, then $Y^o_{i,P} = Y^o_{i,R} = Y_{i,R}$ and $(\delta_{i,P}, \delta_{i,R}) = (0, 1)$.
6. If neither Y_P nor Y_R occurs, then $(\delta_{i,P}, \delta_{i,R}) = (0, 0)$. Since P and R are competing risks, $(\delta_{i,P}, \delta_{i,R}) = (1, 1)$ is not possible.
7. If censoring is due to the fact that $Y_{i,j} > \min\{t - e_i, t_{F,j}\}$ and the patient is alive at trial time t, then censoring is assumed to be independent of $Y_{i,j}$. For $j \neq D$, if censoring of $Y_{i,j}$ is due to death, i.e. $Y_{i,D} < Y_{i,j}$, then the censoring is not independent and $Y^o_{i,j} = Y^o_{i,D}$ and $\delta_{i,j} = 0$.

Index NK cell doses by $d = 1, 2, 3$, and define $Z = 0$ for LBD and $Z = 1$ for HDB. Index $r = 1, 2, 3$, for disease types CLL, ALL, NHL, and denote $\mathbf{Z} = (Z, r)$. For patient $i = 1, \cdots, n(t)$, denote dose by d_i and covariates by $\mathbf{Z}_i = (Z_i, r_i)$. The joint distribution of $\boldsymbol{Y}$ is defined on the set

$$\mathscr{Y}_D = \{\mathbf{y} \subset [0, \infty)^5 : \max(y_P, y_R, y_C, y_T) < y_D\},$$

since death censors any nonfatal event but not conversely. On $\mathscr{Y}_D$, Y_R and Y_P are competing risks, since at most one of the two events can be observed. Thus, h_R and h_P are subhazard functions, that is, cause-specific hazards, with $h_j(y)$, $j = R, P$ interpreted as the hazard of R or P occurring at time y and the outcome being j, with $h_R(y) + h_P(y)$ the overall hazard of either P or R on $\mathscr{Y}_D$.

The regression modeling strategy for the five outcomes as functions of dose and subgroup is to specify a marginal distribution for each $Y_{i,j} \mid d_i, \mathbf{Z}_i$ and use iid multivariate normal patient frailties to induce a joint distribution on $\boldsymbol{Y}_i \mid d_i, \mathbf{Z}_i$. The marginal of the jth event time distribution is assumed to be Weibull,

$$Y_{i,j} \mid \alpha_j, \lambda_{i,j} \overset{indep}{\sim} \text{Weibull}(\alpha_j, e^{\lambda_{i,j}}) \tag{3.5}$$

for $i = 1, \ldots, n(t)$ and $j \in \{P, R, T, C, D\}$, with shape parameter $\alpha_j > 0$ and scale parameter $\exp(\lambda_{i,j})$, where $\lambda_{i,j}$ is a function of $(d_i, \mathbf{Z}_i)$. The hazard and survival functions are

$$h_j(y \mid \alpha_j, \lambda_{i,j}) = \alpha_j e^{\lambda_{i,j}} y^{\alpha_j - 1} \quad \text{and} \quad S_j(y \mid \alpha_j, \lambda_{i,j}) = \exp\{-e^{\lambda_{i,j}} y^{\alpha_j}\}$$

for $y > 0$. Denoting the sample size and data up to trial time t by $n(t)$ and $\mathscr{D}_{n(t)}$ and the parameter vector by $\boldsymbol{\theta}$, the joint conditional likelihood is the product over patients and outcomes given by

$$\mathscr{L}(\boldsymbol{\theta} \mid \mathscr{D}_{n(t)}) = \prod_{i=1}^{n(t)} \prod_{j \in \{P,R,T,C,D\}} h_j(Y_j^o \mid \alpha_j, \lambda_{i,j})^{\delta_j} \, S_j(Y_{i,j}^o \mid \alpha_j, \lambda_{i,j}).$$

Given the form (3.5) of the Weibull marginals, a joint model is formulated to account for the effects of both dose and subgroup on each Y_{ij} by including dose-subgroup parameters in each $\lambda_{i,j}$ and defining $u_{i,j}$ to be a latent frailty associated with patient i for outcome j, with $\boldsymbol{u}_i = (u_{i,P}, \cdots, u_{i,D})$. The relationships between $Y_{i,j}$, d_i, and $\mathbf{Z}_i = (Z_i, r_i)$ are based on the following regression model, given the frailties,

$$\lambda_{i,j}(\mathbf{Z}_i, d_i, u_{i,j}) = \beta_j Z_i + \xi_j \psi_{r_i, d_i} + u_{i,j}, \tag{3.6}$$

requiring $\xi_D \equiv 1$ to ensure identifiability. The parameter $\psi_{r,d}$ is the effect of d on the death rate for disease type r, with a larger value of $\psi_{r,d}$ corresponding to a higher risk of death for patients with disease type r given dose d. The multiplicative parameter ξ_j scales ψ_{r_i,d_i} for outcome $j \neq D$, with $\xi_j > 0$ implying a higher risk and $\xi_j < 0$ a lower risk for Y_j. The parameter β_j is the additive effect of HBD ($Z = 1$), and $\beta_j > 0$ reflects a higher rate, equivalently a smaller mean E(Y_j), for HBD versus LBD. Compared to a saturated model with $6 \times 3 \times 5 = 90$ values of $\psi_{Z,r,d,j}$, the regression model in (3.6) reduces the number of parameters to $5 + 9 + 4 = 18$. The model thus is parsimonious. It provides a practical basis for adaptive subgroup-specific decision-making, given the trial's limited sample size, while still being flexible enough to accommodate a wide variety of possible relationships between d, $\mathbf{Z}$, and $\boldsymbol{Y}$.

Priors for the Frailties and Parameters

For the patient frailties, it is assumed that $\boldsymbol{u}_i \sim iid \ \text{N}_5(\mathbf{0}, \Omega)$ with $\Omega \sim$ inverse-Wishart(ν, Ω^0) for fixed $\nu > J - 1$ and 5×5 positive definite hyperparameter matrix Ω^0. Incorporating $\{\boldsymbol{u}_i, i = 1, \cdots, n\}$ into the five hazard functions accounts for possible heterogeneity between patients beyond that due to the prognostic subgroups, and the correlations among the $u_{i,j}$'s induce association among the outcomes of each patient. Combining (3.5) and (3.6), the conditional hazard function of $[Y_{ij} \mid \mathbf{Z}_i, u_{i,j}]$ is

$$h_j(y \mid \alpha_j, \xi_j, \psi_{r_i,d_i}, Z_i, u_{i,j}) = \alpha_j \, y^{\alpha_j - 1} \, \exp(\beta_j Z_i + \xi_j \psi_{r_i,d_i} + u_{i,j}), \quad y > 0.$$

Conditional independence of the elements of $\boldsymbol{Y}_i$ given $\boldsymbol{u}_i$ and $\boldsymbol{\theta}$ on the set on $\mathscr{Y}_D$ is assumed. Again, each $S_j(y \mid \alpha_j, \xi_j, \psi_{r_i,d_i}, Z_i)$ is not a survival function, due to the competing and semicompeting risks framework. Suppressing the patient index i, the joint survival function for $\mathbf{y} \in \mathscr{Y}_D$ is obtained by averaging over the frailty distribution,

$$S(y', y_T, y_C, y_D \mid \boldsymbol{\alpha}, \boldsymbol{\xi}, \psi_{r,d}, Z, \nu, \Omega^0) =$$

$$\int_{\boldsymbol{u}} \prod_{j \in \{P,R\}} S_j(y' \mid \alpha_j, \xi_j, \psi_{r,d}, u_j) \prod_{j \in \{T,C,D\}} S_j(y_j \mid \alpha_j, \xi_j, \psi_{r,d}, u_j) p(\boldsymbol{u}, \Omega \mid \nu, \Omega^0) d\boldsymbol{u}.$$

Priors for the parameters $\boldsymbol{\alpha} = (\alpha_P, \alpha_R, \alpha_T, \alpha_C, \alpha_D)$, $\beta = (\beta_P, \beta_R, \beta_T, \beta_C, \beta_D)$, $\boldsymbol{\xi} = (\xi_P, \xi_R, \xi_T, \xi_C)$ and $\psi = (\psi_{r,d}, r = 1, \ldots, K, d = 1, \ldots, m)$ are specified as follows. Each $\alpha_j \sim$ *indep* Gamma(a_j, b_j), a gamma distribution with mean a_j/b_j and variance $a_j/b_j{}^2$. Each $\xi_j \sim$ *indep* $\mathrm{N}(\bar{\xi}_j, \omega_j^2)$, with $\xi_D \equiv 1$. Each $\psi_{r,d} \sim$ *indep* $\mathrm{N}(\bar{\psi}_r, \tau_r^2)$ to allow the diseases CLL, ALL, NHL ($r = 1,2,3$) to have different outcome hazards. To reflect higher hazards of adverse outcomes for patients with HBD, their effects are assumed to follow normal distributions truncated below at 0, with $p(\beta_j) \propto \exp\{-(\beta_j - \bar{\beta}_j)^2/(2\sigma_j^2)\}$ for $\beta_j > 0$, for $j = T, C, D$, where $\bar{\beta}_j$ and σ_j^2 denote fixed hyperparameters for all j. The priors express no information on the directions of the HBD effects on the hazards of the sub-distributions of $Y_{i,P}$ and $Y_{i,R}$, so $\beta_j \sim \mathrm{N}(\bar{\beta}_j, \sigma_j^2)$ for $j = P, R$. Denote $\boldsymbol{\theta} = (\boldsymbol{\alpha}, \beta, \boldsymbol{\xi}, \psi, \Omega)$, and the hyperparameter vector

$$\boldsymbol{\theta}^* = (\mathbf{a}, \mathbf{b}, \bar{\boldsymbol{\xi}}, \omega^2, \bar{\psi}, \boldsymbol{\tau}^2, \bar{\beta}, \boldsymbol{\sigma}^2, \nu, \Omega^0)$$

where

$$\begin{aligned}
(\mathbf{a}, \mathbf{b}) &= \{(a_j, b_j), j = P, R, T, C, D\} \\
(\bar{\boldsymbol{\xi}}, \boldsymbol{\omega}^2) &= \{(\bar{\xi}_j, \omega_j^2), j = P, R, T, C\} \\
(\bar{\psi}, \boldsymbol{\tau}^2) &= \{(\bar{\psi}_r, \tau_r^2), r = 1, \ldots, K\} \\
(\bar{\beta}, \boldsymbol{\sigma}^2) &= \{(\bar{\beta}_j, \sigma_j^2), j = P, R, T, C, D\}.
\end{aligned}$$

To establish numerical values of the hyperparameters, $\boldsymbol{\theta}^*$, probabilities of the 12 joint events occurring within 30 days for C and 100 days for all other events $j \neq C$ were elicited for each $j \neq C$ from the clinical investigators, given in Table 3.3. Denote

$$\boldsymbol{t}'_F = (t'_{F,P}, \, t'_{F,R}, \, t'_{F,T}, \, t'_{F,C}, \, t'_{F,D}) = (100, 100, 100, 30, 100).$$

Solving the resulting sets of equations under the assumed model gave the prior means, and dispersion parameters were calibrated to reflect vague prior knowledge.

TABLE 3.3
Probabilities $\pi_{j,Z,r}^{(e)}$ elicited from the clinicians, used to establish values for the prior hyperparameters, $\bar{\psi}$, $\bar{\beta}$, $\bar{\xi}$ and a for the prior in the model underlying the NK cell dose-finding design.

Disease Bulk	Disease Type	Prog	Response	Severe Tox	Death
Low Bulk	CLL ($r=1$)	0.05	0.20	0.25	0.02
Disease	ALL ($r=2$)	0.15	0.50	0.25	0.10
($Z=0$)	NHL ($r=3$)	0.10	0.35	0.25	0.05
High Bulk	CLL ($r=1$)	0.20	0.35	0.25	0.10
Disease	ALL ($r=2$)	0.40	0.60	0.40	0.20
($Z=1$)	NHL ($r=3$)	0.40	0.40	0.25	0.15

Given $\boldsymbol{\theta}^*$ and interim data $\mathscr{D}_{n(t)}$ at trial time t, the joint posterior of all parameters $\boldsymbol{\theta}$ and patient-specific random effects $\boldsymbol{u} = \{\boldsymbol{u}_i, i=1,\ldots,n(t)\}$ is

$$p(\boldsymbol{\theta},\boldsymbol{u} \mid \mathscr{D}_{n(t)},\boldsymbol{\theta}^*) \propto \prod_{i=1}^{n(t)} p(\boldsymbol{y}_i^o,\boldsymbol{\delta}_i \mid \boldsymbol{\theta},\boldsymbol{u}_i) p(\boldsymbol{\theta},\boldsymbol{u} \mid \boldsymbol{\theta}^*) =$$

$$\prod_{i=1}^{n(t)} \prod_{j\in\{P,\ldots,D\}} h_j(y_{i,j}^o \mid \alpha_j,\lambda_{i,j})^{\delta_{i,j}(t)} S_j(y_{i,j}^o \mid \alpha_j,\lambda_{i,j}) p(\boldsymbol{\theta},\boldsymbol{u} \mid \boldsymbol{\theta}^*).$$

MCMC simulation was used to generate posterior samples of $\boldsymbol{\theta}$ and $\boldsymbol{u}$.

Utilities were computed for all combinations of events occurring within 30 days for C and within 100 days for all $j \neq C$, corresponding to the elicited probabilities in Table 3.3. Denote $\boldsymbol{\delta}' = (\delta_P', \delta_R', \delta_T', \delta_C', \delta_D')$, where each $\delta_j' = 1$ if Y_j is observed by follow-up time $t_{F,j}'$, and 0 otherwise. Since $(\delta_P', \delta_R') = (1,1)$ is impossible since P and R are competing risks, there are $3 \times 2^3 = 24$ possible outcome indicator vectors $\boldsymbol{\delta}' \in \Delta$. Denote the subset of patients who survive 100 days ($\delta_D = 0$) by $\Delta^0 \subseteq \Delta$. The key to obtaining a practical utility-based design is that utilities were elicited on the set Δ^0, rather than attempting to do this for all possible values of $\boldsymbol{Y}$. For the elicitation, a minimum utility $U(\boldsymbol{\delta}') = 0$ first was defined if $\delta_D' = 1$, that is, if a patient died before day 100. This gives a total of $|\Delta^0| = 12$ possible early event combinations for the 100-day survivors. Thus, computing the posterior mean utility for each $(d,\boldsymbol{Z})$ only required evaluation of $\pi(\boldsymbol{\delta}' \mid d,\boldsymbol{Z},\boldsymbol{\theta})$ for the 12 indicator vectors $\boldsymbol{\delta}' \in \Delta^0$. The elicited utilities $U(\boldsymbol{\delta}')$ for all $\boldsymbol{\delta}' \in \Delta^0$ are given in Table 3.4.

Computing mean utilities exploits the fact that the distribution of $[\boldsymbol{\delta}' \mid d,\boldsymbol{Z}]$ is induced by the distribution of $[\boldsymbol{Y} \mid d,\boldsymbol{Z}]$. For example,

$$\pi((1,0,1,0,0) \mid d,\boldsymbol{Z},\boldsymbol{\theta}) =$$

$$\mathrm{P}(Y_P \le t_{F,P}', Y_R > Y_P, Y_T \le t_{F,T}', Y_C > t_{F,C}', Y_D > t_{F,D}' \mid d,\boldsymbol{Z},\boldsymbol{\theta}) =$$

$$\int_0^{t_{F,P}'} \int_{Y_P}^{\infty} \int_0^{t_{F,T}'} \int_{t_{F,C}'}^{\infty} \int_{t_{F,D}'}^{\infty} \int_{\mathbb{R}^5} p(\mathbf{y} \mid d,\boldsymbol{Z},\boldsymbol{u},\boldsymbol{\theta}) p(\boldsymbol{u} \mid \boldsymbol{\theta}) d\boldsymbol{u}\, d\mathbf{y} =$$

TABLE 3.4
Elicited utilities $U(\boldsymbol{\delta}')$ of the 12 possible combinations of discrete outcomes for patients who survive 100 days for the NK cell dose finding design. If $\delta_D' = 1$ then $U(\boldsymbol{\delta}') = 0$.

δ_C'	δ_T'	(δ_P', δ_R') (1,0)	(0,0)	(0,1)
0	0	20	50	90
	1	10	30	70

δ_C'	δ_T'	(δ_P', δ_R') (1,0)	(0,0)	(0,1)
1	0	10	30	70
	1	5	20	50

$$\int_0^{100} \int_{Y_P}^{\infty} \int_0^{100} \int_{30}^{\infty} \int_{100}^{\infty} \int_{\mathbb{R}^5} p(\mathbf{y} \mid d, \mathbf{Z}, \mathbf{u}, \boldsymbol{\theta}) p(\mathbf{u} \mid \boldsymbol{\theta}) d\mathbf{u}\, d\mathbf{y}.$$

Given $\boldsymbol{\theta}$, the mean utility of giving dose d to a patient with covariates $\mathbf{Z}$ is

$$\bar{U}(d, \mathbf{Z}, \boldsymbol{\theta}) = \sum_{\boldsymbol{\delta}' \in \Delta^0} U(\boldsymbol{\delta}')\, \pi(\boldsymbol{\delta}' \mid d, \mathbf{Z}, \boldsymbol{\theta}).$$

To estimate $\bar{U}(d, \mathbf{Z}, \boldsymbol{\theta})$, the design uses posterior predictive mean utilities. Given the final data, $\mathscr{D}_{N_{\max}}$, for a future patient with covariates $\mathbf{Z}$, the posterior predictive mean utility of giving dose d to that patient is

$$u(d, \mathbf{Z} \mid \mathscr{D}_{N_{\max}}) = \int_{\boldsymbol{\theta}} \bar{U}(d, \mathbf{Z}, \boldsymbol{\theta}) p(\boldsymbol{\theta} \mid \mathscr{D}_{N_{\max}}) d\boldsymbol{\theta}. \tag{3.7}$$

While utilities were elicited over the early follow-up intervals of 30 or 100 days, all follow-up data on $(\boldsymbol{Y}_i, \boldsymbol{\delta}_i)$, for $i = 1, \cdots, N_{\max}$ up to 365 days were included in $\mathscr{D}_{N_{\max}}$.

For trial design and conduct, the following subgroup-specific safety monitoring rules were used. For each $(d, \mathbf{Z})$, denote $\pi_D(d, \mathbf{Z}, \boldsymbol{\theta}) = \mathrm{P}(Y_D \leq 100 \mid d, \mathbf{Z}, \boldsymbol{\theta}) = \mathrm{P}(\delta_D' = 1 \mid d, \mathbf{Z}, \boldsymbol{\theta})$. Let $\bar{\pi}_D(\mathbf{Z})$ denote an elicited fixed upper limit on $\pi_D(d, \mathbf{Z}, \boldsymbol{\theta})$ for subgroup $\mathbf{Z}$, and let $p_{D,1}$ be a fixed cut-off probability. At trial time t, if

$$\mathrm{P}\{\pi_D(d, \mathbf{Z}, \boldsymbol{\theta}) > \bar{\pi}_D(\mathbf{Z}) \mid \mathscr{D}_{n(t)}\} > p_{D,1} \tag{3.8}$$

then d is considered unsafe for subgroup $\mathbf{Z}$ due to an unacceptably high early death rate, and is no longer administered to patients in that subgroup. Elicited values of $\bar{\pi}_D(\mathbf{Z})$ are given in Table 3.5, where simulation results are summarized. To obtain a design with high subgroup-specific probabilities of stopping a truly unsafe dose and selecting the best safe dose for each $\mathbf{Z}$, based on preliminary simulations the cut-off $p_{D,1} = 0.80$ was used.

The design is defined in terms of the possible actions $\mathscr{A} = \{0, 1, 2, 3\}$, where $d = 1, 2, 3$ are doses and $d = 0$ denotes the action to not administer any NK cells. Let $\mathscr{A}(\mathbf{Z}, \mathscr{D}_{n(t)})$ be the subset of acceptable actions for a patient with covariates $\mathbf{Z}$ at trial time t based on interim data $\mathscr{D}_{n(t)}$. If no doses are safe for $\mathbf{Z}$, i.e. $\mathscr{A}(\mathbf{Z}, \mathscr{D}_{n(t)}) = \{0\}$, then no patient in subgroup $\mathbf{Z}$ is treated. The subgroup-specific nature of the design is reflected by the fact that the acceptable dose sets $\mathscr{A}(\mathbf{Z}_1, \mathscr{D}_{n(t)})$ and $\mathscr{A}(\mathbf{Z}_2, \mathscr{D}_{n(t)})$ may differ for $\mathbf{Z}_1 \neq \mathbf{Z}_2$ at trial time t. Moreover, these sets may change adaptively as data accumulate.

TABLE 3.5
Simulation Results for the NK cell dose-finding design. π_D^{true} = true probability of death within 100 days for each combination of disease type (r), dose level (d), disease bulk (Z). $\bar{\pi}(\mathbf{Z})$ = fixed safety threshold. Death rates for unsafe doses are given in red. $\bar{U}^{true}$ = true mean utility. Optimal doses are underlined.

		LBD ($Z=0$)				HBD ($Z=1$)			
		$d=1$	$d=2$	$d=3$	$\bar{\pi}_D$	$d=1$	$d=2$	$d=3$	$\bar{\pi}_D$
Scenario 1									
CLL ($r=1$)	π_D^{true}	0.02	0.02	0.02	*0.15*	0.10	0.10	0.10	*0.30*
	$\bar{U}^{true}$	<u>46.32</u>	<u>46.32</u>	<u>46.32</u>		<u>44.04</u>	<u>44.04</u>	<u>44.04</u>	
	P(Stop)	0.00	0.00	0.00		0.00	0.00	0.00	
	P(Select)	0.33	0.37	0.30		0.33	0.34	0.33	
ALL ($r=2$)	π_D^{true}	0.10	0.10	0.10	*0.20*	0.25	0.25	0.25	*0.40*
	$\bar{U}^{true}$	<u>50.52</u>	<u>50.52</u>	<u>50.52</u>		<u>37.97</u>	<u>37.97</u>	<u>37.97</u>	
	P(Stop)	0.00	0.01	0.00		0.03	0.03	0.02	
	P(Select)	0.33	0.35	0.32		0.32	0.36	0.33	
NHL ($r=3$)	π_D^{true}	0.05	0.05	0.05	*0.20*	0.15	0.15	0.15	*0.40*
	$\bar{U}^{true}$	<u>49.18</u>	<u>49.18</u>	<u>49.18</u>		<u>38.35</u>	<u>38.35</u>	<u>38.35</u>	
	P(Stop)	0.00	0.00	0.00		0.00	0.00	0.00	
	P(Select)	0.33	0.34	0.33		0.35	0.31	0.34	
Scenario 2									
CLL ($r=1$)	π_D^{true}	0.02	0.45	0.60	*0.15*	0.04	0.70	0.84	*0.30*
	$\bar{U}^{true}$	<u>42.34</u>	22.86	16.00		<u>39.73</u>	9.49	4.63	
	P(Stop)	0.00	0.89	0.97		0.00	0.98	1.00	
	P(Select)	1.00	0.00	0.00		1.00	0.00	0.00	
ALL ($r=2$)	π_D^{true}	0.40	0.60	0.05	*0.20*	0.64	0.84	0.10	*0.40*
	$\bar{U}^{true}$	23.81	15.82	<u>40.41</u>		10.99	4.62	<u>36.02</u>	
	P(Stop)	0.67	0.92	0.00		0.81	0.98	0.00	
	P(Select)	0.00	0.00	1.00		0.00	0.00	1.00	
NHL ($r=3$)	π_D^{true}	0.65	0.05	0.35	*0.20*	0.88	0.10	0.58	*0.40*
	$\bar{U}^{true}$	14.03	<u>40.47</u>	26.31		3.46	<u>36.08</u>	13.26	
	P(Stop)	0.95	0.00	0.55		1.00	0.00	0.68	
	P(Select)	0.00	1.00	0.00		0.00	1.00	0.00	
Scenario 3									
CLL ($r=1$)	π_D^{true}	0.42	0.38	0.37	*0.15*	0.66	0.62	0.60	*0.30*
	$\bar{U}^{true}$	40.33	44.40	<u>44.55</u>		20.71	24.52	<u>24.81</u>	
	P(Stop)	0.88	0.85	0.82		0.96	0.95	0.94	
	P(Select)	0.07	0.11	0.13		0.03	0.05	0.05	
ALL ($r=2$)	π_D^{true}	0.52	0.58	0.65	*0.20*	0.77	0.83	0.88	*0.40*
	$\bar{U}^{true}$	<u>33.99</u>	29.43	24.52		<u>14.34</u>	11.01	7.57	
	P(Stop)	0.93	0.96	0.98		0.99	0.99	1.00	
	P(Select)	0.06	0.03	0.01		0.01	0.01	0.00	
NHL ($r=3$)	π_D^{true}	0.40	0.42	0.45	*0.20*	0.64	0.67	0.70	*0.40*
	$\bar{U}^{true}$	<u>42.49</u>	40.21	38.79		<u>22.61</u>	20.32	18.95	
	P(Stop)	0.73	0.77	0.84		0.85	0.87	0.94	
	P(Select)	0.19	0.16	0.07		0.12	0.11	0.04	

Table 3.5 (continued)

		LBD ($Z = 0$)				HBD ($Z = 1$)			
		$d = 1$	$d = 2$	$d = 3$	$\bar{\pi}_D$	$d = 1$	$d = 2$	$d = 3$	$\bar{\pi}_D$
Scenario 4									
CLL	π_D^{true}	0.01	0.10	0.25	*0.15*	0.01	0.11	0.27	*0.30*
($r = 1$)	$\bar{U}^{true}$	48.99	<u>58.91</u>	45.84		<u>38.03</u>	24.74	14.45	
	P(Stop)	0.00	0.00	0.26		0.00	0.00	0.13	
	P(Select)	0.04	0.95	0.01		1.00	0.00	0.00	
ALL	π_D^{true}	0.01	0.09	0.27	*0.20*	0.01	0.10	0.29	*0.40*
($r = 2$)	$\bar{U}^{true}$	48.82	<u>58.66</u>	43.95		<u>37.54</u>	26.47	14.55	
	P(Stop)	0.00	0.00	0.14		0.00	0.00	0.05	
	P(Select)	0.02	0.98	0.00		1.00	0.00	0.00	
NHL	π_D^{true}	0.01	0.08	0.30	*0.20*	0.01	0.09	0.33	*0.40*
($r = 3$)	$\bar{U}^{true}$	48.87	<u>58.22</u>	40.45		<u>40.28</u>	26.92	11.70	
	P(Stop)	0.00	0.00	0.29		0.00	0.00	0.12	
	P(Select)	0.01	0.99	0.00		1.00	0.00	0.00	
Scenario 5									
CLL	π_D^{true}	0.01	0.09	0.30	*0.15*	0.01	0.12	0.38	*0.30*
($r = 1$)	$\bar{U}^{true}$	<u>44.30</u>	33.82	22.05		<u>41.40</u>	27.18	14.33	
	P(Stop)	0.00	0.00	0.48		0.00	0.00	0.44	
	P(Select)	1.00	0.00	0.00		1.00	0.00	0.00	
ALL	π_D^{true}	0.12	0.03	0.18	*0.20*	0.16	0.04	0.23	*0.40*
($r = 2$)	$\bar{U}^{true}$	30.68	<u>40.15</u>	27.96		23.62	<u>35.88</u>	20.52	
	P(Stop)	0.00	0.00	0.02		0.00	0.00	0.01	
	P(Select)	0.00	1.00	0.00		0.00	0.99	0.00	
NHL	π_D^{true}	0.10	0.15	0.01	*0.20*	0.13	0.20	0.01	*0.40*
($r = 3$)	$\bar{U}^{true}$	32.73	28.73	<u>44.09</u>		26.24	21.60	<u>41.22</u>	
	P(Stop)	0.00	0.01	0.00		0.00	0.00	0.00	
	P(Select)	0.00	0.00	1.00		0.00	0.00	1.00	
Scenario 6									
CLL	π_D^{true}	0.35	0.03	0.13	*0.15*	0.75	0.10	0.37	*0.30*
($r = 1$)	$\bar{U}^{true}$	41.74	<u>59.80</u>	57.69		14.53	<u>55.48</u>	40.90	
	P(Stop)	0.76	0.00	0.09		0.99	0.00	0.34	
	P(Select)	0.00	0.56	0.44		0.00	0.99	0.01	
ALL	π_D^{true}	0.08	0.45	0.02	*0.20*	0.24	0.86	0.06	*0.40*
($r = 2$)	$\bar{U}^{true}$	57.75	34.95	<u>57.83</u>		47.19	7.81	<u>55.07</u>	
	P(Stop)	0.00	0.82	0.00		0.02	0.99	0.00	
	P(Select)	0.80	0.00	0.20		0.04	0.00	0.96	
NHL	π_D^{true}	0.05	0.10	0.30	*0.20*	0.16	0.29	0.70	*0.40*
($r = 3$)	$\bar{U}^{true}$	<u>59.50</u>	57.18	46.28		<u>52.57</u>	44.10	18.84	
	P(Stop)	0.00	0.01	0.51		0.00	0.03	0.91	
	P(Select)	0.53	0.47	0.01		0.89	0.11	0.00	

Unlike traditional dose-finding trials with cytotoxic agents, with NK cell therapy there is no biological reason to assume that any h_j increases monotonically with cell dose. The design thus assigns patients to doses using a random sequential allocation rule, with disease types as blocks. During trial conduct, for each disease type *r*, patients are randomized among the three doses in order of entry to the trial by randomly permuting the integers $(1,2,3)$. Safety monitoring is begun for each disease type *r* when nine patients have been enrolled in *r* and at least five of the nine have either died or been followed for 100 days. While disease bulk *Z* is not used to define cohorts or decide when safety monitoring will begin, it is used with *r* for determining safety. For each disease type *r*, the action sets are $\mathscr{A}((0,r),t)$ and $\mathscr{A}((1,r),t)$, for the two disease bulk subgroups. E.g., suppose that the initial permuted dose blocks are $(3,1,2)$ for $r = 1$, $(3,2,1)$ for $r = 2$, and $(2,1,3)$ for $r = 3$. Thus, the first three patients in the cohort with disease type $r = 1$ are assigned to doses $(3,1,2)$ as they enroll. Once safety monitoring is begun, unsafe doses are eliminated from each block adaptively. For example, if the design gives doses $(3,1,2)$ for a cohort with disease type *r*, the following two possible cases illustrate details of trial conduct.

Case 1. Suppose that the first patient in the cohort has $\mathbf{Z} = (0,r)$. If the updated $\mathscr{A}((0,r),t) = \{0\}$, i.e. no dose is safe for this subgroup, no NK cells are given to the patient. If $3 \in \mathscr{A}(\mathbf{Z},t)$, the patient is treated at $d = 3$. If not, move on to a dose in the permutation that has not been used and is safe for $\mathbf{Z}$.

Case 2. Suppose that $d = 3$ was given to the previous patient in the cohort, and the next patient in the cohort has $\mathbf{Z} = (1,r)$. Update $\mathscr{A}(1,r)$ using the most recent data. If $\mathscr{A}((1,r),t) = \{0\}$, do not give any NK cells to the patient. If $1 \in \mathscr{A}((1,r),t)$, then give $d = 1$ to the patient. If not, proceed to $d = 2$. If $d = 2$ is not safe, then $d = 3$ is the only safe dose for $\mathbf{Z} = (1,r)$, since all doses in $(3,1,2)$ have been used. A new cohort is started by randomly permuting (1, 2, 3). Suppose this gives $(1,3,2)$. Since $d = 1$ must be skipped since it is not safe, $d = 3$ is given to the patient. At this point, only $d = 2$ is left in that block for the next patient with disease type *r*.

An additional rule imposed by a federal regulatory agency (FRAG) also was included in the design. After reviewing the design, the FRAG insisted on a "one size fits all" safety rule that ignores $\mathbf{Z}$ and stops the trial if the estimated probability of death within 30 days at $d = 1$ is too high, compared to a fixed overall death rate. To comply with this federally mandated safety rule requirement, a simplified model was formulated as a basis for constructing this rule. This model assumed that $\delta_D(30) \mid q_D \sim \text{Ber}(q_D)$, where $q_D = \Pr(Y_D < 100 \mid d = 1)$ for all $\mathbf{Z}$, with prior $q_D \sim \text{Be}(0.4, 0.6)$. The FRAG rule says that, if $\text{P}(q_D > .40 \mid \mathscr{D}_{n(t)}) > .90$, then the trial should be stopped with the conclusion that no dose is safe for any patient. Thus, the entire trial could be stopped either by the FRAG rule, or by the subgroup-specific safety rules if all subgroups were found to be unsafe.

To determine a final optimal action for each $\mathbf{Z}$ when $N_{\max} = 60$ at $T_{\max}$, $\mathscr{A}(\mathbf{Z}, T_{\max})$ is identified using the safety rule in (3.8). If $\mathscr{A}(\mathbf{Z}, T_{\max}) = \{0\}$, then no dose is selected for $\mathbf{Z}$, denoted by $d_{\text{sel}}(\mathbf{Z}) = 0$. If $\mathscr{A}(\mathbf{Z}, T_{\max}) \neq \{0\}$, then the optimal dose for a patient in subset $\mathbf{Z}$ is

$$d_{\text{sel}}(\mathbf{Z}) = \underset{d \in \mathscr{A}(\mathbf{Z}, T_{\max})}{argmax} \; u(d, \mathbf{Z} \mid \mathscr{D}_{N_{\max}}).$$

The NK cell dose-finding trial design was simulated under six scenarios to evaluate its performance. For Scenario 1, the prior occurrence probabilities $\{\pi^{(e)}_{j,Z,r}\}$ elicited from the clinicians, given in Table 3.3, were used to simulate data, with fixed "true" parameter values α^{true}_j and $\bar{\lambda}^{true}_{j,Z,r,d}$ determined by solving the equations

$$\pi^{(e)}_{j,Z,r} = 1 - S(t'_{F,j} \mid \alpha^{true}_j, \bar{\lambda}^{true}_{j,Z,r,d}) = 1 - \exp\{-\exp(\bar{\lambda}^{true}_{j,Z,r,d})(t'_{F,j})^{\alpha^{true}_j}\} \qquad (3.9)$$

for (j, Z, r).

In Scenario 1, there is dose has no effect, with the same $\bar{\lambda}^{true}_{j,Z,r,d}$ for all d and no regression relationship in (3.6) for $\bar{\lambda}^{true}$. Data were simulated from the Weibull distribution,

$$y_{i,j} \mid \alpha^{true}_j, \bar{\lambda}^{true}_{j,Z_{1i},r_i,d_i}, u^{true}_{i,j} \overset{indep}{\sim} \text{Weibull}(\alpha^{true}_j, \exp(\bar{\lambda}^{true}_{j,Z_{1i},r_i,d_i} + u^{true}_{i,j})),$$

where $\boldsymbol{u}^{true}_i \overset{iid}{\sim} \text{N}_5(\mathbf{0}, \Omega^{true})$, with $\Omega^{true}_{j,j} = 0.001$, $\Omega^{true}_{j,R} = -0.5 \times 0.001$, $j \neq R$ and $\Omega^{true}_{j,j'} = 0.5 \times 0.001$, $j \neq j'$, $j, j' \neq R$.

Scenarios 2–6 assumed the same Ω^{true} and specified $\boldsymbol{\alpha}^{true}$ and true marginal probability of death by 100o days for LBD ($Z = 0$), disease type r and dose d,

$$\pi^{true}_{D,0,r,d} = \text{P}(Y_D \leq 100 \mid d, Z = 0, r).$$

The survival function in (3.9) with α^{true}_D and $\pi^{true}_{D,0,r,d}$ gives $\bar{\lambda}^{true}_{D,0,r,d}$. For subgroups with $Z = 1$ and the other outcomes, specified parameters were $\bar{\xi}^{true}_j$ with simulation of $\xi^{true}_{j,r,d} \sim indep\, \text{N}(\bar{\xi}^{true}_j, 0.01^2)$ for all combinations of (j, r, d) with $\xi^{true}_{D,r,d} = 1$. Similarly, $\bar{\beta}^{true}_j$ was specified with simulation of $\beta^{true}_{j,r,d} \sim indep\, \text{N}(\bar{\beta}^{true}_j, 0.01^2)$, $j = P, R$, and $\log(\beta^{true}_{j,r,d}) \sim indep\, \text{N}(\log(\bar{\beta}^{true}_j), 0.05^2)$, $j = T, C, D$, to ensure that $\beta^{true}_{j,r,d} > 0$ for adverse outcomes. The rate parameters were

$$\bar{\lambda}^{true}_{j,Z,r,d} = \beta^{true}_{j,r,d} Z + \xi^{true}_{j,r,d} \psi^{true}_{D,r,d}$$

and

$$\lambda^{true}_{i,j} = \bar{\lambda}^{true}_{j,Z_i,r_i,d_i} + u^{true}_{i,j}.$$

The outcome $Y_{i,j}$ for a patient with $\mathbf{Z}_i$ was simulated from Weibull$(\alpha^{true}_j, \lambda^{true}_{i,j})$. Under the model assumed for the simulation truth, β^{true} and ξ^{true} are indexed by j, r and d. This marginal simulation model is more complex, and it includes the design's assumed Weibull model as a special case by letting $\beta^{true}_{j,r,d} = \beta_j$ and $\xi^{true}_{j,r,d} = \xi_j$ for all (r, d). The assumed true probabilities of death, $\pi^{true}_{D,Z,r,d}$ for each (Z, r, d), are shown on the first lines of the simulation scenario boxes in Table 3.5, with the probabilities exceeding the subgroup-specific upper limits $\bar{\pi}_D(\mathbf{Z})$ marked in grey. The second lines give the true expected utilities $\bar{U}^{true}$ for each $(\mathbf{Z},d)$, and the maximum utility for each $\mathbf{Z}$ is underlined. For example, all doses are safe for all $\mathbf{Z}$ under Scenario 1, while all doses unsafe for all $\mathbf{Z}$ under Scenario 3. Under Scenario 1, all doses are equally good, while under Scenarios 2, 4, 5, and 6, the optimal safe doses vary with $\mathbf{Z}$ and using patient subgroup information is critical. Under Scenario 4, doses 2 and 1 are optimal for patients with $Z=0$ and $Z=1$, respectively, regardless of r. Under Scenario 5, the optimal doses vary with disease type r but not with disease bulk Z. Under Scenario 6, the true mean utilities vary with (d,Z,r), and the set of acceptable doses varies with $\mathbf{Z}$. A total of $M=1000$ trials were simulated under each scenario.

Additional simulations examined the design's robustness by generating the $Y_{i,j}$'s from a log-logistic distribution. To obtain fair comparisons, in each scenario, given $\pi^{true}_{j,Z,r,d}$ values under the Weibull distribution, true parameter values under the log-logistic distribution were obtained by matching the $\pi^{true}_{j,Z,r,d}$'s, so the marginal probabilities of occurrence during follow-up were identical for the two models, and truly unsafe doses remained unsafe regardless of the assumed true model used to simulate the data. The rates of occurrence over time under the two models necessarily differed, which in turn caused the mean utilities to change, due to P and R being competing risks and the semi-competing risks structure between D and the other outcomes on $\mathscr{Y}_D$, despite the fact that the marginal probabilities were identical.

As a comparator, a simplified version of the design that does not use covariates or make subgroup-specific inferences was used. This design assumed a simpler model ignoring $\mathbf{Z}$, but it still accounted for the five event times and their competing risks and semi-competing risks relationships. For each j, this model assumes

$$Y_{i,j} \mid \alpha'_j, \lambda'_{i,j} \overset{indep}{\sim} \text{Weibull}(\alpha'_j, \exp(\lambda'_{i,j})),$$

where $\lambda'_{i,j} = \xi'_j \psi'_{d_i} + u'_{i,j}$ with $\xi'_D = 1$ and $\boldsymbol{u}'_i \overset{iid}{\sim} \text{N}_5(\mathbf{0},\Omega)$. Similar to the full model, for the simpler model, we assumed a gamma prior for α'_j, normal priors for ξ'_j and ψ'_d and an inverse Wishart prior for Ω. Under the simpler design, we defined $\mathscr{A}(t)$ as a function of t only, so if a dose was declared unsafe this pertained to all patients. A similar randomization with blocks of size $m=3$ was used for allocating patients to doses in $\mathscr{A}(t)$, and a dose d was declared unsafe if

$$\Pr(\pi'_D(d,\boldsymbol{\theta}') > .25 \mid \mathscr{D}_{n(t)}) > .80,$$

where $\pi'_D(d,\boldsymbol{\theta}')$ is the probability of death within 100 days, with d and $\boldsymbol{\theta}' = (\boldsymbol{\alpha}', \psi', \boldsymbol{\xi}')$. Fixed prior hyperparameters under the simpler model were specified, by using the elicited probabilities in Table 3.3, but ignoring any effects of $\mathbf{Z}$. Posterior mean utility, for each d but ignoring $\mathbf{Z}$, was used as a criterion to choose an optimal dose for any future patient.

The designs were evaluated using two criteria, the probabilities of identifying doses with truly excessive probabilities of death, and of selecting the true optimal safe dose for each $\mathbf{Z}$. For each simulated trial $\ell = 1,\dots,M$, and each $\mathbf{Z}$, each design selects a dose $d_{\text{sel},\ell}(\mathbf{Z})$, with $d_{\text{sel},\ell}(\mathbf{Z}) \equiv d_{\text{sel},\ell}$ for all $\mathbf{Z}$ under the simpler design. Let $\kappa_{1,\ell}(d,\mathbf{Z}) = 1$ if dose d is identified as unsafe for a patient with $\mathbf{Z}$ in simulated trial

ℓ, or 0 if not. Let $\kappa_{2,\ell} = 1$ if trial ℓ is not terminated by the regulator's safety rule, and 0 if not. The simulation results are summarized using the empirical proportions among trials not stopped by the FRAG safety rule, given for each d and $\mathbf{Z}$ by

$$\mathrm{P}(\mathrm{Stop} \mid d, \mathbf{Z}) = \frac{\sum_{\ell=1}^{M} \kappa_{2,\ell} \kappa_{1,\ell}(d, \mathbf{Z})}{\sum_{\ell=1}^{M} \kappa_{2,\ell}}$$

and

$$\mathrm{P}(\mathrm{Select} \mid d, \mathbf{Z}) = \frac{\sum_{\ell=1}^{M} \kappa_{2,\ell} I(d = d_{\mathrm{sel},\ell}(\mathbf{Z}))}{\sum_{\ell=1}^{M} \kappa_{2,\ell}}.$$

Simulation results are summarized in Table 3.5, including the simulation truth to facilitate evaluation, with $\mathrm{P}(\mathrm{Stop} \mid d, \mathbf{Z})$ and $\mathrm{P}(\mathrm{Select} \mid d, \mathbf{Z})$ shown in the third and fourth lines for each $\mathbf{Z}$. Overall, the design reliably identifies unsafe doses and selects optimal safe doses for each subgroup. Large $\mathrm{P}(\mathrm{Stop} \mid d, \mathbf{Z})$ is achieved for $\mathbf{Z}$ and d with large π_D^{true}. When π_D^{true} is substantively greater than $\bar{\pi}_D(\mathbf{Z})$, $\mathrm{P}(\mathrm{Stop} \mid d, \mathbf{Z})$ is particularly high, as in cases with $r = 1$ in Scenario 3. Cases where $\pi_{D,Z,r}^{true}$ is slightly greater than $\bar{\pi}_D(\mathbf{Z})$ tend not to achieve high $\mathrm{P}(\mathrm{Stop} \mid d, \mathbf{Z})$, in part due to the small sub-sample size per subgroup. The design makes more accurate decisions for $Z = 1$ compared to $Z = 0$, due to the prevalences $\Pr(Z = 1) = 2/3$ and $\Pr(Z = 0) = 1/3$. For example, in Scenario 3, π_D^{true} exceeds $\bar{\pi}_D$ by approximately the same difference for $\mathbf{Z} = (0, 1)$ and $\mathbf{Z} = (1, 1)$, but P(Stop) is much larger for $\mathbf{Z} = (1, 1)$ due to there being more data on HBD patients ($Z = 1$). Truly optimal safe doses have large $\mathrm{P}(\mathrm{Select} \mid d, \mathbf{Z})$, shown on the fourth lines for each $\mathbf{Z}$ in the table. In Scenario 1, where doses have the same true mean utilities for each $\mathbf{Z}$, doses are selected with almost equal probabilities for all (Z, r). When there are clearly optimal doses, as in Scenario 2, the design has large $\mathrm{P}(\mathrm{Select} \mid d, \mathbf{Z})$ for those doses. When two doses have similar expected utilities, such as cases with $Z = 0$ in Scenario 6, the design selects both doses with large $\mathrm{P}(\mathrm{Select} \mid d, \mathbf{Z})$. When no dose is safe, as in Scenario 3, $\mathrm{P}(\mathrm{Select} \mid d, \mathbf{Z})$ is small for all d. Scenario 6 is complex in that the pattern of the true utilities varies with both Z and r, but the design captures this pattern quite well and makes correct decisions with high probabilities.

The simulations show that the FRAG safety rule rarely terminates the trial, even for Scenario 3 where π_D^{true} exceeds $\bar{q}_D$ for all $\mathbf{Z}$. This is because the design's subgroup-specific safety rules reliably stop accrual to a dose that is unsafe in a particular subgroup, but not in other subgroups. Under Scenario 3, the trial is terminated earlier by the subgroup-specific safety rule since all doses are unsafe for all subgroups. When only $d = 1$ is unsafe, it is likely that the subgroup-specific safety rule identifies this and stops further allocation of patients to $d = 1$, so no more deaths occur at $d = 1$. This helps to prevent the FRAG "one size fits all" safety rule from incorrectly terminating the entire trial when only $d = 1$ is unsafe, and thus continues accrual for safe doses and improves evaluation of outcomes for those doses.

Stopping and dose selection probabilities under the simpler design that ignores $\mathbf{Z}$ are compared to those under the design with subgroup-specific decisions in Figure 3.7. Panels (a) and (b) of the figure give histograms of the differences, $\mathrm{P}(\mathrm{Stop} \mid d, \mathbf{Z}) - \mathrm{P}(\mathrm{Stop} \mid d)$, for truly safe doses and unsafe doses, respectively, for

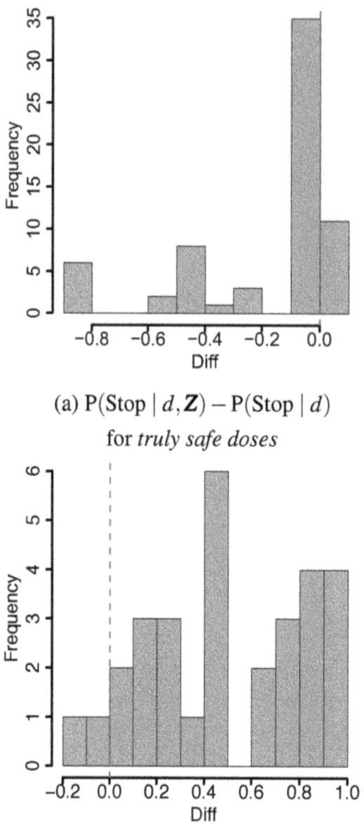

(a) P(Stop | $d, \mathbf{Z}$) – P(Stop | d) for *truly safe doses*

(b) P(Stop | $d, \mathbf{Z}$) – P(Stop | d) for *truly unsafe doses*

(c) P(Select | $d, \mathbf{Z}$) – P(Select | d) for *truly optimal doses*

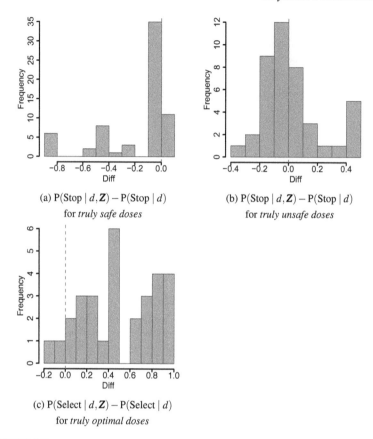

FIGURE 3.7
Comparison to the simpler NK cell dose-finding design that ignores subgroups defined by $\mathbf{Z}$. Histograms of differences in empirical proportions of stopping are given for truly safe doses in panel (a), and truly unsafe doses in panel (b). Panel (c) gives the histogram of differences in empirical proportions of correctly selecting a truly optimal dose. Each histogram is for all $\mathbf{Z}$ and all scenarios combined.

all $(d, \mathbf{Z})$ and all scenarios. Panel (a) shows that the design accounting for $\mathbf{Z}$ often has much smaller P(Stop | $d, \mathbf{Z}$) for truly safe doses (thus negative differences). This advantage is substantial in cases like Scenario 2, where the true safety of a dose varies greatly across subgroups, and most doses are unsafe. The histogram in (b) shows that the simpler design often stops truly unsafe doses with higher probability (thus negative differences). This is mainly due to Scenario 3 where all doses are unsafe for all subgroups. However, when a dose is unsafe only for some subgroups, as in Scenarios 4 and 5, the design accounting for $\mathbf{Z}$ greatly increases the probability of correctly stopping truly unsafe doses, shown by the large cluster above 0.40. Panel (c) gives the histogram of differences in empirical proportions P(select | $d, \mathbf{Z}$) – P(select | d), for truly optimal doses for all $(d, \mathbf{Z})$ across Scenarios 2-6. The design accounting for

$\boldsymbol{Z}$ is much more likely to select truly optimal doses (thus many more positive differences). The simulations showed that when unsafe doses vary between subgroups, as in Scenario 2, more patients are treated at unsafe doses under the simpler design.

To examine how much the design's performance may be improved by a larger sample size, additional simulations were run using $N_{\max} = 120$ rather than $N_{\max} = 60$. While 120 may seem very large for a dose-finding trial, it is important to bear in mind that the trial accounts for six prognostic subgroups with proportions 1/9 to 2/9 each, so six separate trials with overall $N_{\max} = 120$ would have sample sizes of roughly 13 to 27 each. Under all scenarios, with the larger sample size subgroup-specific dose assignments are improved and probabilities of correctly stopping unsafe doses are greatly increased for many combinations of $(d, \boldsymbol{Z})$. For example, in Scenario 6, for patients with NHL, $N_{\max} = 120$ increases $\text{P}(\text{Stop} \mid d, \boldsymbol{Z})$ from 0.51 to 0.67 for LBD, and from 0.91 to 0.99 for HBD, for d = 3. For subgroups with HBD, truly optimal doses are selected with higher rates for $N_{\max} = 120$. An additional simulation investigated the performance of the design with shorter follow-up, reducing the follow-up times (365, 365, 100, 100, 365) to (100, 100, 100, 30, 100). The results showed that the design's performance deteriorates substantially with this shorter follow-up, in terms of both $\text{P}(\text{Stop} \mid d, \boldsymbol{Z})$ and $\text{P}(\text{Select} \mid d, \boldsymbol{Z})$. For example, the probability that dose 1 is correctly identified as unsafe decreases from 0.76 to 0.53 for (CLL, LBD) in Scenario 6. Thus, incorporating outcome data from patients monitored for a longer time period greatly enhances the design's performance.

Computer Software A computer program, *NKcelldosefinding*, for implementing this methodology is available from urlhttps://users.soe.ucsc.edu/ juheelee/.

3.5 Generalized Phase I-II Designs

A pervasive problem in cancer treatment is that, in many settings, an early anti-disease effect, nominally "response", may not be durable. The problem is that cancer patients for whom treatment achieves an early response often have a substantial probability of disease recurrence, i.e. relapse, within a few months. If relapse occurs, the next therapeutic option is to give the patient some form of salvage therapy in an attempt to reduce their disease burden and possibly achieve a second response. While response and relapse rates depend on the disease, each patient's prognostic variables, and their initial treatment, the general pattern is the same across diseases.

Conventionally, nearly all early phase dose-finding designs rely on the implicit assumption that there is a strong positive association between treatment effects on short-term outcomes, such as response, and long-term outcomes, such as progression-free survival (PFS) time. In a phase I-II trial, a dose d is declared to be optimal based on a criterion defined in terms of early response and toxicity, $\boldsymbol{Y} = (Y_E, Y_T)$, evaluated over a short follow-up period $[0,\ t_{F,1}]$. The assumption is that, if the selected d is likely to achieve a response, then it also is likely to maximize the therapeutic success rate over a longer follow-up period, $[0,\ t_{F,2}]$, for $t_{F,2}$ much larger than $t_{F,1}$. In practice, this often is not true. For example, typical early and later disease evaluation times in oncology may be $t_{F,1}$ = 1 to 2 months for short-term (Y_E, Y_T), and $t_{F,2}$ = 6 to 12 months for long term treatment success. To see potential problems with this assumption, let R denote remission duration (RD), recorded among responders starting at the follow-up time $t_{F,1}$ when $\boldsymbol{Y}$ are evaluated. In the sequel, the fact that R is defined conditional on $Y_E = 1$ often will be suppressed in the notation for brevity. Define long-term therapeutic success as $(R > t_{F,2} - t_{F,1})$, the event that the patient will be alive without disease progression at follow-up time $t_{F,2}$. For example, if $t_{F,1}$ = 1 month to evaluate response and $t_{F,2}$ = 6 months, then RD among responders is evaluated 5 months after the response is achieved, at 6 months. This is the event that the time from evaluation of the early outcomes to long-term treatment failure, defined as disease progression or death, is no smaller than 6 –1 = 5 months. Distinguishing between this long-term success event and early response, $Y_E = 1$ evaluated at $t_{F,1}$, is very important. This is a pervasive problem in cancer therapeutics.

To formalize this, given the early and late follow-up times $t_{F,1}$ and $t_{F,2}$, define *response durability* among responders as the probability

$$S_R(t_{F,2} - t_{F,1} \mid d, \boldsymbol{\theta}) = \Pr(R > t_{F,2} - t_{F,1} \mid d, Y_E = 1, \boldsymbol{\theta}),$$

where $S_R(\cdot \mid d, \boldsymbol{\theta})$ denotes the survival function of R, which is defined among responders. Thus, response durability is the probability that a patient who has responded to treatment at or before time $t_{F,1}$ will be alive with disease still in remission at the later time $t_{F,2}$. Response durability has been discussed in many oncology settings, including radiation oncology (Tseng et al., 2015), and donor lymphocyte infusion at disease recurrence after an allogeneic bone marrow transplantation (Dazzi et al., 2000). The key point is that many early phase clinical trial protocols include provisions to follow each patient for a longer time $t_{F,2}$ after $\boldsymbol{Y}$ is evaluated. This allows response durability to be estimated using data on R.

To address the problem of how to choose an optimal dose of a new agent based on both early response and toxicity outcomes and response durability, Thall et al. (2023) proposed a *generalized phase I-II* (Gen I-II) design. The distributions of $\boldsymbol{Y}$ evaluated over $[0,\ t_{F,1}]$ and R evaluated over $[t_{F,1},\ t_{F,2}]$ are used to optimize d. For example, the early evaluation period may be $t_{F,1}$ = 1 month, with R evaluated at month $t_{F,2}$ = 6. The Gen I-II design is practical in settings where investigators conducting a phase I-II dose-finding trial, with early outcomes $\boldsymbol{Y}$ evaluated at $t_{F,1}$, also plan to follow each patient to time $t_{F,2}$ to estimate response durability. Let $\phi(d, \boldsymbol{\theta})$ be an objective function, defined in terms of the early outcome distribution $p(\boldsymbol{Y} \mid d, \boldsymbol{\theta})$, that is used as a criterion to optimize d in a phase I-II trial. Denote the experimental agent by X, and let $X(d)$ denote X administered at d. A central assumption underlying any dose-finding design is that the effects of $X(d)$ and $X(d')$ on $\boldsymbol{Y}$ or R may be different for $d \neq d'$. The common practice of referring to X qualitatively without specifying a dose implies that an optimal dose of X is known, so the statistical problem of determining a dose that is safe and maximizes anti-disease effect already has been addressed. This practice has led to many problems, both scientific and therapeutic, in settings where the dose of a new agent has not been optimized reliably.

Denote a criterion for characterizing long-term treatment success by $\xi(d, \boldsymbol{\theta})$. This may be $S_R(t_{F,2} - t_{F,1} \mid d, \boldsymbol{\theta})$, or possibly the conditional mean $E(R \mid d, \boldsymbol{\theta})$. An implicit assumption underlying phase I-II trials, which seldom is stated explicitly, is that, if a selected dose $d^{sel,\phi}$ maximizes an estimate of $\phi(d, \boldsymbol{\theta})$ at the end of the trial, then $d^{sel,\phi}$ also maximizes $\xi(d, \boldsymbol{\theta})$. The validity of this assumption depends on how the distributions $p(\boldsymbol{Y} \mid d, \boldsymbol{\theta})$ and $p(R \mid d, \boldsymbol{\theta})$ vary with d, and associations between R and $\boldsymbol{Y}$, which depend on how these outcomes are defined in a given clinical setting. Conventionally, using $\phi(d, \boldsymbol{\theta})$ in place of $\xi(d, \boldsymbol{\theta})$ to optimize d is motivated mainly by a desire for logistical convenience when conducting a sequentially adaptive dose-finding trial, because $\boldsymbol{Y}$ is observed in $[0,\ t_{F,1}]$, much sooner than the long-term event $[R > t_{F,2} - t_{F,1}]$ evaluated at $t_{F,2}$.

In many settings, the assumption that a selected dose $d^{sel,\phi}$ based on the early outcome criterion $\phi(d, \boldsymbol{\theta})$ also maximizes $\xi(d, \boldsymbol{\theta})$ is not true. As always, the statistical estimate $d^{sel,\phi}$ may not be the true optimal dose $d^{opt,\phi}$ in terms of the phase I-II criterion $\phi(d, \boldsymbol{\theta})$, due to random variation and the typically limited phase I-II sample size. Aside from the statistical issue of determining $d^{opt,\phi}$ based on the estimate $d^{sel,\phi}$ computed from phase I-II data, a different problem is that $d^{sel,\phi}$ may not necessarily be optimal in terms of a long-term success criterion. It can be shown very easily by example that, for assumed true values $\phi^{true}(d_j, \boldsymbol{\theta})$ and $\xi^{true}(d_j, \boldsymbol{\theta})$ of the two criterion functions evaluated for doses $d_1, \cdots, d_J$, the true optimal doses $d^{opt,\phi}$ and $d^{opt,\xi}$ may be different. Many of the simulation scenarios studied below will reflect this problem. While these functions are hypothetical constructs, there is a very long history of cancers that have been brought into remission but recur within a fairly short timeframe. For dose-finding, even if R depends on $\boldsymbol{Y}$ as well as d, an optimal phase I-II dose $d^{sel,\phi}$ based on $\phi(d, \boldsymbol{\theta})$ may be suboptimal in terms of $\xi(d, \boldsymbol{\theta})$. With targeted agents, immunotherapies, or cellular therapies, differences between how $p(\boldsymbol{Y} \mid d, \boldsymbol{\theta})$ and $p(R \mid d, \boldsymbol{\theta})$ vary with d may be due to direct biological effects of $X(d)$ on R that are not mediated by either Y_E or Y_T.

The fact that $\phi(d, \boldsymbol{\theta})$ and $\xi(d, \boldsymbol{\theta})$ are different criteria based on different outcomes may have very undesirable practical consequences that impact patients, pharmaceutical companies, and health care providers. If the dose $d^{opt,\xi}$ that is truly optimal in terms of $\xi(d, \boldsymbol{\theta})$ is different from $d^{sel,\phi}$ then, on average, $d^{sel,\phi}$ may give substantially smaller R than $d^{opt,\xi}$. In this case, a randomized trial of $X(d^{sel,\phi})$ versus a control, C, with outcome R or OS time is studying a suboptimal version $X(d^{sel,\phi})$ of X. This reduces the probability that the trial will yield a positive result, compared to what would have been obtained if $X(d^{sel,\xi})$ based on the longer-term outcome R had been used. It also reduces patient benefit in terms of R. Another possibility, which may be less obvious, is that a completed phase III trial may conclude that $X(d^{sel,\phi})$ is superior to C, when in fact $X(d^{sel,\phi})$ is inferior to $X(d')$ in terms of $\xi(d, \boldsymbol{\theta})$ for some dose $d' \neq d^{sel,\phi}$, but d' is not known. In this case, future patients treated with $X(d^{sel,\phi})$ based on the trial's results will have stochastically smaller R than they would have had if d' had been chosen prior to phase III. Thus, choosing a dose in phase I-II that is optimal in terms of $\phi(d, \boldsymbol{\theta})$ but suboptimal in terms of $\xi(d, \boldsymbol{\theta})$ reduces the potential long-term effectiveness of X before a phase III trial is even begun. Unfortunately, it is very difficult to determine reliably whether a truly optimal dose different from $d^{sel,\phi}$ was discarded erroneously prior to phase III or, if so, how severely values of R for future patients may have been reduced.

All of the considerations described above may be summarized by saying that the design described here is motivated by the fact that $\boldsymbol{Y}$ is not a surrogate for R. Relationships between short-term and long-term outcomes are a major issue in treatment evaluation. This has been discussed extensively, often with regard to use of an early outcome as a surrogate for a long-term outcome. Common examples are relationships between early response and PFS time, and between the times to progression and death (Anderson et al., 1983; Simon and Makuch, 1984; Buyse and Piebois, 1996; Fleming and Powers, 2012).

An even worse disconnect between a selected dose and the desire to increase R may arise if a phase I trial based on Y_T is used to choose a dose. It is well known that a "maximum tolerated dose (MTD)" based on Y_T in a phase I trial may be very different from $d^{sel,\phi}$ based on $\phi(Y_E, Y_T, \boldsymbol{\theta})$. Explanations and numerous examples are provided by Yuan et al. (2016), Yan et al. (2018), and Gauthier et al. (2019), among others. If a dose of X has been chosen based on Y_T alone, there is little or no relationship between the criterion used for dose selection and either early or late anti-disease effect. The best that one can hope for is that the probability of response or mean PSF time increases with dose. It thus is not surprising that an MTD from phase I is very unlikely to be optimal in terms of $\xi(d, \boldsymbol{\theta})$, since a dose-toxicity probability function $p(Y_T \mid d)$ provides little or no information about the long-term outcome distribution $p(R \mid d)$. For these reasons, the design described below begins with a phase I-II design based on (Y_E, Y_T), and not a phase I design based on Y_T alone.

All of the problems due to a possible disconnect between $p(\boldsymbol{Y} \mid d)$ and $p(R \mid d)$ are addressed by Gen I-II design, which has three stages. In stage 1, any conventional phase I-II design may be used, with $\boldsymbol{Y}$ evaluated over a short follow-up period $[0,\ t_{F,1}]$, using an objective function $\phi(d, \boldsymbol{\theta})$ that characterizes dose desirability. In stage 2, the phase I-II design is modified so that each successive dose is chosen using adaptive randomization (AR), with randomization probabilities defined in terms of $\phi(d, \boldsymbol{\theta})$. This is done to reduce the probability of getting stuck at a suboptimal dose, and to explore $\{d_1, \cdots, d_J\}$ more fully. At the end of stage 2, a set $\mathscr{C}$ of acceptable *candidate doses* having estimated $\phi(d, \boldsymbol{\theta})$ close to the maximum estimate is determined. In stage 3, additional patients are randomized among the selected candidates, and followed up to a longer time, $t_{F,2}$, to obtain data on R. At the end of the trial, the candidate dose maximizing the posterior mean of a long-term success criterion function $\xi(d, \boldsymbol{\theta})$ is selected.

The Gen I-II design is modular in that different phase I-II outcomes $\boldsymbol{Y}$ and designs may be used for stages 1 and 2, and the long-term time-to-event outcome may be R, PFS time, or overall survival time. In this sense, Gen I-II may be considered a paradigm for constructing a dose-finding design. The clinical setting that will be used here to illustrate the design has the following early outcomes, evaluated over the early interval $[0,\ 1]$ month. These are a binary indicator Y_T of toxicity, and a three-level ordinal response variable $Y_E^{\prime}$ taking on the possible values 0 to indicate progressive disease or death (PD), 1 for stable disease (SD), and 2 for response (RES). This 3-level ordinal variable may be reduced to the binary response indicator $Y_E = I[Y_E^{\prime} = 2]$.

The intermediate event, SD = $(RES \cup PD)^c$, is included to accommodate settings where PD and RES are not complementary events, that is, a patient may not respond to early therapy but also not have PD, with this defined as stable disease. The early outcome $\boldsymbol{Y}$ may be defined as either (Y_E, Y_T) or (Y_E', Y_T), depending on the particular phase I-II design being used. More generally, depending on the setting, the phase I-II design may use other early outcomes, including ordinal toxicity with multiple levels of severity, or Y_E' with more than three levels, with appropriate modifications of design parameters.

Using remission duration as the long-term outcome, to evaluate $X(d)$, for any $t_{F,2} > t_{F,1}$ the long-term success criterion may be defined as

$$\xi(d, \boldsymbol{\theta}) = \Pr(R > t_{F,2} - t_{F,1} \mid Y_E = 1, d, \boldsymbol{\theta}).$$

The Gen I-II paradigm is quite general, however. For the long-term outcome, alternatively, a Gen I-II design may define $\xi(d, \boldsymbol{\theta})$ in terms of Y_S = PFS or overall survival time, with $\xi(d, \boldsymbol{\theta}) = \Pr(Y_S > t_{F,2} \mid d, \boldsymbol{\theta})$ or $E(Y_S \mid d, \boldsymbol{\theta})$. Similarly, for the early outcomes, different optimality criteria $\phi(d, \boldsymbol{\theta})$ may be used based on $\boldsymbol{Y} = (Y_E, Y_T)$, depending on the phase I-II design. Denote $\pi_k(d, \boldsymbol{\theta}) = \Pr(Y_k = 1 \mid d, \boldsymbol{\theta})$ for $k = E, T$. Examples of a criterion function $\phi(d, \boldsymbol{\theta})$ based on bivariate binary $\boldsymbol{Y}$ include the response probability $\phi(d, \boldsymbol{\theta}) = \pi_E(d, \boldsymbol{\theta})$ (Thall and Sung, 1998), the odds ratio

$$\phi(d, \boldsymbol{\theta}) = \frac{\pi_E(d, \boldsymbol{\theta})/\{1 - \pi_E(d, \boldsymbol{\theta})\}}{\pi_T(d, \boldsymbol{\theta})/\{1 - \pi_T(d, \boldsymbol{\theta})\}}$$

(Yuan et al., 2016), and the efficacy-toxicity probability trade-off function $\phi(d, \boldsymbol{\theta}) = f\{\pi_E(d, \boldsymbol{\theta}), \pi_T(d, \boldsymbol{\theta})\}$ used by the EffTox design (Thall and Cook, 2004; Thall et al., 2014). A trade-off function may be constructed as follows. Suppressing d and $\boldsymbol{\theta}$ for brevity, given three equally desirable outcome probability pairs $\boldsymbol{\pi}_1^* = (\pi_{1,E}^*, 0)$, $\boldsymbol{\pi}_2^* = (1, \pi_{2,T}^*)$, and $\boldsymbol{\pi}_3^* = (\pi_{3,E}^*, \pi_{3,T}^*)$, the trade-off function is defined as

$$f\{\pi_E, \pi_T\} = 1 - \left\{ \left(\frac{\pi_E - 1}{\pi_{1,E}^* - 1} \right)^p + \left(\frac{\pi_T - 0}{\pi_{2,T}^* - 0} \right)^p \right\}^{1/p}$$

for $p > 0$. Solving $f(\pi_{R,3}^*, \pi_{T,3}^*) = 0$ for p gives a target contour Π_0 in the unit square $[0,1]^2$. Given this target, if the contour is shifted by specifying $f(\pi_E, \pi_T) = \delta$ to obtain Π_δ, then all $\boldsymbol{\pi}$ on this contour have desirability δ, and $f(\pi_E, \pi_T)$ increases as $\boldsymbol{\pi}$ moves along any straight line from a point in $[0,1]^2$ to the optimal pair $(1,0)$. If numerical utilities, $U(\boldsymbol{Y})$, of the early outcomes can be elicited, the early decision criterion may be defined as the mean utility,

$$\phi(d, \boldsymbol{\theta}) = E\{U(\boldsymbol{Y}) \mid d, \boldsymbol{\theta}\}.$$

The utility-based approach is flexible, contains some important trade-off functions as special cases (Zhou et al., 2019), and also is more scalable than the trade-off-based function of the marginal toxicity and efficacy probabilities. The development given here will use $U(Y_E', Y_T)$.

For stage $s = 1,2,3$, denote the sample size by n_s, with overall sample size $N = n_1 + n_2 + n_3$. Values of n_1 and n_2 are specified at the start of the trial, but n_3 is determined adaptively at the end of stage 2, as described below. For the ith patient enrolled in the trial, denote the assigned dose by $d_{[i]}$ and let C_i be the independent right censoring time starting from the time $t_{F,1}$ when $Y'_{i,E}$ is evaluated. The observed time to failure or censoring from $t_{F,1}$ is $R_i^o = \min\{R_i, C_i\}$. Let $\varepsilon_i = 1$ if $R_i^o = R_i$ and $\varepsilon_i = 0$ if $R_i^o = C_i < R_i$, and denote the data from the first n patients enrolled in the trial by

$$\mathscr{D}_n = \{(Y'_{i,E}, Y_{i,T}, R_i^o, \varepsilon_i, d_{[i]}) : i = 1, \cdots, n\}.$$

Stages 1 and 2 of a Gen I-II design include the Bayesian dose acceptability criteria

$$\Pr\{\pi_E(d,\boldsymbol{\theta}) > \underline{\pi}_E \mid \mathscr{D}_n\} > .10 \text{ and } \Pr\{\pi_T(d,\boldsymbol{\theta}) < \overline{\pi}_T \mid \mathscr{D}_n\} > .10, \tag{3.10}$$

where $\underline{\pi}_E$ and $\overline{\pi}_T$ are fixed limits corresponding to the disease and clinical setting. To determine these probabilities, binary versions of Y'_E and Y_T must be defined. These inequalities say that an acceptable dose must not be unlikely to have a response rate at least $\underline{\pi}_R$, or a toxicity rate at most $\overline{\pi}_T$. Denote the set of acceptable doses satisfying (3.10) by $\mathscr{A}_n$. During stages 1 and 2, no patient is treated with an unacceptable dose, and if it is determined that no dose is acceptable, i.e. $\mathscr{A}_n$ is empty, the trial is stopped, stage 3 is not conducted, and no dose is selected.

Steps for conducting stage 1 of a utility-based Gen I-II trial

Step 1. Treat the first cohort of patients at a starting dose chosen by the physician investigator.

Step 2. For each new cohort, update the posterior distribution of $\boldsymbol{\theta}_1$ and compute the admissible dose set $\mathscr{A}_n$ and posterior mean utility $u(d_j, \mathscr{D}_n)$ for each d_j.

Step 3. If $\mathscr{A}_n$ is empty, stop the trial and do not select any dose. If $\mathscr{A}_n$ is not empty, treat the next cohort of patients at the dose in $\mathscr{A}_n$ maximizing $u(d_j, \mathscr{D}_n)$, subject to the constraint that an untried dose may not be skipped when escalating.

Step 4. If the current dose d_j is the highest acceptable dose that has been tried so far, escalate one dose level. This supersedes Step 3.

Step 5. Repeat steps 1-4 until all stage 1 cohorts have been treated and their early outcomes $\boldsymbol{Y}$ evaluated.

Step 4 is included because, due to its simplicity, the Dirichlet-multinomial model assumed below cannot use the available data to estimate $\boldsymbol{p}(d_j)$ for any untried dose d_j. Step 5 reduces the chance getting stuck at a locally optimal but globally sub-optimal dose, and it facilitates exploring untried doses. If, instead of a multinomial model, a parametric dose-outcome model for $p(Y_T, Y_E \mid d)$ that borrows strength between doses is assumed, then step 4 may be dropped.

For brevity, denote $n_{1,2} = n_1 + n_2$. For each dose d_j and $n = 1, \cdots, n_{1,2}$, denote the posterior mean of the early outcome dose selection criterion by $\hat{\phi}_{j,n} = E\{\phi(d_j, \boldsymbol{\theta}) \mid \mathscr{D}_n\}$. In stage 1, doses are chosen to maximize $\hat{\phi}_{j,n}$ for the first n_1 patients, as in a usual phase I-II design. In stage 2, doses are chosen for the next n_2 patients using AR. Given fixed scale parameter ζ between 0 and 1, AR probabilities may be defined as

$$r_{j,n} = \frac{(\hat{\phi}_{j,n})^{\zeta}}{\sum_{l:d_l \in \mathscr{A}_n} (\hat{\phi}_{l,n})^{\zeta}}$$

for $n = n_1 + 1, \cdots, n_{1,2}$ and $d_j \in \mathscr{A}_n$. Values of $\zeta < 1$ shrink $r_{j,n}$ toward the fair randomization probability $1/|\mathscr{A}_n|$ corresponding to $\zeta = 0$, and the most common value used in practice is $\zeta = 1/2$. AR distributes patients more evenly among the doses during stage 2. For the Gen I-II design, simulations described below show that using AR in stage 2 has the advantages that it may reduce the additional stage 3 per-dose sample sizes, and it also increases the probability of making a correct final dose selection based on the long-term outcome. To facilitate trial conduct in stage 2, the AR probabilities may be updated and assigned for cohorts of $c > 1$, rather than for individual patients, and in this case, n_2 should be specified as a multiple of c.

At the end of stage 2, for a given fixed $0 < \rho < 1$, the *candidate dose set* is defined to be all $d_j \in \mathscr{A}_{n_{1,2}}$ having posterior mean desirability that is close to the maximum value,

$$\mathscr{C} = \left\{ d_j \in \mathscr{A}_{n_{1,2}} : \hat{\phi}_{j,n_{1,2}} \geq \rho \max_{d_l \in \mathscr{A}_{n_{1,2}}} \hat{\phi}_{l,n_{1,2}} \right\}.$$

To use this structure in practice, since a given numerical value of ρ has no intuitive meaning, one should not fix a value of ρ arbitrarily. Rather, the trial should be simulated using several different values to determine a ρ that gives a design with good operating characteristics (OCs) under a set of dose-outcome scenarios. In practice, a value between .60 and .90 should give a design with good properties.

The stage 3 sample size, n_3, is determined adaptively using the data $\mathscr{D}_{n_{1,2}}$ from stages 1 and 2 and the per-dose sample sizes $\{n_{1,2}(d_j) : d_j \in \mathscr{C}\}$ at the end of stage 2. For any $n_{1,2}$ and phase I-II design, the $n_{1,2}(d_j)$ values are random because doses are chosen adaptively in stages 1 and 2. Denote the stage 3 sample size of dose $d_j \in \mathscr{C}$ by $n_3(d_j)$. Thus, $n_3 = \sum_{d_j \in \mathscr{C}} n_3(d_j)$ and the per-dose sample sizes from all three stages are $N(d_j) = n_1(d_j) + n_2(d_j) + n_3(d_j)$ for each d_j. To determine $n_3(d_j)$ adaptively, a fixed overall per dose sample size $N(d) = N(d_j)$ for all j may be chosen that ensures a desired level of reliability for selecting an optimal dose from $\mathscr{C}$ at the end of the trial. Since the candidate dose set $\mathscr{C}$ is a random, the value of $N(d)$ should be chosen from a set of feasible values, such as $N(d)$ = 10, 15, or 20. This may be based on simulations of the trial, for given n_1, n_2, ρ, and assumed true values of the long-term success probabilities,

$$\boldsymbol{\xi}^{true} = (\xi^{true}(d_1), \cdots, \xi^{true}(d_J)),$$

and the short-term success probabilities,

$$\boldsymbol{\phi}^{true} = (\phi^{true}(d_1), \cdots, \phi^{true}(d_J)).$$

Each stage 3 per-dose sample size, $n_3(d_j) = N(d_j) - n_{1,2}(d_j)$, depends on $\mathscr{C}$ and the values of $n_{1,2}(d_j)$ for the candidate doses $d_j \in \mathscr{C}$. For example, if $J = 4$, $\mathscr{C} = \{d_3, d_4\}$, $n_{1,2}(d_3) = 12$, and $n_{1,2}(d_4) = 6$, then $N(d) = 20$ requires $n_3(d_3) = 8$ and $n_3(d_4) = 14$. Thus, in stage 3 a total of 22 additional patients would be randomized between d_3 and d_4, restricted to obtain overall per-dose sample sizes of 20.

For the final dose selection, an additional requirement is that each candidate dose $d \in \mathscr{C}$ must satisfy the futility inequality

$$\Pr\{\xi(d, \boldsymbol{\theta}) > \underline{\xi} \mid \mathscr{D}_N\} > .10,$$

where fixed $\underline{\xi}$ is the smallest acceptable value of $\xi(d_j, \boldsymbol{\theta})$. This avoids selecting the best dose from a set of candidate doses that all are unlikely to have a long-term therapeutic success rate that is at least $\underline{\xi}$, which may be the historical mean of ξ with standard therapy. For example, if $\underline{\xi} = .20$, this says that a dose that cannot provide at least a 20% chance of long-term success, characterized by the probability $\xi(d, \boldsymbol{\theta})$, is not worth further consideration. Denoting the final set of acceptable doses in $\mathscr{C}$ by $\mathscr{A}_N^{\xi}$, the final selected optimal dose in $\mathscr{A}_N^{\xi}$ is that having largest posterior mean long term success probability,

$$d_N^{sel,\xi} = \underset{d_j \in \mathscr{A}_N^{\xi}}{\operatorname{argmax}} E\{\xi(d_j, \boldsymbol{\theta}) \mid \mathscr{D}_N)\}.$$

If desired, one may use the alternative definition

$$d_N^{sel,\xi} = \underset{d_j \in \mathscr{A}_N^{\xi}}{\operatorname{argmax}} \Pr\left[\xi(d_j, \boldsymbol{\theta}) = \max_{d_r \in \mathscr{A}_N^{\xi}} \{\xi(d_r, \boldsymbol{\theta})\} \mid \mathscr{D}_N)\right].$$

The design parameters include all values required to specify the phase I-II design and objective function $\phi(d, \boldsymbol{\theta})$ used in stages 1 and 2, including $t_{F,1}$, n_1, n_2, cohort size c, acceptability limits $\underline{\pi}_E$ and $\overline{\pi}_T$, and the exponent ζ used to define the AR probabilities. For stage 3, one must specify the long-term follow-up time $t_{F,2}$, ρ, $\underline{\xi}$, and the overall per-dose sample size $N(d)$ required for each $d_j \in \mathscr{C}$. For the Bayesian model, one must specify hyperparameters $\tilde{\boldsymbol{\theta}}_1$ of the noninformative prior $p(\boldsymbol{\theta}_1 \mid \tilde{\boldsymbol{\theta}}_1)$ in the model for $p(\boldsymbol{Y} \mid d, \boldsymbol{\theta}_1)$, and hyperparameters $\tilde{\boldsymbol{\theta}}_2$ of the noninformative prior $p(\boldsymbol{\theta}_2 \mid \tilde{\boldsymbol{\theta}}_2)$ in the conditional failure time distribution $p(R \mid d, \boldsymbol{Y}, \boldsymbol{\theta}_2)$. A schematic for conducting a Gen 1-II design is given in Figure 3.8.

The Dirichlet-multinomial model for counts of the discrete early outcome pair (Y_E', Y_T) as functions of d is assumed because it has a simple form with a conjugate Dirichlet prior. This assumption is made to simplify computations. If desired, a parametric dose-outcome model $p(\boldsymbol{Y} \mid d, \boldsymbol{\theta})$ may be used to provide a basis for borrowing strength between doses in stages 1 and 2. However, the simulations given below show that the Dirichlet-multinomial model gives a Gen I-II design with good OCs.

The following development corresponds to the case where Y_E' has three possible values and Y_T has two possible values, but it can be applied for any discrete Y_E and Y_T. For each dose d_j, $j = 1, \cdots, J$, denote the joint early outcome probability

$$p_{a,b}(d_j) = \Pr(Y_E' = a, Y_T = b \mid d_j, \boldsymbol{\theta})$$

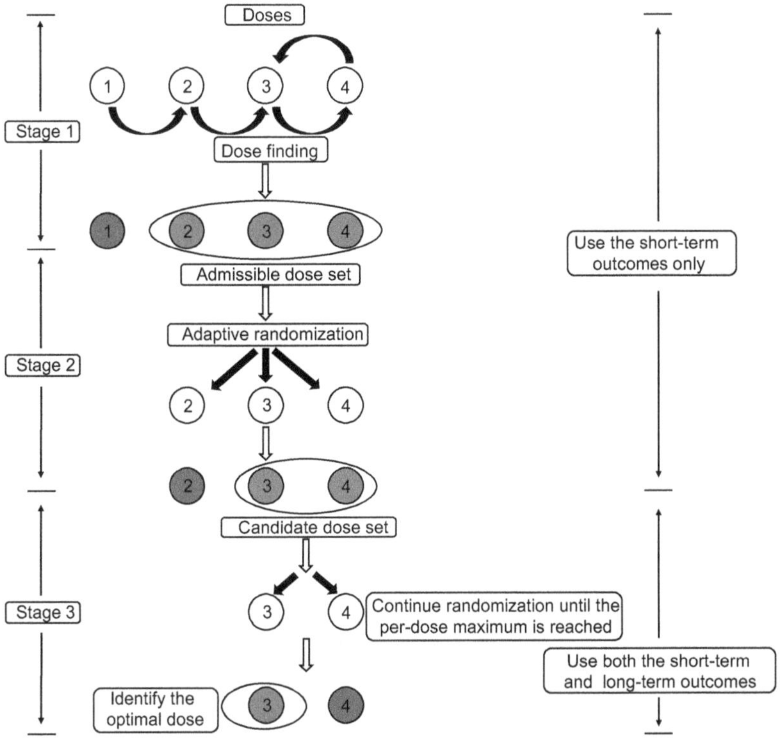

FIGURE 3.8
Schematic for conduct of a generalized phase I-II trial. In the illustration, the dose set is {1,2,3,4}, the candidate dose set is {2,3,4}, and the final selected dose is {3}

for $a = 0$, 1, or 2 and $b = 0$ or 1. The vector of all joint early outcome probabilities in this case is

$$\boldsymbol{p}(d_j) = (p_{0,0}(d_j), p_{0,1}(d_j), p_{1,0}(d_j), p_{1,1}(d_j), p_{2,0}(d_j), p_{2,1}(d_j)),$$

which implies $\sum_{a=0}^{2}\sum_{b=0}^{1} p_{a,b}(d_j) = 1$. Denoting $\boldsymbol{\theta}_1 = (\boldsymbol{p}(d_1), \cdots, \boldsymbol{p}(d_J))$, the dimension of $\boldsymbol{\theta}_1$ is the product of the number of possible values of Y_E', the number of possible values of Y_T, and the number of doses.

The vector of counts for the six elementary outcomes $\{(0,0), \cdots, (2,1)\}$ of $\boldsymbol{Y}$ is defined as follows. For each dose d_j and interim sample size n, the six-dimensional elementary outcome count vector is defined as

$$\boldsymbol{V}_n(d_j) = \sum_{i=1}^{n} \left(I[(\boldsymbol{Y}_i = (0,0)], \cdots, I[\boldsymbol{Y}_i = (2,1)]\right) I[d_{[i]} = d_j].$$

This is assumed to be multinomial with parameters $n(d_j)$ and $\boldsymbol{p}(d_j)$. It is assumed that $\boldsymbol{p}(d_j)$ follows a non-informative conjugate Dirichlet prior with parameter $\frac{1}{6}$ in each cell, denoted by $\boldsymbol{p}(d_j) \sim Dir(\frac{1}{6}, \cdots, \frac{1}{6})$, which has effective sample size 1. This model does not assume dose-outcome functions for $p(Y_E' \mid d)$ or $p(Y_T \mid d)$, and consequently it does not borrow strength between doses. Due to this model's simplicity, it improves robustness and facilitates posterior computation, and it yields a design with good properties. If desired, parametric dose-outcome functions for $p(Y_E' \mid d)$ and $p(Y_T \mid d)$ may be assumed to borrow strength, although this will complicate prior specification and posterior computations. By conjugacy of the multinomial likelihood and Dirichlet prior, the posteriors are

$$\boldsymbol{p}(d_j) \mid \boldsymbol{V}_n(d_j) \sim Dir\big((\frac{1}{6}, \cdots, \frac{1}{6}) + \boldsymbol{V}_n(d_j) \big)$$

for each d_j. The early outcome objective function is the mean utility

$$\phi(d_j, \boldsymbol{\theta}_1) = \overline{U}(d_j, \boldsymbol{\theta}_1) = \sum_{a=0}^{2} \sum_{b=0}^{1} U(a,b)\, p_{a,b}(d_j) \ \text{ for } \ j = 1, \cdots, J.$$

This model and utility structure for $\boldsymbol{Y}$ can easily be generalized to accommodate any dimensions of discrete Y_E' and Y_T.

For the distribution of $[R \mid d]$, a flexible but parsimonious parametric model for R is a Weibull distribution with pdf

$$f_R(t \mid Y_T, d_j, \boldsymbol{\theta}_2) = \frac{\alpha t^{\alpha-1}}{\lambda^{\alpha}} \exp\{-(t/\lambda)^{\alpha}\}, \quad t > 0,$$

where $\alpha > 0$ is the shape parameter. The rate parameter λ is assumed to be a log-linear function of Y_T and dose,

$$\log\{\lambda(Y_T, d_j, \boldsymbol{\theta})\} = \beta_0 + \beta_T\, Y_T + \gamma_j\, I[j > 1], \tag{3.11}$$

so γ_j is the effect of dose d_j versus d_1 on R for $j = 2, \cdots, J$, with $j = 1$ as the baseline dose. This gives parameter vectors

$$\boldsymbol{\theta}_2 = (\alpha, \beta_0, \beta_T, \gamma_1, \cdots, \gamma_J)$$

and $\boldsymbol{\theta} = (\boldsymbol{\theta}_1, \boldsymbol{\theta}_2)$. Non-informative $N(0, 10^2)$ priors may be assumed for elements of $\boldsymbol{\theta}_2$, and a $Gamma(.01,\ .01)$ prior may be assumed for α. The joint likelihood from n patients with data $\mathscr{D}_n$ is

$$\mathscr{L}(\mathscr{D}_n, \boldsymbol{\theta}) = \prod_{i=1}^{n} p(\boldsymbol{Y}_i \mid d_{[i]}, \boldsymbol{\theta}) \left\{ f_R(R_i^o \mid \boldsymbol{Y}_i, d_{[i]}, \boldsymbol{\theta})^{\varepsilon_i}\, S_R(R_i^o \mid \boldsymbol{Y}_i, d_{[i]}, \boldsymbol{\theta})^{(1-\varepsilon_i)} \right\}^{I[Y_{i,R}'=2]}.$$

The Gen I-II design is illustrated by a trial of CD70 CAR-NK cells as a targeted immunotherapy for patients with recurrent or treatment-resistant B-cell hematologic malignancies. This trial was the original motivation for the Gen I-II design. The trial enrolled patients with recurrent or relapsed disease following frontline treatment with chemotherapy or an allogeneic stem cell transplant. Treatment began with lymphodepleting chemotherapy for three days, followed by one CAR-NK cell infusion at the patient's selected dose. This therapeutic strategy was motivated by reverse signaling of CD70 that enhances the anti-disease activity of the CAR-NK cells against B-cell malignancies (Al Sayed et al., 2017; Liu et al., 2020). The trial's goal was to determine an optimal dose among the four values $5.0 \times (10^6, 10^7, 10^8, 10^9)$ cells, with standardized doses denoted by $(d_1, d_2, d_3, d_4) = (1, 2, 3, 4)$. The early outcomes were the indicators Y_E of response, defined as complete remission at $t_{F,1}$ = 1 month, and Y_T of grade 3 or 4 non-hematologic toxicity or cytokine release syndrome within 1 month. The fixed early outcome dose acceptability limits were $\overline{\pi}_T$ = .30 for the probability of toxicity and $\underline{\pi}_R$ = .50 for the probability of response.

For stages 1 and 2 of the CD70 NK Gen I-II trial design, a maximum of 48 patients are treated in cohorts of size 3 each, with the first $n_1 = 15$ patients treated by choosing doses to optimize the posterior mean of the function $\phi(d, \boldsymbol{\theta})$, and the next $n_2 = 33$ patients' doses chosen using AR. The first cohort is treated at d_1, no untried dose may be skipped when escalating, and no unacceptable doses may be used to treat a patient. The candidate dose set $\mathscr{C}$ is defined as all acceptable d at the end of stage 2 satisfying

$$E\{\phi(d, \boldsymbol{\theta}) \mid \mathscr{D}_{48}\} \geq .70 \times E\{\phi(d^{sel}, \boldsymbol{\theta}) \mid \mathscr{D}_{48}\},$$

so $\rho = .70$. For long-term therapeutic success, R = PFS time is defined starting at $t_{F,1} = 1$ month given that $Y_E' = 2$, monitored for up to 5 more months, so $t_{F,2} = 6$, and the long term success probability is defined as

$$\xi(d, \boldsymbol{\theta}) = Pr(R > 5 \mid Y_E = 1, d, \boldsymbol{\theta}_2).$$

The fixed lower limit on the long-term success probability acceptability criterion is $\underline{\xi}$ = .40, so the futility rule is quite demanding.

In stage 2 of the CD70 NK cell trial, $n_2 = 33$, and 11 cohorts of 3 patients each are randomized sequentially among the doses in $\mathscr{A}_{n_{1,2}}$ using the AR probabilities $r_{j,n}$, defined with shrinkage parameter $\zeta = 0.5$. The admissible dose set $\mathscr{A}_n$ is updated after each cohort's outcomes $\boldsymbol{Y}$ have been evaluated at $t_{F,1}$ = 1 month of follow-up. In stage 3, additional patients are randomized among the doses in the candidate set $\mathscr{C}$. The per-dose stage 3 sample sizes are chosen adaptively to ensure that a three-stage total of $N(d_j) = 15$ patients are treated at each $d_j \in \mathscr{C}$, where the overall per-dose sample size 15 is chosen based on simulations to ensure a reasonably high reliability for the final dose selection. Thus, the stage 3 sample sizes $n_3(d_j) = N(d_j) - n_{1,2}(d_j)$ are random. Denoting the final overall sample size by N, a dose $d^{opt,\xi}$ is chosen at the end of stage 3 to maximize the posterior mean of the long-term success probability,

$$E\{\xi(d_j, \boldsymbol{\theta}) \mid \mathscr{D}_N\} = E\{S_R(5 \mid d_j, \boldsymbol{\theta}) \mid \mathscr{D}_N\}.$$

Because the patients in the CD70 CAR-NK cell trial have active disease at enrollment, to define Y_E' for the 1-month evaluation, PD is defined as worsening of disease, and RES is defined as complete remission (CR), with $SD = (PD \cup RES)^c$. Thus, a patient with SD does not have worse disease at $t_{F,1}$ = 1 month than at trial entry,

TABLE 3.6
Numerical utilities for possible values of the outcome $\boldsymbol{Y} = (Y_E^{\prime}, Y_T)$ of the CD70 CAR-NK cell dose-finding trial.

		$Y_E^{\prime}$		
		$2 = RES$	$1 = SD$	$0 = PD$
Y_T	0 = No DLT	100	50	20
	1 = DLT	60	30	0

but does not have CR. To obtain a utility, the two boundary values $U(1,0) = 0$ for the worst possible outcome and $U(0,2) = 100$ for the best possible outcome first are fixed, and the remaining four intermediate values then are determined, subject to the admissibility constraints

$$0 < U(a,0) \leq U(a,1) \leq U(a,2) < 100,$$

and

$$0 < U(1,b) \leq U(0,b) < 100$$

for $a = 0,1$ or b = 0, 1, or 2. These inequalities require that either a worse disease status or toxicity must reduce the utility. Table 3.6 gives the numerical utility used for the simulations. The short-term outcome objective function is defined as the mean utility (3.5). During stages 1 and 2, given interim data $\mathscr{D}_n$, the posterior mean utility is defined as $u(d_j, \mathscr{D}_n) = E\{\overline{U}(d_j, \boldsymbol{\theta}) \mid \mathscr{D}_n\}$.

The following simulation study was conducted to evaluate the operating characteristics of the utility-based Gen I-II design, using the CD70 CAR-NK cell trial as the basis for constructing simulation settings. The scenarios ranged over different patterns of $\phi(d_j)^{true}$, $\xi(d_j)^{true}$, and outcome distributions. Figure 3.9 shows the assumed true dose-outcome curves $\pi_T^{true}(d_j)$ and $\pi_E^{true}(d_j) = \Pr^{true}(RES \mid d_j)$, and the assumed true long-term success probability curves, $\xi^{true}(d_j)$. As comparators, two conventional phase I-II designs, Conv 1 and Conv 2, were considered. The Conv 1 design consists of stages 1 and 2 of the Gen I-II design, but not stage 3, with an optimal dose selected to maximize the posterior mean utility $u(d_j, \mathscr{D}_n)$ based on the short term outcomes. The Conv 2 design is almost identical to the Conv 1 design, with the one difference that more patients are randomized in stage 2 in order to match the sample size of the Gen I-II design. The fixed dose acceptability limits used in the simulations were the upper limit $\overline{\pi}_T$ = .30 for the probability of toxicity, the lower limit $\underline{\pi}_R$ = .50 for the probability of response, and the lower limit $\underline{\xi}$ = .40 for long-term success probability. A total of 5,000 trials were simulated under each scenario using each of the three designs. To examine the Gen I-II design's robustness in the simulations, $(Y_E^{\prime}, Y_T)$ were generated by first simulating bivariate normal latent variables $(W_{R^{\prime}}, W_T)$, and defining each of the observed outcomes using dose-dependent intervals for the latent variables. The RD outcome R was simulated from a piecewise exponential distribution.

Table 3.7 summarizes the OCs of the Gen I-II, Conv 1, and Conv 2 designs, including dose selection percentages, mean number of patients treated at each dose,

TABLE 3.7
Dose selection % and mean number of patients treated at each dose level, and mean sample size under the Gen I-II, Conv 1, and Conv 2 designs. $\xi^{true}(d_j) = \Pr^{true}(R > 5 \mid Y_E = 1, d_j)$. Dose level 0 denotes the % of trials terminated early with no dose selected. Fixed dose acceptability limits were $\overline{\pi}_T$ = .30, $\underline{\pi}_E$ = .50, and $\underline{\xi}$ = .40. Boldface indicates results at the true optimal decision.

Designs		Dose levels 0	1	2	3	4	Sample size	R %
				Scenario 1				
	$\pi_T^{true}(d_j)$		0.10	0.20	0.40	0.50		
	$\pi_E^{true}(d_j)$		0.30	0.40	0.50	0.55		
	$\overline{U}^{true}(d_j)$		59.1	62.5	60.9	59.9		
	$\xi^{true}(d_j)$		0.05	0.1	0.15	0.3		
Gen I-II	Selection %	**93.5**	0	0.50	2.6	3.3	35.6	NA
	Patients		9.5	13.5	8.8	3.7		
Conv 1	Selection %	**56.2**	6.5	22.6	12.4	2.2	34.9	NA
	Patients		9.4	13.2	8.7	3.6		
Conv 2	Selection %	**57.7**	5.6	22.1	12.4	2.2	35.4	NA
	Patients		9.4	13.4	8.8	3.7		
				Scenario 2				
	$\pi_T^{true}(d_j)$		0.04	0.06	0.08	**0.10**		
	$\pi_E^{true}(d_j)$		0.40	0.50	0.60	**0.60**		
	$\overline{U}^{true}(d_j)$		61.2	67.0	74.2	**75.0**		
	$\xi^{true}(d_j)$		0.20	0.40	0.50	**0.70**		
Gen I-II	Selection %	2.9	0.3	7.0	19.3	**70.5**	52.7	91.0
	Patients		10.6	13.0	14.9	**14.3**		
Conv 1	Selection %	2.2	4.6	13.5	39.7	**39.9**	47.4	79.1
	Patients		9.6	11.7	13.3	**12.7**		
Conv 2	Selection %	2.2	3.6	13.2	40.2	**40.8**	52.8	79.9
	Patients		10.5	13.1	14.9	**14.3**		
				Scenario 3				
	$\pi_T^{true}(d_j)$		0.01	0.03	**0.06**	0.08		
	$\pi_E^{true}(d_j)$		0.6	0.7	**0.70**	0.70		
	$\overline{U}^{true}(d_j)$		72.1	80.9	**81.3**	80.6		
	$\xi^{true}(d_j)$		0.4	0.45	**0.65**	0.50		
Gen I-II	Selection %	0.1	4.3	9.2	**69.1**	17.2	58.1	91.5
	Patients		13.9	15.1	**14.8**	14.4		
Conv 1	Selection %	0	8.7	33.0	**31.1**	27.7	48.0	80.1
	Patients		11.5	12.7	**12.1**	11.6		
Conv 2	Selection %	0.1	7.0	33.2	**32.3**	27.3	58.1	80.6
	Patients		13.8	15.3	**14.8**	14.1		

Table 3.7 (continued)

Designs		Dose levels 0	1	2	3	4	Sample size	R %
				Scenario 4				
	$\pi_T^{true}(d_j)$		0.03	0.05	0.1	**0.15**		
	$\pi_E^{true}(d_j)$		0.4	0.6	0.75	**0.65**		
	$\overline{U}^{true}(d_j)$		63.1	75.2	82.3	**72.7**		
	$\xi^{true}(d_j)$		0.3	0.45	0.5	**0.65**		
Gen I-II	Selection %	1.2	1.4	14.2	24.1	**59.1**	54.3	89.1
	Patients		10.5	14.7	15.8	**13.3**		
Conv 1	Selection %	1.0	2.2	22.8	61.3	**12.8**	47.7	77.1
	Patients		9.1	13.0	14.1	**11.4**		
Conv 2	Selection %	1.1	1.6	20.1	64.6	**12.6**	54.3	77.5
	Patients		10.2	14.9	16.2	**13.1**		
				Scenario 5				
	$\pi_T^{true}(d_j)$		0.02	**0.04**	0.07	0.10		
	$\pi_E^{true}(d_j)$		0.4	**0.60**	0.65	0.75		
	$\overline{U}^{true}(d_j)$		54.4	**72.6**	72.5	77.8		
	$\xi^{true}(d_j)$		0.45	**0.70**	0.5	0.45		
Gen I-II	Selection %	0.6	5.4	**68.4**	15.7	9.8	54.3	90.0
	Patients		10.0	**14.4**	14.7	15.2		
Conv 1	Selection %	0.9	1.4	**25.1**	25.3	47.2	47.7	75.2
	Patients		8.8	**12.9**	12.8	13.2		
Conv 2	Selection %	1.1	1.0	**23.9**	23.6	50.5	54.2	74.6
	Patients		9.9	**14.6**	14.5	15.1		
				Scenario 6				
	$\pi_T^{true}(d_j)$		0.03	0.05	**0.1**	0.15		
	$\pi_E^{true}(d_j)$		0.4	0.5	**0.6**	0.50		
	$\overline{U}^{true}(d_j)$		66.1	70.3	**73.5**	68.6		
	$\xi^{true}(d_j)$		0.3	0.45	**0.65**	0.4		
Gen I-II	Selection %	3.3	2.0	15.0	**70.9**	8.8	52.2	90.6
	Patients		11.2	13.7	**15.1**	12.1		
Conv 1	Selection %	2.8	11.4	26.2	**42.9**	16.6	47.4	78.9
	Patients		10.1	12.5	**14.0**	10.7		
Conv 2	Selection %	2.8	10.2	26.2	**44.6**	16.3	52.2	79.7
	Patients		11.1	13.9	**15.3**	11.8		

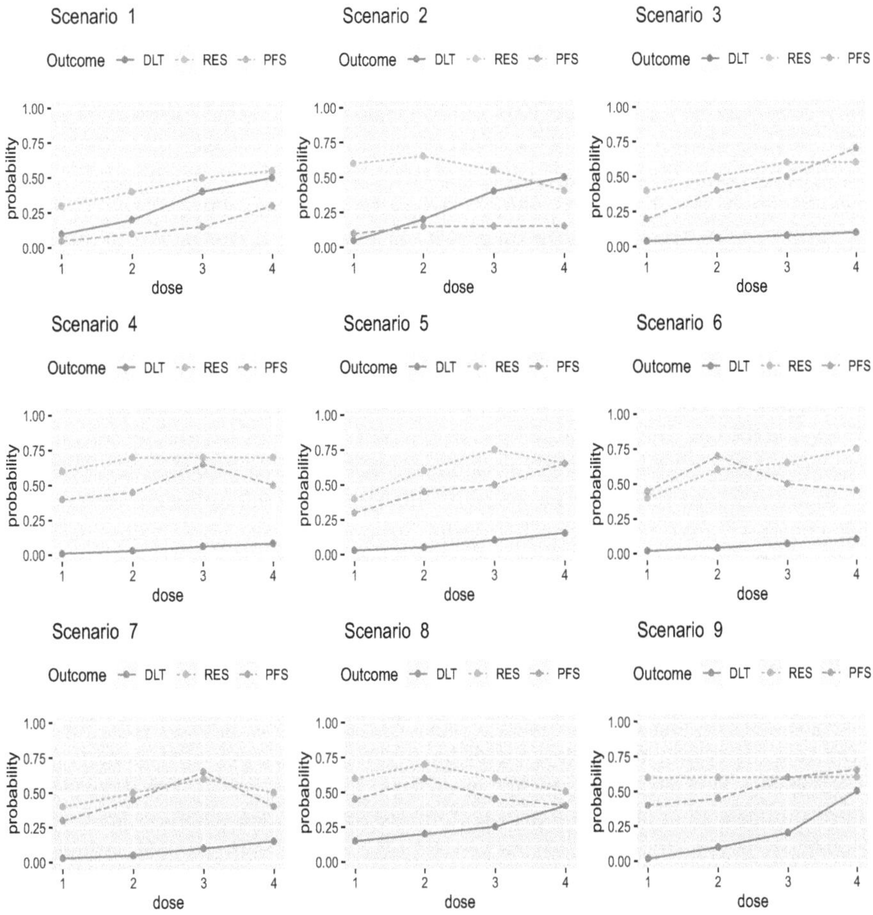

FIGURE 3.9
Dose-outcome curves for the scenarios in the Gen I-II design simulation study. The red, green, and blue lines represent $\pi_T^{true}(d_j)$, $\pi_E^{true}(d_j)$, and $\xi^{true}(d_j) = \Pr^{true}(R > t_{F,2} - t_{F,1} \mid Y_E = 1, d_j)$ for cell doses d_j = 1, 2, 3, 4.

and mean overall sample size. The number given under dose 0 is the percentage of trials terminated early with no dose selected. Scenario 1 is a null case where no dose has both acceptable $\pi_T^{true}(d_j)$ and acceptable $\xi(d)^{true}$. In scenario 1, the Gen I-II design is much safer in that it correctly terminates the trial early 93.5% of the time compared to 56.2% for the Conv 1 design and 57.7% for the Conv 2 design.

Recall that the conventional phase I-II utility criterion, $\overline{U}^{true}(d_j)$, is based on $\boldsymbol{Y}$ only, and ignores the long-term success rate that determines the truly optimal dose. Table 3.7 shows that, in scenarios 2 and 3, multiple doses are nearly optimal in terms of mean utilities $\overline{U}^{true}(d_j)$, but only one dose is optimal in terms of the long-term success rate $\xi^{true}(d_j)$. In scenario 2, d_3 and d_4 have similar mean utilities near 75, but d_4 is truly optimal with the highest $\xi^{true}(d_4) = 0.70$, compared to $\xi^{true}(d_3) = 0.50$. In scenario 3, the Gen I-II design has a 69.1% chance of correctly selecting d_3, while the Conv 1 and Conv 2 designs have only 31% and 32% chances of selecting d_3, and are about as likely to incorrectly select d_2 or d_4, again because both conventional designs ignore R. In scenario 3, d_2, d_3 and d_4 all have similar mean utilities $\overline{U}^{true}(d_j)$ near 81, but d_3 is truly optimal with $\xi^{true}(d_3) = 0.65$, compared to $\xi^{true}(d_2) = .45$ and $\xi^{true}(d_4) = .50$. The same large superiority of the Gen I-II design over Conv 1 and Conv 2 is seen in this scenario. Scenarios 4 and 5 are important cases where the truly optimal dose in terms of $\xi^{true}(d_j)$ and the dose with highest mean utility $\overline{U}^{true}(d_j)$ differ. In scenario 4, d_4 is truly optimal with the highest $\xi^{true}(d_4) = 0.65$, whereas d_3 has the highest mean utility $\overline{U}^{true}(d_3) = 82.3$. In scenario 6, d_2 is truly optimal with the highest $\xi^{true}(d_2) = 0.70$, while d_4 has highest mean utility $\overline{U}^{true}(d_4) = 77.8$. The Conv 1 and Conv 2 designs both have over a 60% chance of incorrectly selecting d_3 as optimal in scenario 5, about a 50% chance of incorrectly selecting d_4 as optimal in scenario 6, and they have 15% to 25% chances of correct optimal dose selection in scenarios 4 and 5. The Gen I-II design has much higher correct dose selection rates of 59.1% in scenario 4 and 68.4% in scenario 5. In scenario 6, where the truly optimal dose and the dose with highest mean utility are identical, the Gen I-II still outperforms Conv 1 and Conv 2, with a 25% higher correct optimal dose selection percentage.

In summary, the Gen I-II design outperforms the two conventional phase I-II designs substantially in all scenarios, especially in terms of optimal dose selection. The Gen I-II design also always yields the highest R value in all the scenarios, with a value that is at least 10% higher in each of scenarios 3 – 7. The three designs have similar patient allocation distributions, essentially because all designs allocate patients based on short-term outcomes, while R is only used in the final optimal dose selection of the Gen I-II design.

While the basic idea underlying a Gen I-II design is simple, it addresses a severe and pervasive problem with conventional dose-finding designs. As noted earlier, the Gen I-II design paradigm is modular in that $\boldsymbol{Y}$ can be any early outcome used for dose finding, any phase I-II design with early outcome criterion $\phi(d, \boldsymbol{\theta})$ can be used for stages 1 and 2, and any long term event time may be used for stage 3, This allows a Gen I-II design to be tailored for a wide variety of different clinical settings. A major practical advantage of the Gen I-II design is that, if the clinical, computational, and administrative structure for conducting a phase I-II trial is in place, and the investigators plan to follow patients long enough to assess response durability, then implementing a Gen I-II design is straightforward. Given the strong interest in assessing response durability among clinicians, this should be the case in most oncology settings.

Computer Software A computer program written in R code for implementing the Gen I-II design is available from https://github.com/ yongzang2020.

4

Basket Trials

4.1 Patient Heterogeneity and Baskets

Targeted agents may be engineered to kill cancer cells in a variety of different ways. They may disrupt the process of cancer cell reproduction, overcome a cancer cell's ability to protect itself from immune cells via a checkpoint blockade, cut off the blood supply to a solid tumor, or activate apoptosis, the natural process of programmed cell death, in the cancer cells. In settings where patient subgroups correspond to different diseases, disease subtypes, or histologies that share a common biological target, a class of Bayesian clinical trial designs known as *basket trials* may be used (Redig and Jänne, 2015). The idea is that the subgroups characterize patent heterogeneity, but they are tied together by the fact that they share the agent's biological target. Consequently, the agent's effects within subgroups should be qualitatively similar, while the magnitudes of the effects may differ between subgroups. A logistical advantage of running a basket trial is that the targeted agent may be studied in one trial under a "master protocol", rather than running separate trials in the different subgroups. For G subgroups, the key statistical assumption is that subgroup-specific response probabilities $\boldsymbol{\pi} = (\pi_1, \cdots, \pi_G)$ of a binary response outcome, or the event rates $\boldsymbol{\lambda} = (\lambda_1, \cdots, \lambda_G)$ of a time-to-event outcome, are positively associated. Consequently, it is appropriate to assume a Bayesian model in which this is the case in order to borrow strength between subgroups. Including different diseases or disease subtypes in the same trial is motivated by the dual goals of estimating the targeted agent's effect within each subgroup in order to make subgroup-specific decisions of whether or not the agent is promising ("active"), or whether it is safe, while also borrowing strength between subgroups.

Under a Bayesian model that accounts for treatment-subgroup interactions, a response observed in one subgroup of a basket trial should increase the posterior probability of response in each other subgroup. The underlying model may include structures to facilitate adaptively combining subgroups into clusters that the interim data show similar response or failure rates, and set the parameters to be identical for all subgroups in a cluster, thus reducing the overall model dimension and improving reliability. A basket trial design may be considered to be a compromise between the two conventional approaches of conducting a separate trial within each subgroup, or instead including all subgroups while ignoring subgroup effects entirely by assuming homogeneity.

DOI: 10.1201/9781003474258-4

Two general classes of Bayesian regression models for outcome as a function of treatment and subgroup can be used to account for heterogeneity due to disease subgroups. Which type of model is appropriate depends on what is appropriate to assume about treatment effects within subgroups on biological grounds. The first type of model is a Bayesian analysis of covariance or regression model that includes treatment-subgroup interactions with non-exchangeable subgroups, with a level 1 prior but no level 2 prior. This type of Bayesian model is used by the single-arm phase II design of Wathen et al. (2008), which will be reviewed below. The second is a Bayesian hierarchical model in which, for example, if the outcome is a binary response indicator, the response probabilities $\pi_1, \cdots, \pi_G$ of G subgroups are conditionally independent under an assumed hyperprior, with dependence induced by the hyperprior. Under this model, the π_g's are exchangeable, which says that, if the subgroups are re-indexed as $r(1), \cdots, r(G)$, the joint distributions of $(\pi_{r(1)}, \cdots, \pi_{r(G)})$ and $(\pi_1, \cdots, \pi_G)$ are the same. Since an extreme example of exchangeability is independence, in practice the hyperprior is constructed so that the π_g's have the same marginal prior but are positively associated. An example is a setting where the subgroups are histological subtypes of a disease. Such a hierarchical model is not appropriate, however, if the subgroups are not exchangeable. This is the case, for example, if subgroups correspond to ordinal prognostic levels, such as {Good, Intermediate, Poor}, or if subgroups are defined by two binary biomarkers, as $(+,+)$, $(+,-)$, $(-,+)$, $(-,-)$. A more appropriate model in such settings might assume that $\text{logit}(\pi_g) = \mu + \alpha_g$ with $\alpha_1 = 0$. While it may be argued that these cases are not basket trials because the subgroups may not necessarily share the same biological target, they do have response probabilities that may be assumed to be positively associated for a given experimental agent. Both model classes account for association among the outcome probabilities or event rates in different subgroups. These models may be refined by defining a third class of models to account for the more complex possibility that a set of G predefined subgroups may be overly refined, for example in the case $G = 3$ where in fact $\pi_1 = \pi_2$ but π_3 is a distinct value. In this case, the subgroups should be clustered as $\{1,2\}$ and $\{3\}$ and there are two distinct response probabilities rather than three. In such settings, additional model components that use interim data during a clinical trial to adaptively identify subgroup clusters are needed.

Basket trial designs may be applied in a broad variety of different settings. Patient subgroups may be defined by a disease's histological subtypes, biomarkers, different diseases that share the same biological target of an agent being evaluated, or some combination of these variables. Compared to traditional early-phase trial designs, a basket trial has the advantages that it allows different decisions, actions, and conclusions within subgroups, while also borrowing strength between subgroups, which improves the trial's efficiency in terms of sample size and trial duration (Simon et al., 2016). This approach also has the advantage that it may allow inclusion of patients with a rare cancer that is biologically related to the other cancer subtypes in the trial, but it is not feasible to run a separate trial in the rare cancer.

A variety of different basket trial designs have been proposed. Practical advances have been obtained from the use of multi-level hierarchical models, utilities to accommodate multiple outcomes, and rules, as noted above, that adaptively combine similar subgroups during trial conduct to borrow strength and obtain more reliable decisions. In this sense, the dose-finding designs proposed by Chapple and Thall (2018) and Lee et al. (2021), described in Chapter 3, that use latent subset membership variables to cluster subgroups, may be considered basket trials. A basket trial design proposed by Simon et al. (2016) used a Bayesian model that includes a parameter to quantify the heterogeneity of treatment effects across disease subtypes. Cunanan et al. (2017) proposed an efficient two-stage basket trial design. Trippa and Alexander (2017) proposed using AR in a Bayesian basket trial design. Chu and Yuan (2018a) proposed a method for calibrating a Bayesian hierarchical model to improve design performance. Chu and Yuan (2018b) proposed a Bayesian design that uses latent variables representing treatment sensitivity to account for subtype heterogeneity, including a rule to adaptively cluster disease subtypes. Psioda et al. (2019) proposed a Bayesian design for a basket trial with binary endpoints that relies on adaptive model averaging to identify sets of subgroups having similar response rates. This approach has the same purpose as using latent subgroup membership variables to adaptive subgroup clustering. Lin et al. (2021c) proposed a phase I-II basket design to jointly optimize the dose and schedule of an agent within prognostic disease subgroups, based on late-onset toxicity and efficacy. The following sections will describe several of these designs.

4.2 A Phase II Design for Non-Exchangeable Subgroups

The following example illustrates what may go wrong in a phase II trial if patient heterogeneity is ignored in a setting where subgroups are not exchangeable. Suppose that patients are classified as being biomarker positive (*pos*) or negative (*neg*) with probability .50 each. Suppose that the true response probabilities for a standard control treatment C are $\pi^{true}_{C,pos} = 0.50$ in the biomarker *pos* subgroup, and $\pi^{true}_{C,neg} =$ 0.60 in the biomarker *neg* subgroup, and the corresponding values for experimental treatment E are $\pi^{true}_{E,pos} = 0.70$ and $\pi^{true}_{E,neg} = 0.40$. Since the overall response probability is the biomarker subgroup weighted average $.50\ \pi^{true}_{X,pos} + .50\ \pi^{true}_{X,neg} = .55$ for both $X = C$ and $X = E$, in this case a single futility rule assuming homogeneity would be very likely to stop accrual for all patients. This would be a correct decision in the biomarker-negative patients, where E decreases the response rate of C from .60 to .40, but a false-negative decision in the biomarker-positive patients, where E increases the response rate of C from .50 to .70. This shows what can happen if heterogeneity is ignored when there is a qualitative treatment-subgroup interaction, since

$$\Delta^{true}_{pos} = \pi^{true}_{E,pos} - \pi^{true}_{C,pos} = .70 - .50 = .20$$

and

$$\Delta^{true}_{neg} = \pi^{true}_{E,neg} - \pi^{true}_{C,neg} = .40 - .60 = -.20$$

have different signs. While this illustration is an extreme case, similar problems also arise if the $\Delta_{X,g}^{true}$'s have the same sign but their magnitudes are substantively different.

Suppose that, instead, two separate sets of Bayesian futility rules are used, one within each subgroup, based only on that subgroups's data. Each rule may stop accrual and declare E not promising within its subgroup. Applying the phase II design of Thall and Simon (1994) in each subgroup g, with targeted improvement δ_g, the rule stops accrual and declares E not promising in g if $\Pr(\Delta_g \geq \delta_g \mid \mathscr{D}_g)$ is unacceptably small. However, there is a flaw with this approach that may not be obvious. It ignores the fact that, if the data $\mathscr{D}_g$ show that E does not provide an improvement in subgroup g then, because the response probabilities are different but not independent across subgroups, this should decrease the posterior probability that E provides an improvement over C in the other subgroup. That is, for distinct subgroups g and r, the two rules should borrow strength from each other, based on an assumed positive association between $\pi_{X,g}$ and $\pi_{X,r}$, rather than operating independently. This also suggests that each subgroup-specific rule should be constructed so that it is based on the combined data from all subgroups. That is the basic idea underlying basket trials.

Wathen et al. (2008) proposed a Bayesian design that does subgroup-specific futility monitoring for single-arm phase II trials, based on either a binary response or time-to-event endpoint. The design addresses settings where the subgroups are not exchangeable, so a hierarchal model on subgroup-specific response probabilities is not appropriate. This is the case, for example, when subgroups are determined by prognostic levels, so the subgroups are ordered by prognosis, or by disease subtypes having different historical response rates, so the subgroups are not ordered. In such settings, because the outcome rates are not exchangeable across subgroups, it is not appropriate to assume a hierarchical model for the subgroups effects, as in the basket trial design given by Thall et al. (2003). The design proposed by Wathen et al. (2008) is intermediate between ignoring subgroups or doing a separate trial within each subgroup. The design includes subgroup-specific monitoring rules that account for non-exchangeable subgroups and borrow strength between subgroups. It may be based on the following Bayesian regression model.

Index subgroups by $g = 1, \cdots, G$, and let

$$\pi_{X,g}(\boldsymbol{\theta}) = \Pr(Y = 1 \mid X, g, \boldsymbol{\theta})$$

denote the response probability for treatment $X = E$ or C in subgroup g. Denote the linear term

$$\eta_{X,g}(\boldsymbol{\theta}) = \text{logit}\{\pi_{X,g}(\boldsymbol{\theta})\} = \alpha + \beta_g + \xi_g I(X = E), \tag{4.1}$$

with all parameters real-valued. Using subgroup $g = 1$ as the baseline with $\beta_1 \equiv 0$, the between-subgroup effects are $\beta_2, \cdots, \beta_G$, and the E-versus-C treatment-subgroup interactions are $\xi_1, \cdots, \xi_G$, with α the baseline real-valued response rate. Denoting $\boldsymbol{\beta} = (\beta_2, \cdots, \beta_G)$ and $\boldsymbol{\xi} = (\xi_1, \cdots, \xi_G)$, the model parameter vector is $\boldsymbol{\theta} = (\alpha, \boldsymbol{\beta}, \boldsymbol{\xi})$. This parameterization facilitates construction of futility monitoring rules that borrow strength between subgroups. Denoting the response indicators for n patients by $Y_1, \cdots, Y_n$ and their subgroups by $g_1, \cdots, g_n$, the data are

$$\mathscr{D}_n = \{(Y_1, g_1), \cdots, (Y_n, g_n)\}$$

and the likelihood is

$$\mathscr{L}(\mathscr{D}_n, \boldsymbol{\theta}) = \prod_{i=1}^{n} \pi_{E,g_i}(\boldsymbol{\theta})^{Y_i}\{1 - \pi_{E,g_i}(\boldsymbol{\theta})\}^{1-Y_i}.$$

If Y is a time-to-event outcome, such as the time to response or PFS time, a flexible model is the Weibull with pdf

$$f_X(y \mid g, \boldsymbol{\theta}) = \frac{\zeta y^{\zeta-1}}{(\lambda_{X,g})^{\zeta}} \exp\{-(y/\lambda_{X,g})^{\zeta}\}, \quad y > 0.$$

In this model, ζ is a shape parameter, $\{\lambda_{X,g} : X = E, C, \ g = 1, \cdots, G\}$ are scale parameters, the means are $\lambda_{X,g}\Gamma(1 + 1/\zeta)$, and the survival functions are

$$S_X(y \mid g, \boldsymbol{\theta}) = \Pr(Y > t \mid g, \boldsymbol{\theta}) = \exp\{-(y/\lambda_{X,g})^{\zeta}\}.$$

In this case, the linear term (4.1) may take the same form to account for subgroup effects, but it is defined as $\eta_{X,g}(\boldsymbol{\theta}) = \log(\lambda_{X,g})$. For this model, $\boldsymbol{\theta} = (\zeta, \alpha, \boldsymbol{\beta}, \boldsymbol{\xi})$. Denoting Y^o = time to the event or right censoring and $\delta = I(Y^o = Y)$, the data are

$$\mathscr{D}_n = \{(Y_1^o, \delta_1, , g_1), \cdots, (Y_n^o, \delta_n, g_n)\}$$

and the likelihood for time-to-event data is

$$\mathscr{L}(\mathscr{D}_n, \boldsymbol{\theta}) = \prod_{i=1}^{n} f_E(Y_i^o \mid E, g_i, \boldsymbol{\theta})^{\delta_i} S(Y_i^o \mid E, g_i, \boldsymbol{\theta})^{1-\delta_i}.$$

Subgroup-specific futility stopping rules based on binary or time-to-event data may be specified as follows, as a simple generalization of Thall and Simon (1994). For probabilities for binary outcomes with targeted improvement δ_g, accrual to subgroup g will be terminated early if

$$\Pr\{\Delta_g(\boldsymbol{\theta}) > \delta_g \mid data\} < p_g.$$

The probability cut-off p_g must be calibrated to obtain a small stopping probability, such as .10, if the targeted improvement is achieved in subgroup g, that is, if the true response probability is $\pi_{E,g}^{true} = \mu_{C,g} + \delta_g$. This is a property of the within-subgroup rule. For time-to-event data, $\pi_{E,g}(\boldsymbol{\theta})$ and $\pi_{C,g}(\boldsymbol{\theta})$ are replaced by the corresponding subgroup-specific mean event times, with the δ_g's defined in the time domain. Given a maximum overall sample size N_{max}, if enrollment to a subgroup is terminated early then, to improve overall design reliability, the sample sizes of the remaining subgroups should be enriched while maintaining N_{max}.

For the Bayesian futility monitoring rule to work, the prior is formulated following the general approach of Thall and Simon (1994) in the homogeneous case, by specifying an informative prior on $\boldsymbol{\theta}_C$ and a non-informative prior on $\boldsymbol{\theta}_E$. The idea is that there must be prior knowledge about the distribution of Y with standard therapy C, which provides the basis for specifying an informative prior on $\boldsymbol{\theta} - \boldsymbol{\xi}$. However, the prior on $\boldsymbol{\xi}$, which are the additional subgroup-specific effects of E, must be non-informative, since before the trial nothing is known about the effects of E. Otherwise, a trial of E would not be needed. The informativeness of each prior may be quantified in terms of the effective sample sizes (ESS's) of the prior probabilities $\pi_{X,g}(\boldsymbol{\theta})$. This approach uses the idea that a Beta(a,b) distribution has ESS $= a+b$ and mean $a/(a+b)$. One may exploit this fact by approximating the prior of each $\pi_{X,g}(\boldsymbol{\theta})$ with a beta distribution. This is done matching the prior mean and ESS of $\pi_{X,g}(\boldsymbol{\theta})$ with those of a beta, and then solving for the prior hyperparameters. Since $\mu_{X,g} = \mathrm{E}\{\pi_{X,g}(\boldsymbol{\theta})\}$ for $t = E,S$, this is done subject to the constraint that the prior means for E and C are the same, $\mu_{E,g} = \mu_{C,g}$, for each $g = 1,\cdots,G$.

Taking this approach, the hyperparameters may be determined as follows. A practical approach is to assume normal priors on all parameters, where priors on elements of $\boldsymbol{\theta} - \boldsymbol{\xi}$ have suitably small variances and priors on elements of $\boldsymbol{\xi}$ have large variances. For each treatment X and subgroup g, let $p_{X,g}(q)$, for $0 \le q \le 1$, denote the prior on $\pi_{X,g}(\boldsymbol{\theta})$ induced by the prior on $\boldsymbol{\theta}$. Let $p^*_{X,g}(q)$ be the approximating beta distribution, beta$(a_{X,g}, b_{X,g})$, which has mean $a_{X,g}/(a_{X,g}+b_{X,g})$ and ESS $= a_{X,g}+b_{X,g}$. Assume that the historical estimates $\hat{\pi}_{C,g}(\boldsymbol{\theta})$ of the mean and $N_{C,g}$ = ESS with C have been obtained either from data or by elicitation, and let $N_{E,g}$ be suitably small specified values, between .5 and 2.

Algorithm for Establishing Prior Hyperparameters

Step 1: For each $g = 1,\cdots,G$, set $a_{C,g} = N_{C,g}\hat{\pi}_{C,g}(\boldsymbol{\theta})$ and $b_{C,g} = N_{C,g}(1-\hat{\pi}_{C,g}(\boldsymbol{\theta}))$.

Step 2: Solve for normal hyperparameters $(\mu_\alpha, \sigma^2_\alpha)$ to minimize $\int_0^1 |p_{C,1}(q) - p^*_{C,1}(q)|dq$ subject to $\mathrm{E}\{\pi_{C,1}(\boldsymbol{\theta})\} = \hat{\pi}_{C,1}(\boldsymbol{\theta})$.

Step 3: Given $(\mu_\alpha, \sigma^2_\alpha)$ from Step 2, for each $g = 2,\cdots,G$, solve for normal hyperparameters $(\mu_\beta, \sigma^2_\beta)$ to minimize $\int_0^1 |p_{C,g}(q) - p^*_{C,g}(q)|dq$ subject to $\mathrm{E}\{\pi_{C,g}(\boldsymbol{\theta})\} = \hat{\pi}_{C,g}(\boldsymbol{\theta})$ for each $g = 2,\cdots,G$.

Step 4: Set $a_{E,g} = N_{E,g}\hat{\pi}_{E,g}(\boldsymbol{\theta})$ and $b_{E,g} = N_{E,g}(1-\hat{\pi}_{E,g}(\boldsymbol{\theta}))$ for each $g = 1,\cdots,G$.

Step 5: Given all hyperparameters from Steps 1 – 4, solve for (μ_ξ, σ^2_ξ) to minimize $\int_0^1 |p_{C,g}(q) - p^*_{C,g}(q)|dq$ subject to $\mathrm{E}\{\pi_{E,g}(\boldsymbol{\theta})\} = \hat{\pi}_{C,g}(\boldsymbol{\theta})$ for each $g = 1,\cdots,G$.

The following simulation results, from Table II of Wathen et al. (2008), give the OCs of this design, denoted by S-TI, for a trial with binary outcomes, $G = 4$ prognostic subgroups with equal prevalences of .25 each, and overall maximum sample size $N_{max} = 200$. The first comparator is a design, S-NTI, that assumes there are subgroup effects $\beta_2,\cdots,\beta_G$, but no treatment-subgroup interactions, with one E-versus-C effect, $\xi = \xi_1 = \cdots = \xi_G$, and the linear term is

$$\eta_{X,g}(\boldsymbol{\theta}) = \alpha + \beta_g + \xi I(X = E).$$

The second comparator, SEP, conducts G separate trials, one within each subgroup, without assuming a regression model. The third comparator, NS, completely ignores subgroups and uses one monitoring rule. For comparability, all four designs were assumed to have $N_{max} = 200$, with all $\delta_g = .15$. Each SEP trial was conducted with a maximum of 50 patients per subgroup. For all designs, the priors were calibrated to have ESS = 100 for C and ESS = 1 for E, and the subgroup-specific null values were $\hat{\pi}_{C,1} = .25$, $\hat{\pi}_{C,2} = .45$, $\hat{\pi}_{C,3} = .55$, $\hat{\pi}_{C,4} = .60$, corresponding to prognostic levels that improved with g. For NS, since it ignores subgroups, the null probability was set to the average value .46, with prior

$$\pi_C \sim Beta(46, 54), \quad \pi_E \sim Beta(.46, .54).$$

For SEP, the following four priors for the subgroup-specific response probabilities were assumed:

$$\pi_{C,1} \sim Beta(25, 75), \quad \pi_{E,1} \sim Beta(.25, .75),$$

$$\pi_{C,2} \sim Beta(45, 55), \quad \pi_{E,2} \sim Beta(.45, .55),$$

$$\pi_{C,3} \sim Beta(55, 45), \quad \pi_{E,3} \sim Beta(.55, .45),$$

$$\pi_{C,4} \sim Beta(60, 40), \quad \pi_{E,4} \sim Beta(.60, .40).$$

The simulation results are summarized in Table 4.1. In Scenario 1, E achieves the targeted improvement in subgroups 3 and 4 but not in subgroups 1 or 2. In this case, it is desirable to have small early stopping probabilities for E in subgroups and 4 and large early stopping probabilities in subgroups 1 and 2. Both S-TI and SEP maintain stopping probabilities ≤ 0.10 within subgroups 3 and 4, S-NTI incorrectly stops early 64% and 70% of the time in subgroups 3 and 4, and NS incorrectly stops early 64% of the time in both subgroups 3 and 4. In subgroup 1, S-TI has the largest probability of correctly stopping, 75%, compared to 74%, 64%, and 65% for S-NTI, NS, and SEP, respectively. In subgroup 2, all designs have approximately the same chance of rejecting E. In this case, S-TI reliably terminates patient accrual to these subgroups, and consequently allows more patients in subgroups 3 and 4 to be treated with an effective treatment due to the enrichment rule.

In Scenario 2, E achieves the targeted improvement in subgroups 1 and 4 but not in subgroups 2 or 3, and the simulation results are very similar to those of Scenario 1. In Scenario 3, E only achieves the targeted improvement in subgroup 4. While both S-TI and SEP maintain stopping probability 0.10 in subgroup 4, S-NTI and NS incorrectly stop in this subgroup 92% and 90% of the time, respectively. Because the stopping probabilities in subgroups 1, 2, and 3 with S-TI uniformly dominate those of SEP, this scenario provides a strong motivation for borrowing strength while accounting for treatment–subgroup interactions. In Scenario 4, E achieves the targeted improvement in all subgroups and thus, by design, all four methods have a stopping probability of 0.10. In Scenario 5, E fails to achieve the targeted improvement in all subgroups and thus it is desirable for a design to stop in all subgroups with high probability. Because S-NTI and NS do not include treatment-subgroup interactions, both procedures combine the information for the subgroups and thus, in this case, they are nearly certain to stop early. S-TI has subgroup-specific stopping probabilities ranging from 0.76% to 0.83% in this case.

TABLE 4.1
Simulations comparing the S-TI design with subgroup-specific stopping rules based on a model with treatment-subgroup interactions. The S-NTI design assumes no treatment-subgroup interactions, the NS design assumes homogeneity with no subgroup effects, and the SEP design uses a separate monitoring rule in each subgroup. All $\delta_g = .15$. Each $\pi_{E,g}^{true} = \mu_{C,g} + .15$ is given in boldface. Each entry in columns 5 – 8 is the simulated Pr(Stop Early) (Mean sample size).

Scenario	g	$\mu_{C,g}$	$\pi_{E,g}^{true}$	S-TI	S-NTI	NS	SEP
1	1	.25	.25	.75 (35)	.74 (35)	.64 (30)	.65 (33)
	2	.45	.45	.67 (36)	.67 (31)	.64 (30)	.65 (34)
	3	.55	**.70**	.10 (64)	.64 (35)	.64 (30)	.10 (47)
	4	.60	**.75**	.10 (64)	.70 (35)	.64 (30)	.10 (47)
2	1	.25	**.40**	.10 (65)	.67 (35)	.64 (30)	.10 (47)
	2	.45	.45	.69 (35)	.67 (31)	.64 (30)	.65 (34)
	3	.55	.55	.73 (34)	.65 (32)	.64 (30)	.73 (32)
	4	.60	**.75**	.10 (65)	.66 (31)	.64 (30)	.10 (47)
3	1	.25	.25	.74 (39)	.94 (21)	.90 (20)	.65 (33)
	2	.45	.45	.73 (38)	.92 (22)	.90 (20)	.65 (34)
	3	.55	.55	.74 (37)	.91 (22)	.90 (20)	.73 (32)
	4	.60	**.75**	.10 (83)	.92 (2)	.90 (20)	.10 (47)
4	1	.25	**.40**	.10 (50)	.10 (47)	.10 (46)	.10 (47)
	2	.45	**.60**	.10 (50)	.10 (47)	.10 (46)	.10 (47)
	3	.55	**.70**	.10 (50)	.10 (47)	.10 (46)	.10 (47)
	4	.60	**.75**	.10 (50)	.10 (47)	.10 (46)	.10 (47)
5	1	.25	.25	.83 (42)	.99 (16)	.99 (14)	.65 (33)
	2	.45	.45	.76 (46)	.99 (16)	.99 (14)	.65 (34)
	3	.55	.55	.81 (41)	.99 (16)	.99 (14)	.73 (32)
	4	.60	.60	.82 (41)	.99 (16)	.99 (14)	.75 (31)

4.3 An Early Basket Trial

Thall et al. (2003) proposed what now is considered a "basket trial" design, although this terminology did not exist at the time. The design was for a phase II activity trial of imatinib for the treatment of sarcomas. It was constructed to address the complication that patients with 10 different sarcoma histologies would be enrolled, and the aims were to construct a futility stopping rule for each subtype while borrowing strength between subtypes. At the time the trial was designed, only two conventional approaches for dealing with this type of setting were available. The first approach was to ignore the sarcoma subtypes by assuming homogeneity, with one response rate for all 10 subtypes, and construct a single "one size fits all" futility rule that would stop or continue accrual for all subtypes. The second approach was to conduct 10 separate trials, one for each subtype. While neither of these approaches was considered fully satisfactory, it did not occur to the medical investigators that any other approach was possible. Consequently, the investigators' discussions were focused on weighing the advantages and disadvantages of these two approaches. Assuming homogeneity does not accommodate the possibility that the agent may be active in some subtypes but not others. For example, if the agent has substantive anti-disease activity in subtype 1 but not subtypes 2 – 10, then ignoring subtypes and using a single stopping rule carries the risk of failing to detect activity in subtype 1, because its outcome data are averaged with data from other subtypes where the agent is ineffective and thus no responses are observed. While, in concept, the alternative approach of conducing 10 separate trials allows the possibility of different response rates, this is logistically very cumbersome, produces unreliable results in subtypes with small sample sizes, and is not feasible for rare subtypes.

The following model and design formalize the ideas that the subtypes might have different response rates, but observing anti-disease activity in one subtype should increase the posterior probability of activity in the other subtypes. Thall et al. (2003) constructed a phase II design with subtype-specific futility stopping rules that does this. In terms of its decision rules, the design was very similar to the precision phase II design of Wathen et al. (2008). The fundamental difference between the two designs is that, before the trial was conducted, it was assumed that the response rates were exchangeable across the 10 sarcoma subtypes. This motivated a Bayesian hierarchical model, which was used as the basis for constructing 10 subtype-specific futility stopping rules. Each patient's disease was evaluated at enrollment and at months 2 and 4 following treatment, with disease status at each evaluation classified as complete response (CR), partial response (PR), stable disease (SD) or progressive disease (PD) compared to baseline. For the purpose of treatment evaluation, "response" was defined as either (1) CR or PR at month 2, or (2) SD at month 2 followed by CR, PR, or SD at month 4.

The assumed model is given in general as follows. For G exchangeable subgroups, denote the response probabilities by $\pi_1, \cdots, \pi_G$. For subtype g, denoting the number of responders by R_g and number of patients evaluated by n_g at any point in the trial, $R_g \mid \pi_g, n_g \sim Binom(\pi_g, n_g)$. A conditionally independent hierarchical model (CIHM) was constructed as follows. Transforming the response probabilities to the real-valued parameters $\theta_g = \log\{\pi_g/(1-\pi_g)\}$, for a level 1 prior it was assumed that

$$\theta_1, \cdots, \theta_G \mid \mu, \tau \sim iid\ N(\mu, \tau^{-1}).$$

The assumed level 2 priors (hyperpriors) were

$$\mu \sim N(\tilde{\mu}, \tilde{\sigma}^2) \quad \text{and} \quad \tau \sim Gamma(\tilde{\alpha}_1, \tilde{\alpha}_2).$$

Numerical hyperparameter values were obtained by elicitation. To represent the prior belief that the average response rate was midway between the targeted value 0.30 and the nominally uninteresting value 0.10, the prior mean of μ was set to equal to $\tilde{\mu}$ = log{0.20/(1-.20)} = -1.386. Setting $\tilde{\sigma}^2 = 10$ ensured that $\tau = 1/\sigma^2$ had prior mean .10 and variance .005. The remaining numerical hyperprior parameter values were obtained by eliciting the following probabilities. Three elicited prior probabilities were the unconditional value, $\Pr(\pi_1 > .30) = 0.45$, and the conditional probabilities $\Pr(\pi_1 > .30 \mid R_1 = 2, n_1 = 6) = 0.525$, and $\Pr(\pi_1 > .30 \mid R_2 = 2, n_2 = 6) = 0.47$. This says the prior belief was that, if 2/6 responses were observed for any given subtype, then this should increase the probability of at least a 30% response rate from .45 to .525. But if this was seen in another histology, then it should increase this probability from .45 to .47.

Denote $\boldsymbol{\theta} = (\theta_1, \cdots, \theta_G)$, the G-vector of 1's by $\mathbf{1}_G$ and the $G \times G$ matrix with all entries 1 by $\mathbf{J}_G$, the level 1 prior of the CIHM may be given as

$$\boldsymbol{\theta} \mid \mu, \sigma^2 \sim N_G(\mu \mathbf{1}_G, \sigma^2 \mathbf{J}_G).$$

To see how this induces association among the θ_g's, for subtypes $g = 1$ and $g = 2$,

$$\theta_1 \mid \theta_2, \sigma^2 \sim N\left(\frac{\sigma^2\tilde{\mu} + \tilde{\sigma}^2\theta_2}{\sigma^2 + \tilde{\sigma}^2}, \frac{\sigma^4 + 2\sigma^2\tilde{\sigma}^2}{\sigma^2 + \tilde{\sigma}^2}\right).$$

Thus, the model shrinks the mean of θ_1 toward θ_2, since the conditional mean $E(\theta_1 \mid \theta_2)$ is a weighted average of the hypermean $E(\mu) = \tilde{\mu}$ and θ_2. Since the distribution of each π_g thus is associated with those of all other π_j's for $j \neq g$, the early stopping criterion for each subtype depends on the data from all 10 subtypes. The rules were to stop accrual to subtype g if

$$Pr(\pi_g > .30 \mid \mathscr{D}_n) < .005,$$

with the cutoff .005 determined based on preliminary simulations. Table 4.2 illustrates how association among the π_g's may change the early stopping decision for each subtype. The 10-separate-trial rules are based on assuming conventional independent beta(.20, .80) priors.

Table 4.3 gives the design's operating characteristics, applying the rule for each subtype after at least eight patients are accrued in that subtype. To avoid treating a large number of patients in a subtype with a high accrual rate but a low response rate, the following additional rule was applied. During the first three months of the trial, if subtype g had accrued 15 patients but not all 15 had been evaluated, then accrual would be suspended temporarily in subtype g unless the stopping criterion would not be met even if all patients not yet evaluated were to fail. The slowly accruing subtypes had smaller early stopping probabilities because there is less information per unit time compared to the quickly accruing subtypes. In simulation scenario 1, all 10 subtypes have true response rate $\pi_g^{true} = 0.10$, with a rule applied assuming conventional beta distributions, the early stopping probability would be 0.67 for each of subtypes 1–5 and 0.42 for each of subtypes 6 – 10. The higher stopping probabilities

TABLE 4.2
Comparison of stopping decisions within disease subtypes based on either a conventional or a hierarchical Bayesian model.

		Decisions Within Sarcoma Subtypes	
Case	Outcomes	Conventional	Hierarchical
1	5 subtypes with 0/8	Stop	Stop
	5 subtypes with 1/8	Continue	Continue
2	3 subtypes with 0/8	Stop	Continue
	2 subtypes with 1/8	Continue	Continue
	5 subtypes with 2/8	Continue	Continue
3	2 subtypes with 1/17	Stop	Continue
	3 subtypes with 5/17	Continue	Continue
	5 subtypes with 7/23	Continue	Continue
4	3 subtypes with 0/8	Stop	Stop
	2 subtypes with 1/8	Continue	Stop
	5 subtypes with 2/23	Stop	Stop
5	3 subtypes with 1/8	Continue	Continue
	2 subtypes with 2/22	Continue	Stop
	5 subtypes with 3/30	Stop	Stop

0.77 and 0.47 under the hierarchical model are due to its ability to borrow strength. Scenarios 2 and 3 show that, while a quickly accruing subtype with target response probability $\pi_g^{true} = 0.30$ has a 9.5% false negative rate, if two subtypes have $\pi_g^{true} = 0.30$ then each has false negative rate 8.0% due to borrowing strength.

This simple hierarchical model for borrowing strength between disease subtypes is better than the two simple competitors of either ignoring subtypes or conducting completely separate trials. However, it is limited by the fact that all posterior means are shrunk toward the same hyperparameter $\tilde{\mu}$. Moreover, there is no method for adaptively combining subtypes that have similar estimated response rates in order to borrow strength. More recently proposed basket trial designs, discussed below, address these limitations by including refinements of both the hierarchical model and the decision-making scheme.

4.4 A Basket Trial Design with Bayesian Model Averaging

For basket trials based on a binary response, Psioda et al. (2019) proposed a design that uses Bayesian model averaging (BMA), rather than latent subgroup membership variables, to address settings where some subgroups have the same response rates

TABLE 4.3
Operating characteristics of the phase II sarcoma trial design for 10 sarcoma subtypes. Assumed accrual rates are 3 to 6 per month for subtypes 1 – 5 and .5 to 2 per month for subtypes 6 – 10. # patients and stopping percentages are per subtype.

		Sarcoma Subtype		
Scenario		1 – 5		6 – 10
1	π_g^{true}	.30		.10
	# Patients	23.2		12.6
	% Stop	77		47
		1	2 – 5	6 – 10
2	π_g^{true}	.30	.10	.10
	# Patients	29.2	23.2	12.8
	% Stop	9.5	76	45
		1 –2	3 – 5	6 – 10
3	π_g^{true}	.30	.10	.10
	# Patients	29.3	23.9	13.1
	% Stop	8.0	73	42
		1 – 5	6	7 – 10
4	π_g^{true}	.10	.30	.10
	# Patients	23.4	14.2	12.9
	% Stop	75	7	49

while others have distinct values. For example, if $\pi_1 = \pi_2$ and this common value is distinct from π_3, then there actually are two distinct response probabilities rather than three. In such cases, a problem with assuming a CIHM is that it is symmetric in $\pi_1, \cdots, \pi_G$, and consequently it shrinks all posterior estimates toward one average value. In cases of the sort described above, this may produce decision rules with undesirably large false positive or false negative rates for some subgroups. BMA addresses this problem by putting prior probabilities on the possible models for $\boldsymbol{\pi}$, computing posterior model probabilities, and averaging over this posterior. Like the phase II basket trial designs of Thall et al. (2003) and Wathen et al. (2008), the design of Psioda et al. (2019) includes Bayesian subgroup-specific stopping rules that may terminate accrual to some subgroups but not others, while borrowing strength between subgroups.

Index baskets (patient subgroups) by $g = 1, \cdots, G$ and denote models by $M_1, \cdots, M_J$. Let $\pi_{j,p}$ denote the pth distinct response probability under M_j. Each model is characterized by a clustering of the G baskets and distinct response probabilities, one for each cluster. As an illustration, Table 4.4 presents all possible models for the response probabilities if $G = 3$. Clustering baskets with the same $\pi_{j,p}$'s reduces the model dimension and thus provides a basis for borrowing strength between

TABLE 4.4
Illustration of the possible models for $G = 3$ baskets. Under model M_j the pth distinct response probability is denoted by $\pi_{j,p}$.

Model	Basket Clusters	Response Probabilities $g = 1$	$g = 2$	$g = 3$	Dimension($\boldsymbol{\pi}$)
M_1	{1,2,3}	$\pi_{1,1}$	$\pi_{1,1}$	$\pi_{1,1}$	1
M_2	{1,2}, {3}	$\pi_{2,1}$	$\pi_{2,1}$	$\pi_{2,2}$	2
M_3	{1,3}, {2}	$\pi_{3,1}$	$\pi_{3,2}$	$\pi_{3,1}$	2
M_4	{1}, {2,3}	$\pi_{4,1}$	$\pi_{4,2}$	$\pi_{4,2}$	2
M_5	{1}, {2}, {3}	$\pi_{5,1}$	$\pi_{5,2}$	$\pi_{5,3}$	3

baskets. The Bayesian model averaging approach does this by placing probabilities on the M_j's, computing their posteriors based on the interim data, and averaging over this model posterior. For the example in Table 4.4, models M_2, M_3, and M_4 are intermediate cases where it would be useful to know the cluster structure and construct stopping rules accordingly. For example, under M_2, subgroups 1 and 2 should have the same rule, based on the common response probability $\pi_{2,1}$.

The design includes subgroup-specific rules to stop enrollment to subgroup g if it is determined that either π_g is too small for the agent to be promising (futility) or if π_g is large enough so that its activity level motivates study of the agent in that basket in a future trial (superiority). In subgroup g, at the ith interim analysis, let $n_{i,g}$ denote the number of patients and $Y_{i,g}$ the number of responses, with data

$$\mathscr{D}_i = \{(Y_{i,g}, n_{i,g}) : g = 1, \cdots, G\}.$$

Let p_0 be a fixed small response rate that corresponds to an agent being not promising, $p_1 > p_0$ a larger response rate that corresponds to an agent being promising, and denote $\delta = (p_0 + p_1)/2$. For a futility rule, accrual to subgroup g is stopped with the agent declared not promising in that subgroup if

$$P(\pi_g > p_0 + \delta \mid \mathscr{D}_i) \leq \phi_0. \tag{4.2}$$

For a superiority rule, accrual to subgroup g is stopped with the agent declared promising in that subgroup if

$$P(\pi_g > p_0 \mid \mathscr{D}_i) > \phi_1. \tag{4.3}$$

The rules (4.2) and (4.3) are subgroup-specific versions of the rules given by Thall and Simon (1994) in the first Bayesian phase II design, which assumed that patients were homogeneous. Given a prespecified maximum number of analyses, I, accrual to a subgroup is terminated if either stopping criterion is met, or if I is reached. If all subgroups are stopped the trial ends. However, it is important to note that, in practice, clinicians very seldom use superiority stopping rules in phase II trials. This is because they almost invariably want to continue to treat patients with a new agent that has been found to be successful. Thus, a practical version of the BMA design would not include superiority stopping rules.

Temporarily suppressing the index i, it is assumed that $Y_g \mid \pi_g, n_g \sim \text{Binomial}(\pi_g, n_g)$ for each g. Under model M_j, let P_j denote the number of distinct response probabilities and $\Omega_{j,r}$ the set of all subgroup labels for subgroups having the rth distinct response probability, $r = 1, \cdots, P_g$. Thus, each $\Omega_{j,r}$ is a cluster of subgroup labels. For example, under model M_2 in Table 4.4, the subgroup clusters are $\Omega_{2,1} = \{1,2\}$ and $\Omega_{2,2} = \{3\}$. Conditional on model M_j, denoting $\boldsymbol{\pi}_j = (\pi_{j,1}, \cdots, \pi_{j,P_j})$, the likelihood is

$$\mathscr{L}(\boldsymbol{\pi}_j \mid \mathscr{D}, M_j) \propto \prod_{r=1}^{P_j} \prod_{g \in \Omega_{j,r}} \binom{n_g}{Y_g} \pi_{j,r}^{Y_g} (1 - \pi_{j,r})^{n_g - Y_g}.$$

For G subgroups, denote the set of all models given that there are at most P distinct response probabilities by $\mathscr{M}_{G,P}$, so necessarily $P \le G$. The complete model space $\mathscr{M}_{G,G}$ includes the two extreme cases of complete homogeneity where all $\pi_g = \pi$ and the case where there are G distinct probabilities.

To specify priors, a default prior on $\mathscr{M}_{G,P}$ assumes that $p(M_j) \propto P_j^{\alpha}$, where the tuning parameter $\alpha \ge 0$. A uniform prior is obtained of $\alpha = 0$, and $\alpha > 0$ gives larger prior weight to models with more parameters, which produces less borrowing. The response probability priors under each model are assumed to be simple betas,

$$\pi_{j,r} \mid M_j \sim Beta(a_0, b_0).$$

A practical hyperparameter choice sets the prior mean $a_0/(a_0 + b_0) = p_1$ and effective sample size $a_0 + b_0 = 1$, which is somewhat optimistic but uninformative. By conjugacy of the binomial distribution and beta prior, the posteriors are

$$\pi_{j,r} \mid M_j, \mathscr{D}_i \sim Beta\left(a_0 + \sum_{g \in \Omega_{j,r}} Y_g,\ b_0 + \sum_{g \in \Omega_{j,r}} (n_g - Y_g) \right) = Beta(a_{j,r}, b_{j,r}),$$

which are used to compute the subgroup-specific stopping probabilities. The marginal likelihood under M_j is

$$p(\mathscr{D} \mid M_j) = \prod_{g=1}^{G} \binom{n_g}{y_g} \prod_{r=1}^{P_j} \frac{\mathscr{B}(a_{j,r}, b_{j,r})}{\mathscr{B}(a_0, b_0)}$$

where $\mathscr{B}(a,b)$ denotes the complete beta function. Applying Bayes' Law, and following the BMA framework of Madigan and Raftery (1994), this can be used to compute the posterior model probabilities,

$$p(M_j \mid \mathscr{D}) = \frac{p(\mathscr{D} \mid M_j) p(M_j)}{\sum_r p(\mathscr{D} \mid M_r) p(M_r)}.$$

This structure is used to obtain the subgroup-specific stopping criteria (4.2) and (4.3) as the model weighted averages

$$P(\pi_g > x \mid \mathscr{D}) = \sum_j P(\pi_g > x \mid \mathscr{D}, M_j) p(M_j \mid \mathscr{D})$$

for $x = p_0$ or $p_0 + \delta$. A nice feature of this construction is that the computations are straightforward.

To evaluate the BMA design and compare it to other basket trial designs, Psioda et al. (2019) conducted a simulation study, based on tests of the hypotheses H_0 : $\pi_g \leq p_0$ versus $H_1 : \pi_g > p_0$ for each subgroup $g = 1, \cdots, G = 5$. They used the numerical values $p_0 = .15$ (an "inactive" subgroup) with target $p_1 = .45$ (an "active" subgroup) for computing power. The design parameters α, ϕ_0 and ϕ_1 were calibrated by simulation so that the false positive rate for each subgroup, FPR(g), was controlled to be below a fixed limit. The subgroup-specific power was characterized by the "true positive rate" for each basket, TPR(g).

The proposed value $\alpha = 2$ provides a balance between information borrowing between subgroups and design stability for various accrual rates, and cut-offs in the range $.2 \leq \phi_0 \leq .4$ work well. The simulation study was conducted for two-stage designs, $L = 2$, to facilitate comparison to Simon's two-stage design (Simon, 1989), under a homogeneity assumption. It is important to note that, in practice, many phase II designs require must more frequent monitoring. While this is especially important for monitoring toxicity probabilities to protect patient safety, the design of Psioda et al. (2019) ignores toxicity, which limits its practicality. The BMA design was implemented with its futility rule but not its superiority stopping rule, which is more realistic since oncologists seldom stop a phase II trial for superiority. Additional comparators were the CIHM-based designs of Chu and Yuan (2018a), CBH who used a calibrated hierarchical model, and Cunanan et al. (2017), CUN, who used the interim data to adaptively choose between the two extremes of complete homogeneity and complete heterogeneity. Three accrual patterns were evaluated, Poisson, slow to all active subgroups, and fast to all active subgroups. Poisson accrual corresponded to patients arriving according to a homogeneous Poisson process with a rate of two patients per subgroup per month, so the subgroups were assumed to be equally likely. For comparability, all four designs were calibrated to meet the following criteria:

1. If all $\pi_g = p_0$ then the family-wise FPR $\leq .05$
2. If $\pi_5 = .45$ and all other $\pi_g = .15$, under Poisson accrual, then TPR(5)$\geq .78$.
3. If $\pi_5 = .45$ and all other $\pi_g = .15$, under slow accrual, then TPR(5) $\geq .60$.

The most prominent result in Table 4.5 is that the CUN design does a very poor job of controlling the family-wise FPR $\leq .05$ when the number of active subgroups, namely those with $\pi_g^{true} = .45$, is larger. This is worst for active subgroups with fast accrual. For basket-specific TPR(g) rates, the BMA, CBH, and CUN designs all provide larger subgroup-specific TPR(g) than the Simon design for both Poisson and fast active accrual. The results are mixed for slow active accrual, with basket-specific TPR(g) values for the information borrowing designs larger than those of the Simon design only when there are at least three (CUN) or four (BMA and CBH) baskets active. In this case, the CUN design has the smallest reduction in TPR(g) compared to the Simon design, but this advantage is negated by the fact that the CUN design has greatly inflated family-wise FPR in the cases of 2, 3, or 4 active subgroups. General

TABLE 4.5
Operating characteristics of two-stage BMA, CBH, CUN, and Simon designs, for $G = 5$ subgroups with response rates either $\pi_g^{true} = .15$ (agent is inactive) or $\pi_g^{true} = .45$ (agent is active). FPR = false positive rate, TPR = true positive rate.

	#	Family-Wise FPR				Subgroup Specific TPR(g)			
Accrual	Active	BMA	CBH	CUN	Simon	BMA	CBH	CUN	Simon
Poisson	0	.05	.05	.05	.05	–	–	–	–
	1	.05	.05	.06	.04	.78	.78	.78	.82
	2	.05	.04	.07	.03	.81	.83	.83	.81
	3	.04	.03	.10	.02	.83	.84	.85	.81
	4	.02	.03	.16	.01	.85	.86	.86	.81
	5	–	–	–	–	.87	.89	.89	.81
Fast Active	0	.05	.05	.05	.05	–	–	–	–
	1	.06	.05	.08	.04	.91	.91	.84	.81
	2	.05	.04	.10	.03	.90	.91	.86	.81
	3	.05	.04	.14	.02	.90	.90	.87	.81
	4	.03	.07	.22	.01	.89	.89	.88	.81
	5	–	–	–	–	.87	.89	.89	.81
Slow Active	0	.05	.05	.05	.05	–	–	–	–
	1	.05	.05	.05	.04	.65	.62	.74	.81
	2	.04	.04	.05	.03	.72	.71	.80	.81
	3	.03	.03	.06	.02	.78	.77	.83	.81
	4	.02	.02	.09	.01	.82	.82	.85	.81
	5	–	–	–	–	.87	.89	.89	.81

conclusions are that (1) the BMA, CBH, and Simon designs perform similarly good jobs of controlling the family-wise FPR, (2) the BMA and CBH designs provide increases in subgroup-specific TPR(g) up to .10 compared to Simon, for both Poisson and fast active accrual, but (3) the Simon design actually has larger subgroup-specific TPR(g) for slow active accrual when 1 or 2 subgroups are active.

Computer Software A computer package *Bayesian Model Averaging for Basket Trials* is available from it https://github.com/ethan-alt/bmabasket.

4.5 Monitoring Response and a Longitudinal Biomarker

Chu and Yuan (2018b) proposed a Bayesian latent subgroup trial (BLAST) basket trial design for phase II settings where a cancer has been classified into disease subtypes, or where several different diseases all have the same targeted biomarker. BLAST addresses an information-rich setting where each patient's clinical outcome consists of both a binary response Y and a biomarker process, Z, that is observed longitudinally over time. To exploit this data structure, a model for Z, Y, and cancer subtype is specified in terms of regression of Z on time and Y, also including a latent disease subtype variable, C, that is used to cluster subtypes having similar distributions. The mean of Z as a function of time is characterized robustly by using penalized splines. Like any basket trial, the design makes adaptive group sequential decisions of whether or not the new agent is active for each disease subtype, while exploiting the clustering to reduce model dimension.

The BLAST design was motivated by a phase II basket trial enrolling patients with advanced solid tumors having CDKN2A deletions, including bladder cancer, brain tumors, melanoma, pancreatic cancer, esophageal cancer, and gastric cancer. The targeted agent was an aurora kinase checkpoint inhibitor targeting the CDKN2A signalling pathway. The goal of the trial was to evaluate the efficacy of the aurora kinase inhibitor in each of these cancers showing the CDKN2A deletion. The trial was designed to enroll up to 20 patients with each type of cancer. Treatment efficacy was scored using the 'Response evaluation criteria in solid tumors', version 1.1, coded as 'response' if the patient achieved CR or PR. The targeted agent was considered unpromising for a disease if the response rate was $< .20$, and promising if the response rate was $> .30$. The longitudinal biomarker was immunofluorescent intensity of phospho-aurora-A (Thr288), which quantifies the inhibitor's ability to hit the target and inhibit the CDKN2A pathway.

For the jth patient with disease subtype $g = 1, \cdots, G$, let $Y_{g,j}$ denote the binary response indicator and $Z_{g,j,l}$ the biomarker observed at time t_l, for $l = 1, \cdots, L$. The response probability for disease subtype g is $\pi_g = \Pr(Y_{g,j} = 1)$. Based on the observed data, the G cancer subtypes may be classified adaptively into $K \leq G$ subgroups, with patients in each subgroup having the same response probability. An important case is that where it can be determined that each disease subtype either is sensitive to the treatment in that patients may respond, or is insensitive with patients not responsive, so $K = 2$. While K is not known and is determined adaptively, given K, for each disease subtype g, ζ_g is a latent cluster membership indicator. It is assumed that

$$\zeta_g \mid K \sim Multinomial(\pi_1, \cdots, \pi_K), \quad \text{for} \quad g = 1, \cdots, G \tag{4.4}$$

where each $\pi_k = \Pr(\zeta_g = k)$ is the probability that the gth cancer subtype belongs to the kth latent subgroup, so $\pi_1 + \cdots + \pi_G = 1$. Since the latent subgroup indicators $\boldsymbol{\zeta} = (\zeta_1, \cdots, \zeta_G)$ are not observed, their posteriors are computed along with the model parameter posterior via MCMC.

For the distribution of the observed outcomes (Y, Z), denoting $\theta_g = \log\{\pi_g/(1 - \pi_g)\}$, it is assumed that

$$\theta_g \mid \zeta_g = k \sim N(\theta_{(k)}, \tau^2_{(k)}).$$

This is a key assumption, since for $K < G$ it reduces the model dimension by assuming that subtypes in the same cluster have the same response rate, and consequently the stopping rules borrow strength between subtypes in the same cluster. If $Y_{g,j}$ is real-valued rather than binary, this model can be modified quite easily by assuming that $Y_{g,j} \sim N(\theta_g, \sigma^2_y)$ with prior $\theta_g \mid \zeta_g = k \sim N(\theta_{(k)}, \tau^2_{(k)})$.

A longitudinal model for the biomarker given the observed response indicator and cluster variable is given by

$$Z_{g,j,l} = \mu_{(k)}(t_l) + v_g + w_{g,j} + \beta Y_{g,j} + \varepsilon_{g,j,l}.$$

The latent variable $v_g \sim iid\ N(0,\sigma_v^2)$ is the cancer subtype-specific random effect, which accounts for the fact that the mean biomarker trajectory for a disease subtype may deviate from the mean trajectory for its subgroup. The latent variable $w_{g,j} \sim N(0,\sigma_w^2)$ is a patient random effect that accounts for the biomarker trajectory of an individual patient deviating from the mean trajectory of their disease subtype, and $\varepsilon_{g,j,l} \sim iid\ N(0,\sigma_\varepsilon^2)$ is measurement error. The parameter β is the effect of response on the biomarker trajectory.

The mean biomarker path $\mu_{(k)}(t_l)$ as a function of time is modeled using penalized splines. For prespecified knots $\kappa_1 < \kappa_2 < \cdots < \kappa_S$ that partition the observation interval $[t_1,\ t_L]$ into $S+1$ subintervals, define

$$(t_l - \kappa_s)_+^d = (t_l - \kappa_s)^d \quad \text{if} \quad t_l > \kappa_s$$

and 0 otherwise. The penalized spline is given by

$$\mu_{(k)}(t_l) = \gamma_{0(k)} + \gamma_{1(k)} t_l + \gamma_{2(k)} t_l^2 + \cdots + \gamma_{d(k)} t_l^d + \sum_{s=1}^{S} a_{s(k)} (t_l - \kappa_s)_+^d,$$

where $\gamma_{0(k)}, \cdots, \gamma_{d(k)}$ are parameters, and $a_{1(k)}, \cdots, a_{S(k)}$ are $N(0,\sigma_{a(k)}^2)$ random effects. This penalized spline is robust to the choice of knots and basis functions.

The longitudinal biomarker process $\{Z_{g,j,l}\}$ plays a key role in determining the posterior distribution of $(\zeta_1, \cdots, \zeta_G)$, since the posterior of $\Pr(\zeta_g = k)$ is a weighted average of multivariate normal distributions of $\mathbf{Z}_g = (\mathbf{Z}_{g,1}, \cdots, \mathbf{Z}_{g,Q_g})$ for Q_g biomarker measurements in subtype g. Since the latent subgroup membership variables play a central role in reducing the model dimension and borrowing strength among the disease subtypes in each cluster, the methodology exploits information in the longitudinal biomarker process to obtain more reliable subtype-specific decision rules. Computational details are given in Appendix A of Chu and Yuan (2018b).

To deal with the fact that the number K of latent subgroups is random, Chu and Yuan (2018b) use the deviance information criterion (DIC) (Spiegelhalter et al., 2002) to choose K adaptively in order to obtain the best fit to the data. While this might be done by fitting the model for each value $K = 1, \cdots, G$ and choosing K to minimize the DIC, in practice it may suffice to do this at the start for $K = 1$, 2, or 3, and later update K during the trial as more data are obtained. A different approach would be to treat K as a parameter and apply reversible jump MCMC (Green, 1995) to accommodate the change in model dimension.

For priors, Chu and Yuan (2018b) assume that, For each $k = 1, \cdots, K$,

$$\theta_{a(k)} \sim N(m_k, c) \ \text{ and } \ \gamma_{d(k)} \sim iid\ N(0,c) \ \text{ for } d = 0,1,2,$$

$$\sigma_{a(k)}^2 \sim iid\ IG(\tilde{a}_a, \tilde{b}_a,) \quad \tau_{(k)}^2 \sim iid\ IG(\tilde{a}_\tau, \tilde{b}_\tau,), \quad (\tau_1, \cdots, \tau_K) \sim Dirichlet(\lambda_1, \cdots, \lambda_K)$$

with $\sigma_v^2, \sigma_w^2, \sigma_\varepsilon^2$ also inverse gamma, and $\beta \sim N(0,c)$. These were specified with hyperparameter values $c = 10^4$, all IG parameters = .001, and all $\lambda_k = 2$.

A BLAST trial is conducted with M planned interim analyses. The objective is to test whether the targeted agent is effective in each subtype, based on a small fixed probability q_0 for which the agent is ineffective, and a larger fixed probability $q_1 > q_0$ for which the agent is effective. The hypotheses for disease subtype g are $H_{g,0} : p_g \leq q_0$ and $H_{g,1} : p_g \geq q_1$. Denoting the data at the mth analysis by

$$\mathscr{D}_m = \{(Y_{g,j}, Z_{g,j,l}), g = 1, \cdots, G, \ j = 1, \cdots, n_{g,m}, \ l = 1, \cdots, L\},$$

the trial is conducted as follows.

Step 1: For each disease subtype $g = 1, \cdots, G$, enroll up to $n_{g,1}$ patients.

Step 2: For the mth interim decision, fit the model to the current data $\mathscr{D}_m$. For each disease subtype $g = 1, \cdots, G$, if

$$\Pr(p_g > (q_0 + q_1)/2 \mid \mathscr{D}_m) < Q_f$$

then stop accrual of patients with disease subtype g and accept $H_{g,0}$. Otherwise, continue to accrue patients with that disease subtype.

Step 3: When the maximum overall sample size has been reached, or accrual has been stopped for all subtypes, evaluate treatment efficacy based on the final data $\mathscr{D}$, as follows. For each subtype g, if

$$\Pr(p_g > q_0 \mid \mathscr{D}) > Q$$

then reject $H_{g,0}$ and conclude that the treatment is effective in subtype g. Otherwise, accept $H_{g,0}$ and declare it ineffective in that subtype. The values of the decision cutoff probabilities Q_f and Q are calibrated to obtain a design with desirable type I and type II error rates.

The BLAST design was studied by simulation based on the motivating trial of CDKN2A deficient solid tumors, for $q_0 = .20$ and $q_1 = .30$, $G = 6$ disease subtypes, and two subgroups, sensitive and insensitive, determined by the response probabilities $\{p_g\}$. The mean biomarker trajectories were assumed to be one of two functions. Biomarker trajectory A was

$$\begin{aligned} \mu_k(t) &= 18 - 12\exp\{-6(t + .05)\} && \text{for g in a sensitive subgroup} \\ &= 9.11 && \text{for g in an insensitive subgroup} \end{aligned}$$

and biomarker trajectory B was a very different shape

$$\begin{aligned} \mu_k(t) &= 8.86 + \frac{8}{1 + \exp\{-8(t - .05)\}} && \text{for g in a sensitive subgroup} \\ &= 4 + 5\exp(-t^{.5}) && \text{for g in an insensitive subgroup} \end{aligned}$$

For trajectory A, the biomarker increases and then plateaus in the sensitive subgroup, and remains constant in the insensitive group. For trajectory B, in the sensitive subgroup, the biomarker first increases slowly and then increases more rapidly, and in the insensitive group the biomarker slowly decays over time, reflecting disease progression. The remaining parameters were set to $\beta = 1$, with σ_ε^2 = 1.5 and $\sigma_v^2 = \sigma_w^2$

TABLE 4.6
Operating characteristics of the designs assuming independent disease subtypes (Indep), a Bayesian hierarchical model (BHM), and BLAST, under four response probability scenarios for biomarker trajectory A.

Scenario	Design		Disease Subtypes 1	2	3	4	5	6	$\bar{N}$
1		p_g^{true}	.20	.20	.20	.20	.20	.20	
	Indep	% Reject H_0	9.9	10.1	10	10.1	10	9.9	133
		% Stop	27.2	27.5	25.9	26.6	26.5	24.7	
	BHM	% Reject H_0	9.8	10.2	9.9	9.9	9.8	9.8	129
		% Stop	30.4	30.1	30.7	28.8	30.6	29.2	
	BLAST	% Reject H_0	9.9	10.1	9.8	9.8	10.2	9.9	129
		% Stop	31.2	30.8	29.1	31	29	29.7	
2		p_g^{true}	.30	.30	.30	.30	.20	.20	
	Indep	% Reject H_0	46.5	45.4	45.9	41.4	9.2	11.6	142
		% Stop	27.2	27.5	25.9	26.6	26.5	24.7	
	BHM	% Reject H_0	69.6	68.6	72.2	70.8	45.8	42.3	147
		% Stop	2.9	2.7	2.9	3.2	6.1	4.8	
	BLAST	% Reject H_0	90.7	91.1	92.3	91.4	11.3	11.9	140
		% Stop	1	1	.6	.8	36.1	37	
3		p_g^{true}	.30	.30	.20	.20	.20	.20	
	Indep	% Reject H_0	45.4	43.4	10	9.4	10.6	10.2	137
		% Stop	5.7	6.4	26.2	25.3	27	27.2	
	BHM	% Reject H_0	46.5	47.4	26.3	26.5	25.2	23.9	141
		% Stop	7.8	7.3	13.6	14.7	15.8	14.3	
	BLAST	% Reject H_0	82.1	85.8	9.9	9.3	7.6	9	133
		% Stop	2.1	1.2	34.6	34.9	32.4	33.3	
4		p_g^{true}	.35	.30	.30	.20	.20	.20	
	Indep	% Reject H_0	69	44.5	46.6	9.7	9.9	10.5	140
		% Stop	2.2	6.4	6.8	25.6	27.	25.2	
	BHM	% Reject H_0	74.9	62.8	66.6	39	36.4	36.4	146
		% Stop	2.6	3.6	4	7.2	8.2	7.1	
	BLAST	% Reject H_0	94.7	89.1	91.4	8.6	9	7.6	138
		% Stop	.4	1.5	.6	36.5	31.7	33.7	

= 4 for trajectory A, and $\sigma_v^2 = \sigma_w^2 = 7$ for trajectory B. Four equally spaced knots were used in the penalized spline. Each disease subtype had a maximum of n_g= 25 patients, so the maximum total sample size was 150. In each disease subtype, here were three interim analyses at 10, 15, and 20 patients. The decision cut-offs were Q_f = .05 with Q calibrated to maintain type I error rate .10 for each subtype.

Simulations for biomarker Trajectory A are summarized in Table 4.6. The two comparators either assumed that the subtype response probabilities were independent, or were based on a Bayesian hierarchical model (BHM). In general, the Independence design does a very poor job of controlling type I error and has poor power, the BHM design performs much better, but the BLAST design is far better than BHM. The worse performance of the BHM design is due to the fact that it treats the p_g's as exchangeable and thus shrinks the posterior estimates of all treatment effects toward the same common mean. In the null case, scenario 1, the treatment is ineffective with a response rate of .20 for all six subtypes, so the percentage for rejecting $H_{g,0}$ is the type I error rate. Compared to the Independent design, because they borrow information across subtypes, the BHM and BLAST designs are more likely to stop the trial early, and thus they have smaller expected total sample sizes. In scenario 2, subtypes 1 - 4 are sensitive to the treatment and subtypes 5 and 6 are insensitive. In scenario 2, ideally, the latent clustering variable vector should be $\boldsymbol{\zeta} = (1,1,1,1,2,2)$, so it is desirable for the posterior to put a high probability on this vector. The BLAST design outperforms both the Independent and BHM designs, with much higher probabilities of correctly rejecting $H_{g,0}$ for each subtypes 1 - 4, while doing a good job of controlling type I error for subtypes 5 and 6. The BHM design fails to control the type I error rate for subtypes 5 and 6, with rates of 48.5% and 42.3%. In scenario 3, only subtypes 1 and 2 are sensitive. The BLAST design had high power figures of 82.1% and 85.8% for these subtypes, and controlled type I error rates well for subtypes 3 - 6. Similar results were seen for Scenario 4, where subtypes 1- 3 were sensitive and subtypes 4 - 6 were insensitive. Simulation results for trajectory *B* showed similar superiority for BLAST over the two competing designs,

In general, compared to the Independent and BHM designs, the BLAST design is more likely to stop treating insensitive disease cancer subtypes and less likely to stop treating the sensitive disease subtypes. These superior properties are due to the fact that the latent variables $\boldsymbol{\zeta}$ adaptively form separate clusters of sensitive and insensitive subtypes, so the model borrows strength in a more precise way than BHM.

Computer Software A computer program for implementing the BLAST design is available from https://odin.mdacc.tmc.edu/~yyuan/index_code.html.

5

Precision Randomized Phase II Designs

5.1 A Two-Stage Phase II Design with Adaptive Matching

Basket trial designs address the issue that, when predefined patient subgroups are determined by prognostic or possibly predictive biological variables, it is not appropriate to assume that parameters characterizing treatment effects are homogeneous across the subgroups. The single-arm phase II design of Wathen et al. (2008) provides a great improvement over phase II designs that assume homogeneity when patient subgroups have been predefined. This design addresses settings where a single-arm trial of an experimental treatment E is planned, so patients are not randomized between E and a historical standard treatment, C. For each subgroup $g = 1, \cdots, G$, an estimate of $\Delta_g = \pi_{E,g} - \pi_{C,g}$ may be the empirical response rate $\hat{\pi}_{E,g}$ computed using the data from the trial of E and the empirical response rate $\hat{\pi}_{C,g}$ computed from historical data. A major problem with this, as discussed earlier, is that the comparative estimator $\hat{\Delta}_g = \hat{\pi}_{E,g} - \hat{\pi}_{C,g}$ in each subgroup g is biased due to confounding of trial-to-historical differences with actual E-versus-C effects within subgroups.

To correct for bias after the design of Wathen et al. (2008) has been used, it may be possible to match trial participants to historical controls. In some settings, however, propensity score matching after a single-arm phase II trial is completed may not be feasible. This occurs if the characteristics of patients treated with E are so different from those of patients treated with C in a historical dataset $\mathscr{H}$ that few matched pairs can be identified. A major practical problem is that the likelihood of this cannot be determined when designing a trial, because the patient data for E are not yet available. In some cases, after the trial has been completed, it is determined that there are too few matched controls that can be identified from $\mathscr{H}$ to provide an adequately powered comparison of E to C. This was the case for several single-arm studies of E = vemurafenib + irinotecan + cetuximab for treating BRAFV600E mutated colorectal cancer. The studies of E all enrolled patients with indolent disease, good performance status, and longer prior survival compared to historical patients treated with C = irinotecan + cetuximab (Kopetz et al., 2021). Consequently, it was not possible to obtain a sufficient number of matched pairs to do a reliable bias-corrected comparison to assess the effect of adding vemurafenib to irinotecan + cetuximab.

Jiang et al. (2023) addressed this problem by proposing a two-stage *Bayesian adaptive synthetic control* (BASIC) design that is a hybrid between a single-arm trial of E and a randomized controlled trial (RCT) of E versus C. The name BASIC is based on the idea that a patient from $\mathscr{H}$ treated with C who is matched with a patient

DOI: 10.1201/9781003474258-5

in the trial treated with E may be called a *synthetic control*. BASIC exploits $\mathscr{H}$ by starting as a single-arm trial of E and doing pair matching during the trial. Based on historical $\mathscr{H}$ for C and interim data on E, the BASIC design predicts the future number of $\mathscr{H}$ patients that can be matched to patients treated with E at the end of the trial. If this number is large enough to compare E to C with a prespecified power, the single-arm trial is continued. If not, the trial is switched to an RCT, with the randomization proportion chosen so that, at the end of the trial, the E and C sample sizes will be balanced. Matching is done using propensity scores.

Deciding whether to make the switch at the end of stage 1 requires predicting future outcomes, matching, and decisions in the trial. This includes predicting the outcomes of future patients who will be enrolled in the single-arm trial if it is continued, determining the likelihood that a match can be found for each patient, and if so predicting the outcome of a final comparison of E to C. Consequently, the interim decision requires the three-stage process of (1) simulating future patients' outcomes, (2) applying propensity score matching to the sample of simulated patients, and (3) doing Bayesian posterior prediction. Deciding whether $\mathscr{H}$ can provide a pseudo-sample of matched control patients for making comparative inferences plays a central role in this process. Related methods are given by Lin et al. (2018), who described a case study using propensity score matching to select patients from $\mathscr{H}$ to augment active controls, and Thorlun et al. (2020), who provided guidelines to assess the validity and quality of trials that use matched control methods.

The usefulness of $\mathscr{H}$ is quantified by the matching efficiency, MatchEff $= N_m/N$, the predicted proportion of matched controls that can be obtained ("synthesized") from $\mathscr{H}$. MatchEff $= 1$ if N matched controls can be obtained from $\mathscr{H}$, while MatchEff $= 0$ if no matched controls can be obtained. Given a fixed threshold π between 0 and 1, if MatchEff $< \pi$ then it is considered unlikely that $\mathscr{H}$ will provide N matched controls, so the BASIC design switches to an RCT. The randomization ratio is chosen to obtain approximately N patients in each of the E and C arms at the end of the trial. In this case, the stage 2 randomization produces a hybrid control arm, with N_m nonrandomized matched controls and $N - N_m$ concurrent randomized controls. If MatchEff $> \pi$ then the single-arm trial of E with N patients is continued, at the end up to N matched controls are identified, and a pair-matched estimator of the E-versus-C effect is computed. The interim decisions of BASIC are as follows:

Singe-Arm to Singe-Arm Decision If MatchEff $\geq \pi$, continue the single-arm trial of E during stage 2 and enroll $N - n$ additional patients.

Singe-Arm to RCT Decision If MatchEff $< \pi$, switch to an RCT in which $2N - n - N_m$ future patients are enrolled during stage 2 and randomized between E and C in the ratio $(N-n):(N-N_m)$.

The threshold π that controls whether the trial is switched to an RCT may be chosen by conducting preliminary computer simulations of the trial using several different values, based on the design's operating characteristics, including power and type I error rate. A practical approach is to choose π to give power within a prespecified margin (e.g., 5%) of a targeted power (e.g., 80%). A value of π between 0.8 and 0.9 typically yields good operating characteristics for a BASIC design.

After completing stage 2 of the BASIC design, the logistic propensity score model is updated by fitting it to the final trial data and $\mathscr{H}$, and propensity score matching is used to construct a final set of matched controls. Depending on the interim decision, this can be either a fully matched control arm, or a hybrid control arm as described above. A standard frequentist test, such as a t-test or chi-square test, or statistical estimates, such as means, proportions, or regression model-based estimates, with confidence intervals, can be used to evaluate the E-versus-C treatment effect. Alternatively, Bayesian posterior probabilities and credible intervals can be used for final inferences. While the randomization and adaptive interim decisions may give final sample sizes of matched or hybrid controls not exactly equal to N, this deviation typically is small and has a negligible impact on the design's operating characteristics, as shown in the simulation study given below.

Jiang et al (2023) evaluated the operating characteristics of the BASIC design by computer simulation, including comparisons to three alternative designs, an RCT with fair randomization between E and C, a conventional single-arm design with $\mathscr{H}$ used as a comparator without matching, and a single-arm design with matched controls generated at the end of the trial using propensity score matching. The matched-control design may be considered a special case of BASIC with $\pi = 0$. The simulations considered both a binary response endpoint and a continuous endpoint, such as biomarker level, with four covariates, including two binary confounders Z_1, Z_2 and two continuous confounders Z_3, Z_4. They assumed that $\mathscr{H}$ included 160 patients, and simulated baseline covariates from a mixed population including patients both similar and dissimilar to the E patients, as follows:

Simulated Covariate Distributions

1. The covariate data of n_h comparable historical patients were generated as $\mathbf{Z} = (Z_1, Z_2, Z_3, Z_4) \sim \text{MVN}_4(\mu_1, \Sigma_1)$ with $\mu_1 = (0,0,0,0)$, the diagonal of Σ_1 given by $(1, 1, 0.25^2, 0.25^2)$ and off-diagonal elements all equal to 0.1. Z_1 and Z_2 were converted to binary covariates using the cut point 0, so that $\text{E}(Z_1) = \text{E}(Z_2) = 0.5$.

2. The covariates of $N_h - n_h$ non-comparable historical patients were generated as $\mathbf{Z} \sim \text{MVN}_4(\mu_0, \Sigma_0)$ with $\mu_0 = (0, 0, 0.8, 1.5)$, the diagonal of Σ_0 given by $(1, 1, 0.25^2, 0.5^2)$ and off-diagonal elements 0.1. Z_1 and Z_2 were converted to binary covariates using cut points $\Phi^{-1}(0.2)$ and $\Phi^{-1}(0.8)$, so that $\text{E}(Z_1) = .2$ and $\text{E}(Z_2) = 0.5$.

In each simulated trial, MatchEff $= N_m/N$ was fixed by setting the value of n_h. For the binary response endpoint, each Y_i was generated using the logit model

$$\text{logit}\ \{\Pr(Y_i = 1 \mid X_i, \mathbf{Z}_i)\} = \beta X_i + \sum_{k=1}^{4} \alpha_k Z_{ik}, \tag{5.1}$$

where $X_i = 1$ for E and 0 for C. Continuous outcomes were generated from a normal linear model with a mean of the above form. Confounding effects were

$\alpha_1 = 0.12, \alpha_2 = -2.6, \alpha_3 = -0.96, \alpha_4 = 2$. Trials had a planned sample size of $N = 80$ per arm for the RCT, and power .80 to detect an improvement of δ of 0.19 ($\beta = 1.21$) for the response probabilities of E versus C, or 0.41 ($\beta = 0.45$) for the standardized difference between the means of the continuous endpoint. For the single-arm design, design parameters were estimated from $\mathscr{H}$, e.g., the historical response rate, and an improvement δ, to estimate the sample size, obtain power .80, and type I error rate .05.

Matching efficiency values MatchEff = 1.0, 0.8, 0.5, 0.3, 0.1, and 0 were considered to represent a wide range of degrees of usefulness of $\mathscr{H}$ for obtaining matched controls. The simulations included BASIC designs with one interim decision based on $\pi = 0.9$ after $n = 40$ of the $N = 80$ patients per arm were enrolled. For all designs, at the end of the trial, a one-sided approximately normal test for binomial proportions was used for the null hypothesis of no E-versus-C effect versus the one-sided alternative that E provides an improvement, with a significance level of 0.05. A total of 5000 trials using each design in each scenario were simulated, calculating the type I error rate, power, mean total sample size, and relative bias $|\hat{\delta} - \delta|/\delta$, where $\hat{\delta}$ is the estimated effect size.

Figure 5.1 illustrates the simulation results for binary endpoints. As expected, the benchmark RCT has high power, low bias, and a type I error rate near 0.05, but it requires the largest sample size. The single-arm design has the smallest sample size, but has by far the lowest power (Fig 5.1B) and the largest bias (Fig 5.1C) of all four designs, especially when patients in $\mathscr{H}$ differ substantially from those in the trial, that is, if MatchEff = 0, 0.1 or 0.3. The single-arm design also fails to control the type I error rate at the nominal level, with values substantially lower than 0.05 (Fig 5.1A). Because it is based on controls matched to the E patients, the BASIC design has much higher power and much lower bias than the single-arm design. In the case where an insufficient number of controls can be synthesized due to large differences between $\mathscr{H}$ patients and trial patients, the matched-control design has lower power and higher bias than the RCT.

BASIC has the best overall performance among the four designs. Compared to the RCT, BASIC has similar power, bias, and type I error, but on average it requires a much smaller sample size (Fig 5.1D). For example, in the ideal case where matched controls for all E patients can be synthesized from $\mathscr{H}$, that is, MatchEff = 1, Fig 5.1D shows that the sample size of BASIC is roughly half that of the RCT. BASIC has much higher power and much smaller bias than the conventional single-arm design. Because BASIC adaptively determines whether to randomize patients to C, depending on the usefulness of $\mathscr{H}$, BASIC avoids the loss of power seen with the matched control design in the case where an inadequate number of controls can be synthesized from $\mathscr{H}$, specifically when MatchEff = 0.1 or 0.3 in Fig 5.1 B). When MatchEff = 1, BASIC has slightly higher power than an RCT. This is because, in this case, each patient in the trial has a matched control and propensity score matching gives better covariate balance than complete randomization, leading to higher power than an RCT. This phenomenon is well known, and has been reported by Joffee and Rosembaum (1999) and Ali et al. (2019). For the same reason, the type I error of

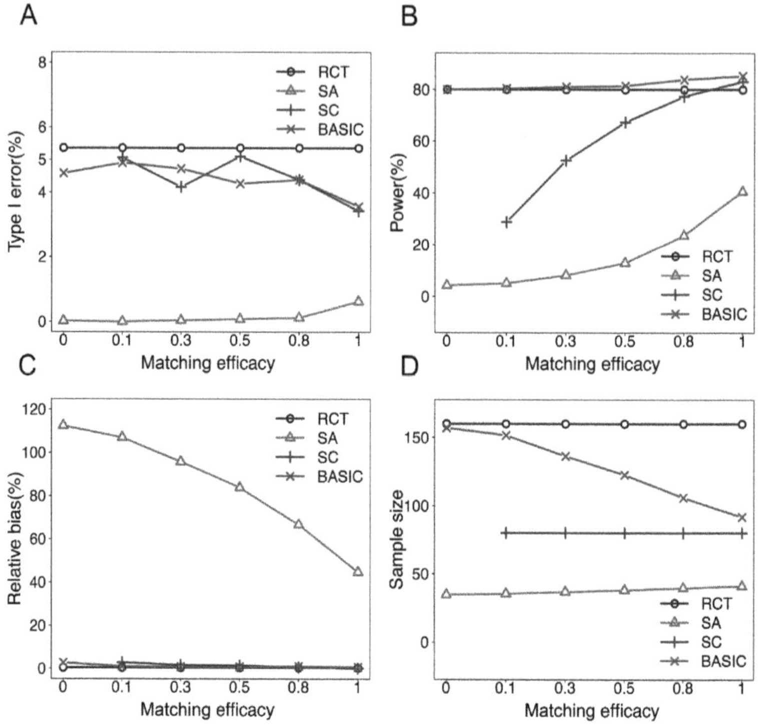

FIGURE 5.1
Simulation results for the RCT, single-arm design (SA), single-arm design with matched controls (SC) and BASIC design, for a binary endpoint. If MatchEff = 0, no historical controls are chosen, and SC is not feasible.

BASIC is slightly lower than the nominal value. Qualitatively similar results were seen for normally distributed outcomes.

In some trials, the baseline patient covariate distribution may drift over time and become different between stages 1 and 2 of the design. To evaluate the performance of BASIC in the presence of drift, the case of a binary endpoint with four covariates and an interim analysis when $t = 50\%$ patients are enrolled was simulated. For patients enrolled after the interim analysis, the mean of Z_4 drifted higher by 0.2 standard deviation. The treatment effect size was adjusted accordingly, so that the RCT had 80% power. In the presence of this population drift, BASIC still had the best overall performance, with power and bias similar to the RCT, but smaller sample size, higher power, and lower bias than the single-arm and matched-control designs. It thus appears that BASIC is robust to patient drift because matching with $\mathscr{H}$ largely eliminates its impact.

A design similar to BASIC was proposed by Gotte et al (2022), who decided whether to switch to a fixed-ratio RCT if a preference score that measures the comparability of covariates between patients receiving E and $\mathscr{H}$ is lower than a fixed threshold. There are two main differences between BASIC and this adaptive fixed threshold two-stage design. BASIC makes an interim decision by predicting the number of matched historical controls that will be obtained at the end of the trial, and switches to an RCT if there are an insufficient number of predicted matched historical controls. This is a more direct approach than using a preference score, which does not have an intuitive interpretation, so choosing a numerical switching threshold is problematic. Gotte et al. (2022) used the arbitrary threshold of 0.5, switching to an RCT if the preference score < 0.5. This may not make sense if, for example, the preference score < 0.5 but a sufficient number of matched controls can be found in $\mathscr{H}$, so there is no need to switch to an RCT. Another problem is that, if the interim data satisfy the switching criterion, the adaptive two-stage design switches to a fixed-ratio RCT. In contrast, BASIC chooses the randomization ratio adaptively based on the effective sample size of $\mathscr{H}$, and thus is more flexible and far more efficient because it only randomizes the number of patients to C that actually are needed.

5.2 Precision Treatment Screening and Selection

Lee et al. (2023) proposed a precision Bayesian design for a randomized clinical trial to evaluate experimental treatments, $E_1, \cdots, E_K$, compared to an active control treatment, C, based on binary or ordinal categorical toxicity Y_T and response Y_R. Including C was motivated by simulation results of Wathen and Thall (2017), who showed that randomized evaluation of multiple experimental treatments can go very wrong if a control arm is not included. The design also accounts for patient heterogeneity, characterized by ordered prognostic subgroups, indexed by $g = 1, \cdots, G$, where $g = 1$ has the best and $g = G$ the worst prognosis. During trial conduct, the randomization is restricted to achieve balance among sample sizes of the treatments within each subgroup. All decisions are based on posterior quantities characterizing E_k-versus-C effects on $\boldsymbol{Y} = (Y_R, Y_T)$ for each g and $k = 1, \cdots, K$. The design allows different subgroups to have different treatment effects, and makes subgroup-specific decisions, including interim screening for safety and futility, and final selection of optimal treatments. To be practical, in most applications $K = 2$ or 3, since there are a total of $(K+1)G$ combinations of (treatment, subgroup) to be evaluated. To reduce the model dimension when the data show that the prespecified subgroups are more refined than necessary, the design adaptively clusters subgroups having similar estimated distributions of $[\boldsymbol{Y} \mid E_k, g]$ for all k. This is done using latent subgroup membership variables. When no meaning is lost, treatments will be identified using the integers $k = 0$ for C and $k = 1, \cdots, K$ for $E_1, \cdots, E_K$. The design elements may be summarized as follows:

Elements of the Screen-and-Select Design

1. Randomize among $E_1, \cdots, E_K$ and C from the start

2. Accommodate platform trials with bivariate binary or ordinal outcomes

3. Account for prognostic subgroups

4. Adaptively combine similar subgroups to borrow strength

5. Interimly drop unsafe or ineffective treatments within subgroups, and enrich the sample sizes of the remaining subgroups

6. Base final subgroup-specific treatment selections on posterior predictive mean utilities

The design was motivated by a randomized phase II trial of first-line therapies for metastatic renal cell carcinoma (mRCC). The trial compared two immunotherapies, E_1 = nivolumab plus ipilimumab (N+I) (Motzer et al., 2019a) and E_2 = pembrolizumab plus lenvatinib (P+L) (Motzer et al., 2021) with the control targeted agent C = pazopanib (Pa). Patients with mRCC were classified using their International Metastatic Renal-Cell Carcinoma Database Consortium (IMDC) prognostic risk scores, defined using clinical variables and biomarkers, including anemia, thrombocytosis, neutrophilia, hypercalcemia, Karnofsky performance status, and time from diagnosis to treatment. Based on their IMDC risk scores, patients were classified into three prognostic subgroups: favorable (IMDC = 0), intermediate (IMDC = 1 or 2) and poor (IMDC $\geq$ 3) (Heng et al., 2013). The ability of these three prognostic subgroups to account for outcome heterogeneity in mRCC has led to their common use as stratification factors in randomized phase III trials of mRCC (Rini et al., 2019; Motzer et al., 2021). IMDC risk score also is recommended by the National Comprehensive Cancer Network (NCCN) guidelines for selecting mRCC treatments (Msaouel et al., 2021). There were no previous randomized trials comparing N+I, P+L, and Pa to determine which of these three regimens would be the most appropriate first-line mRCC therapies for each IMDC subgroup (Adashek et al., 2020). This motivated the randomized phase II trial.

In the mRCC trial, the outcomes were a binary toxicity indicator Y_T, and ordinal Y_R taking the four possible values 0 for progressive disease (PD), 1 for stable disease (SD), 2 for partial response (PR), and 3 for complete response (CR). Interim subgroup-specific screening rules dropped an E_k for any subgroup where the data showed that, compared to C, E had an unacceptably high rate of either toxicity or PD. The design used adaptive model-based clustering to combine two or more adjacent subgroups if the interim or final data showed that they had similar estimated outcome distributions. For example, if the interim data showed that distributions of $[\boldsymbol{Y} \mid k,g]$ and $[\boldsymbol{Y} \mid k,g+1]$ were similar for all treatments $k = 0,1,\cdots,K$, then subgroups g and $g+1$ were combined to form the cluster $\{g,g+1\}$, and the model's parameters are simplified accordingly. Treatment selection was based on posterior estimates of subgroup-specific utility functions, $\{U_g(\boldsymbol{Y}), g = 1,\cdots,G\}$, each quantifying the risk-benefit trade-off between Y_R and Y_T for it's subgroup. At the end of the trial, for each g, a best acceptable E_k was chosen by maximizing the posterior predictive mean of $U_g(\boldsymbol{Y})$.

In general, let $n(t)$ denote the number of patients accrued up to trial time t, and index patients by $i = 1,\dots,n(t)$. For the ith patient, denote treatment by $X_i \in \{0,1,\cdots,K\}$, subgroup by $g_i \in \{1,\dots,G\}$, and outcomes by $\boldsymbol{Y}_i = (Y_{i,T}, Y_{i,R})$, where $Y_{i,j} \in \{0,\dots,M_j - 1\}$ for $j = T$ and R. In the mRCC trial, $Y_{i,\mathrm{T}} = 1$ if severe toxicity occurs and 0 otherwise, so $Y_{i,\mathrm{T}} \in \{0,1\}$ and $M_T = 2$. For response, $Y_{i,\mathrm{R}} \in \{0,1,2,3\}$ and $M_R = 4$. The probability model used by the clustering algorithm may be constructed as follows. First, a model for regression of bivariate ordinal $\boldsymbol{Y}_i$ on treatment and subgroup is obtained by exploiting the construction of Ashford and Sowden (1970). Each observed $\boldsymbol{Y}_i$ is defined in terms of unobserved latent real-valued variables $\mathbf{Z}_i = (Z_{i,T}, Z_{i,R})$ that follow a bivariate normal distribution. This is done by first introducing real-valued correlated latent patient-specific frailties $\boldsymbol{\varepsilon}_i = (\varepsilon_{i,T}, \varepsilon_{i,R})$, and assuming that $\boldsymbol{\varepsilon}_i \mid \Omega \overset{iid}{\sim} \mathrm{N}_2(\mathbf{0}, \Omega)$, where $\mathbf{0} = (0,0)'$ and the 2×2 variance-covariance Ω is random. Conditional independence of $Z_{i,\mathrm{T}}$ and $Z_{i,\mathrm{R}}$ given $\boldsymbol{\varepsilon}_i$ is assumed, with

$$Z_{i,j} \mid g_i = g, X_i = k, \varepsilon_{i,j} \overset{indep}{\sim} \mathrm{N}(\mu_{j,k,g} + \varepsilon_{i,j},\ \sigma^2),$$

for $i = 1,\dots,n(t)$ and $j = T, R$, where the variance σ^2 is fixed. Priors for $\boldsymbol{\mu}_{k,g} = (\mu_{T,k,g}, \mu_{R,k,g})$ and Ω will be given below. For each outcome j and treatment arm k, the observed variable $Y_{i,j}$ is defined using the cutoffs $u^k_{j,0} < u^k_{j,1} < \cdots < u^k_{j,M_j}$, with

$$Y_{i,j} = y_j \text{ if and only if } u^k_{j,m} < Z_{i,j} \le u^k_{j,m+1} \text{ for } y_j = 0,\cdots,M_j - 1.$$

The induced marginal distribution of each observed $Y_{i,j}$ is given by

$$\begin{aligned} \mathrm{P}(Y_{i,j} = y_j \mid g_i = g, X_i = k) &= \Phi_1\left(u^k_{j,m+1} \mid \mu_{j,k,g} + \varepsilon_{i,j}, \sigma^2\right) \\ &- \Phi_1\left(u^k_{j,m} \mid \mu_{j,k,g} + \varepsilon_{i,j}, \sigma^2\right), \end{aligned}$$

with $y_j \in \{0,\dots,M_j - 1\}$ for $j = T$ and R. Φ_d denotes the cumulative distribution function of a d-variate normal distribution. Priors for $\boldsymbol{u} = \{u^k_{j,m}, j = T, R,\ \ m = 2,\dots,M_j - 1, k = 0,\dots,K\}$ for outcomes with $M_j > 2$ are defined below. The cutoffs $\{u^k_{j,m}, m = 2,\dots,M_j - 1\}$ are random for flexibility, and $u^k_{j,1} = 0$ for all (j,k) to avoid non-identifiabilty of the model with fixed σ^2. Setting $u^k_{j,0} = -\infty$ and $u^k_{j,M_j} = \infty$ ensures that

$$\sum_{y_j=0}^{M_j-1} \mathrm{P}(Y_{i,j} = y_j \mid g_i = g, X_i = k) = 1.$$

The joint distribution of $\mathbf{Z}_i$ is obtained by integrating over the distribution of $\boldsymbol{\varepsilon}_i$, which gives

$$[\mathbf{Z}_i \mid g_i = g, X_i = k] \overset{indep}{\sim} \mathrm{N}_2(\boldsymbol{\mu}_{k,g}, \Sigma)$$

with $\Sigma = \Omega + \sigma^2 \mathrm{I}_2$. The distribution of $\mathbf{Z}_i$ induces a bivariate joint distribution for $\boldsymbol{Y}_i$ = $(Y_{i,\mathrm{T}}, Y_{i,\mathrm{R}})$, given by

$$\mathrm{P}(\boldsymbol{Y}_i = \mathbf{y} \mid g_i = g, X_i = k) = \qquad (5.2)$$

$$\Phi_2\left(u^k_{\mathrm{T},y_\mathrm{T}+1}, u^k_{\mathrm{R},y_\mathrm{R}+1} \mid \boldsymbol{\mu}_{g,k}, \Sigma\right) - \Phi_2\left(u^k_{\mathrm{T},y_\mathrm{T}+1}, u^k_{\mathrm{R},y_\mathrm{R}} \mid \boldsymbol{\mu}_{g,k}, \Sigma\right)$$
$$- \Phi_2\left(u^k_{\mathrm{T},y_\mathrm{T}}, u^k_{\mathrm{R},y_\mathrm{R}+1} \mid \boldsymbol{\mu}_{g,k}, \Sigma\right) + \Phi_2\left(u^k_{\mathrm{T},y_\mathrm{T}}, u^k_{\mathrm{R},y_\mathrm{R}} \mid \boldsymbol{\mu}_{g,k}, \Sigma\right),$$

for $\boldsymbol{y} = (y_T, y_R)$. The frailties $\{\varepsilon_{i,1}, \cdots, \varepsilon_{i,n(t)}\}$ account for between-patient heterogeneity not explained by g_i and τ_i, and they induce association between $Y_{i,\mathrm{R}}$ and $Y_{i,\mathrm{T}}$ for patient i.

Regression of each ordinal outcome $Y_{i,T}$ and $Y_{i,\mathrm{R}}$ on subgroup and treatment is done by defining the means of the latent variable distribution as

$$\mu_{j,k,g} = \eta_{j,k} + \alpha_{j,k,g},$$

where $\eta_{j,k}$ is the intercept for treatment k and $\alpha_{j,k,g}$ is an additive treatment-subgroup interaction. This model accommodates a wide variety of different shapes of treatment-subgroup curves for each outcome, and it provides a flexible basis for subgroup-specific treatment screening and selection. Since the prognostic subgroups are ordinal, the constraints

$$\alpha_{T,k,1} \leq \ldots \leq \alpha_{T,k,G} \quad \text{and} \quad \alpha_{\mathrm{R},k,1} \geq \ldots \geq \alpha_{\mathrm{R},k,G}$$

are imposed. These constraints induce a stochastic ordering of the distribution of each Y_j in g for each k, given by

$$\mathrm{P}(Y_T \leq y_T \mid g, k) \geq \mathrm{P}(Y_T \leq y_T \mid g', k)$$

and

$$\mathrm{P}(Y_R \leq y_R \mid g, k) \leq \mathrm{P}(Y_R \leq y_R \mid g', k)$$

for $g < g'$. For settings with non-ordinal subgroups, such as disease subtypes, these constraints should be dropped.

Model-based clustering is used to adaptively combine adjacent subgroups that the data show have similar estimated treatment-subgroup interactions, $\{\alpha_{j,k,g}\}$, assuming that treatment effects are identical within each cluster. This is implemented using a vector of latent cluster membership variables, $\boldsymbol{\zeta} = (\zeta_1, \cdots, \zeta_G)$, where each $\zeta_g \in \{1, \cdots, G\}$. If $\zeta_g = \zeta_{g'}$ for subgroups $g \neq g'$, then the subgroups g and g' belong to a cluster. Similarly to Lee et al. (2021), since the predefined subgroups are ordinal, the method sets $\zeta_1 = 1$, requires $\zeta_1 \leq \ldots \leq \zeta_G$, and defines a prior on $\boldsymbol{\zeta}$ by proceeding sequentially for $g = 2, \ldots, G$. Subgroup $g \geq 2$ is combined with subgroup $g-1$ in a cluster with fixed probability ξ, and is not combined with subgroup $g-1$ with probability $1-\xi$, given by

$$\mathrm{P}(\zeta_g = \zeta_{g-1} \mid \zeta_{g-1}) = \xi \quad \text{and} \quad \mathrm{P}(\zeta_g = \zeta_{g-1} + 1 \mid \zeta_{g-1}) = 1 - \xi.$$

The prior on $\boldsymbol{\zeta}$ is

$$p(\boldsymbol{\zeta} \mid \xi) = \prod_{g=2}^{G} p(\zeta_g \mid \xi, \zeta_{g-1}) = \prod_{g=2}^{G} \xi^{I(\zeta_g = \zeta_{g-1})} (1-\xi)^{1 - I(\zeta_g = \zeta_{g-1})}.$$

This construction allows only neighboring subgroups to be combined, because subgroups are ordinal to represent prognostic level. Let $H \leq G$ denote the number of distinct clusters, so $\zeta_G = H$. The motivating trial has $G = 3$ risk subgroups, so there are four possible cluster configurations, $\boldsymbol{\zeta}$ = (1, 1, 1), (1, 1, 2), (1, 2, 2), or (1, 2, 3), which define, respectively, H = 1, 2, 2, and 3 clusters. For example, in the cluster configuration $\boldsymbol{\zeta} = (1,2,2)$, subgroup 1 has its own cluster $\{1\}$, with $\zeta_1 = 1$, and subgroups 2 and 3 are combined as the cluster $\{2,3\}$, with $\zeta_2 = \zeta_3 = 2$. If the subgroups are not ordinal, such as disease subtypes, the ordering constraint on $\boldsymbol{\zeta}$ should be dropped, and any clustering approach may be used, such as clustering via a Gaussian mixture model with spike-and-slab priors as in Chapple and Thall (2018), or using a random partition as in Xu et al. (2016b).

Borrowing strength using clusters is done using cluster-specific treatment effects, $\alpha^\star_{j,k,h}$, assuming that, given $\boldsymbol{\zeta}$, all subgroups in a cluster have the same parameter, $\alpha_{j,k,g} = \alpha^\star_{j,k,h}$ for all g with $\zeta_g = h$. This implies that, for each treatment k, the distribution of $\boldsymbol{Y}$ is the same for all subgroups in a cluster, so clustering reduces the model's dimension. To illustrate how this works, suppose that $\boldsymbol{\zeta} = (1,2,2)$, giving the clusters $\{1\}$ and $\{2,3\}$. Subgroup 1 has its own distribution of $\boldsymbol{Y}$ with $\mu_{j,k,1} = \eta_{j,k} + \alpha^\star_{j,k,1}$ since $\zeta_1 = 1$, and for subgroup 1 the likelihood is

$$\begin{aligned} P(Y_{i,j} = y_j \mid \boldsymbol{\zeta}, g_i = 1, X_i = k) &= \Phi_1\left(u^k_{j,y_j+1} \mid \eta_{j,k} + \alpha^\star_{j,k,1} + \varepsilon_{i,j}, \sigma^2\right) \\ &- \Phi_1\left(u^k_{j,y_j} \mid \eta_{j,k} + \alpha^\star_{j,k,1} + \varepsilon_{i,j}, \sigma^2\right). \end{aligned}$$

The outcomes for two subgroups in the cluster $\{2,3\}$ have the same means, $\mu_{j,k,2}$ = $\mu_{j,k,3}$ =$\eta_{j,k} + \alpha^\star_{j,k,2}$ since $\alpha_{j,k,1} = \alpha_{j,k,2} = \alpha^\star_{j,k,2}$, for j = T or R. Outcomes in subgroups g = 2 or 3 thus also have the same conditional likelihood,

$$\begin{aligned} P(Y_{i,j} = y_j \mid \boldsymbol{\zeta}, g_i = g, X_i = k) &= \Phi_1\left(u^k_{j,y_j+1} \mid \eta_{j,k} + \alpha^\star_{j,k,2} + \varepsilon_{i,j}, \sigma^2\right) \\ &- \Phi_1\left(u^k_{j,y_j} \mid \eta_{j,k} + \alpha^\star_{j,k,2} + \varepsilon_{i,j}, \sigma^2\right). \end{aligned}$$

This adaptive clustering method uses a distribution over $\boldsymbol{\zeta}$ that stochastically combines adjacent prognostic subgroups found to have similar treatment effects. Optimal treatments are chosen for subgroups by marginalizing over the posterior distribution of $\boldsymbol{\zeta}$. The practical motivation for clustering in this way is that borrowing information through clustering improves estimation of the conditional distribution of $\boldsymbol{Y}$ given k and g, which in turn improves the reliability of subgroup-specific decision-making.

Prior Specification and Posterior Computation

For identifiability, $\alpha^\star_{j,k,1} = 0$ for all j and k, so $\mu_{j,k,g} = \eta_{k,j}$ if $\zeta_g = 1$. A prior for the vector of cluster-specific treatment effects

$$\boldsymbol{\alpha}^\star_T = (\alpha^\star_{T,k,2}, \ldots, \alpha^\star_{T,k,H}) \quad \text{and} \quad \boldsymbol{\alpha}^\star_R = (\alpha^\star_{R,k,2}, \ldots, \alpha^\star_{R,k,H})$$

conditional on $\boldsymbol{\zeta}$ is defined as follows. Given $H > 1$ clusters, normal priors are assumed with ordering constraints on $\boldsymbol{\alpha}_T^\star$ and $\boldsymbol{\alpha}_R^\star$,

$$p(\boldsymbol{\alpha}_T^\star \mid \boldsymbol{\zeta}, \bar{\alpha}_T, v_T^2) \propto \prod_{h=2}^{H} \phi_1(\alpha_{T,k,h}^\star \mid \bar{\alpha}_{T,h}, v_T^2)\mathscr{I}(\alpha_{T,k,h}^\star > \alpha_{T,k,h-1}^\star),$$

$$p(\boldsymbol{\alpha}_R^\star \mid \boldsymbol{\zeta}, \bar{\alpha}_R, v_R^2) \propto \prod_{h=2}^{H} \phi_1(\alpha_{\mathrm{R},k,h}^\star \mid \bar{\alpha}_{\mathrm{R},h}, v_R^2)\mathscr{I}(\alpha_{\mathrm{R},k,h}^\star < \alpha_{\mathrm{R},k,h-1}^\star).$$

The ordering constraints on $\boldsymbol{\zeta}$ and $\boldsymbol{\alpha}_j^\star$, imply that $\alpha_{T,k,1} \leq \ldots \leq \alpha_{T,k,G}$ and $\alpha_{\mathrm{R},k,1} \geq \ldots \geq \alpha_{\mathrm{R},k,G}$ for each treatment k.

Priors for the remaining random model parameters, $\{u_{j,m}^k\}$, $\{\eta_{j,k}\}$ and Ω, are specified as follows. For outcome j with $M_j > 2$, assume $u_{j,m+1}^k = u_{j,m}^k + e_{j,m}^k$, $m = 1, \ldots, M_j - 2$, with $e_{j,m}^k \sim$ *indep* $\text{Gamma}(\bar{e}_{j,m}\kappa_j, \kappa_j)$, with fixed prior mean $\bar{e}_{j,m}$ and prior variance $\bar{e}_{j,m}/\kappa_j$. For treatment and outcome specific intercepts, assume $\eta_{j,k} \sim$ *indep*$\text{N}(\bar{\eta}_j, w_j^2)$, with $\bar{\eta}_j$ and w_j^2 fixed, and let the 2×2 covariance matrix $\Omega \sim \text{inv-Wishart}(\nu, \Omega_0)$ with $\text{E}(\Omega) = \Omega_0/(\nu - 3)$.

Collecting terms, aside from the random subgroup partition $\boldsymbol{\zeta}$, the model parameter vector is $\boldsymbol{\theta} = (\boldsymbol{\eta}, \boldsymbol{\alpha}^\star, \boldsymbol{e}, \Omega)$, where $\boldsymbol{\eta} = \{\eta_{j,k}\}$, $\boldsymbol{\alpha}^\star = \{\alpha_{j,k,h}^\star\}$ and $\boldsymbol{e} = \{e_{j,m}^k\}$. For the renal cancer trial design, the fixed hyperparameters characterizing the priors were established using historical data from Tannir et al. (2020) and Motzer et al. (2013, 2019b, 2021), and prior probabilities also were elicited from the clinical investigators.

The interim dataset $\mathscr{D}_{n(t)}$ at trial time t includes all outcomes and treatment assignments from previously enrolled patients. Given $\mathscr{D}_{n(t)}$, the joint posterior of $\boldsymbol{\theta}$, the latent subgroup variables $\boldsymbol{\zeta}$, and latent patient frailties $\boldsymbol{\varepsilon} = \{\boldsymbol{\varepsilon}_i, i = 1, \ldots, n(t)\}$ is

$$p(\boldsymbol{\theta}, \boldsymbol{\zeta}, \boldsymbol{\varepsilon} \mid \mathscr{D}_{n(t)}) \propto \prod_{i=1}^{n(t)} p(\boldsymbol{\varepsilon}_i \mid \Omega) \prod_{j=T,R} p(Y_{i,j} \mid g_i, \tau_i, \varepsilon_{i,j}, \boldsymbol{\theta}, \boldsymbol{\zeta})\, p(\boldsymbol{\theta} \mid \boldsymbol{\zeta})\, p(\boldsymbol{\zeta} \mid \xi).$$

Assuming that $Y_{i,T}$ and $Y_{i,R}$ are conditionally independent given $\boldsymbol{\varepsilon}$, the likelihood $p(Y_{i,j} \mid g_i, \tau_i, \varepsilon_{i,j}, \boldsymbol{\theta}, \boldsymbol{\zeta})$ given in (5.2) facilitates the posterior computation. MCMC simulation is used to generate posterior samples by iteratively drawing $(\boldsymbol{\zeta}, \boldsymbol{\theta}, \boldsymbol{\varepsilon})$, with each conditional on the values of the others at each iteration through $\boldsymbol{\zeta}$, $\boldsymbol{\theta}$ and $\boldsymbol{\varepsilon}$. The likelihood evaluation in (12.3) depends on $\boldsymbol{\zeta}$, and the joint posterior of $\boldsymbol{\zeta}$ and $\boldsymbol{\theta}$ determines all decision criteria used by the design.

Decision Criteria and Trial Conduct

The utility function accommodates the possibility that clinicians may be more willing to accept a higher risk of toxicity if disease status is more likely to be improved for patients in a poor-risk subgroup. This is formalized by a utility function that allows risk-benefit preferences between Y_T and Y_R to differ between subgroups, with $U_g(\boldsymbol{Y})$ the utility assigned to outcome $\boldsymbol{Y} = (Y_T, Y_R)$ for subgroup g.

In practice, numerical utilities of the $M_T \times M_R$ elementary outcomes should be elicited from clinical collaborators, with specific numerical values reflecting the

TABLE 5.1
Subgroup-specific utilities of $\boldsymbol{Y} = (Y_T, Y_R)$ for the metastatic renal cancer trial.

		PD	SD	PR	CR
$g = 1$ (Favorable)	no Tox	40	60	80	100
	Tox	0	45	65	85
$g = 2$ (Intermediate)	no Tox	30	65	85	100
	Tox	0	50	70	90
$g = 3$ (Poor)	no Tox	20	70	90	100
	Tox	0	55	75	95

physicians' relative preferences. In the motivating trial, there are $2 \times 4 = 8$ possible (Y_T, Y_R) elementary outcomes. For convenience, first fix $U_g(0,3) = 100$ and $U_g(1,0) = 0$ for all g, which are the respective utilities for the best and worst possible outcomes, and elicit the intermediate values for the remaining outcomes. The numerical utilities are given in Table 5.1. Any convenient function satisfying the consistency conditions

$$U_g(Y_T, Y_R) < U_g(Y_T, Y_R + 1) \;\text{ and }\; U_g(0, Y_R) > U_g(1, Y_R)$$

may be used. In addition, it is accepted in practice that patients with favorable IMDC ($g = 1$) generally have more indolent disease and more time to test other subsequent therapies than patients with more aggressive mRCC ($g = 2$ or 3), and they are less willing to tolerate toxicity even if CR is achieved. The utility can be constructed to reflect this by calibrating it across subgroups, by requiring that $U_g(Y_T, Y_R)$ for each (Y_T, Y_R) with $Y_R > 0$ should not decrease with g. For the same reason, even when $Y_R = 3$, having $Y_T = 1$ is penalized more for a favorable risk subgroup, that is, $U_g(1,3)$ increases in g, while $U_g(0,3) = 100$ for all g. However, $U_g(0,0)$ decreases in g, while $U_g(1,0) = 0$ for all g, implying having PD is less desirable for a higher-risk group. Thus, a treatment with a high toxicity probability is more likely to be optimal for higher-risk subgroups if the treatment has good efficacy. The three functions $\{U_g, g = 1,2,3\}$ elicited for the renal cancer trial are illustrated in Figure 5.2.

The mean utility of treating a patient with E_k in subgroup g is

$$\overline{U}_g(k \mid \boldsymbol{\zeta}, \boldsymbol{\theta}) = \sum_{y_T=0}^{M_T-1} \sum_{y_R=0}^{M_R-1} U_g(\boldsymbol{y}) \times p(\boldsymbol{y} \mid k, g, \boldsymbol{\zeta}, \boldsymbol{\theta}).$$

where $p(\boldsymbol{y} \mid k, g, \boldsymbol{\zeta}, \boldsymbol{\theta})$ is given in (5.2). Given data $\mathscr{D}_{n(t)}$ at trial time t, the posterior predictive mean utility of giving treatment k to a future patient in subgroup g is

$$\begin{aligned} u_g(k \mid \mathscr{D}_{n(t)}) &= E\{\overline{U}_g(k \mid \boldsymbol{\zeta}, \boldsymbol{\theta}) \mid \mathscr{D}_{n(t)}\} \\ &= \int_{\boldsymbol{\theta}} \sum_{\boldsymbol{\zeta}} \overline{U}_g(k \mid \boldsymbol{\zeta}, \boldsymbol{\theta}) p(\boldsymbol{\zeta}, \boldsymbol{\theta} \mid \mathscr{D}_{n(t)}) d\boldsymbol{\theta}. \end{aligned}$$

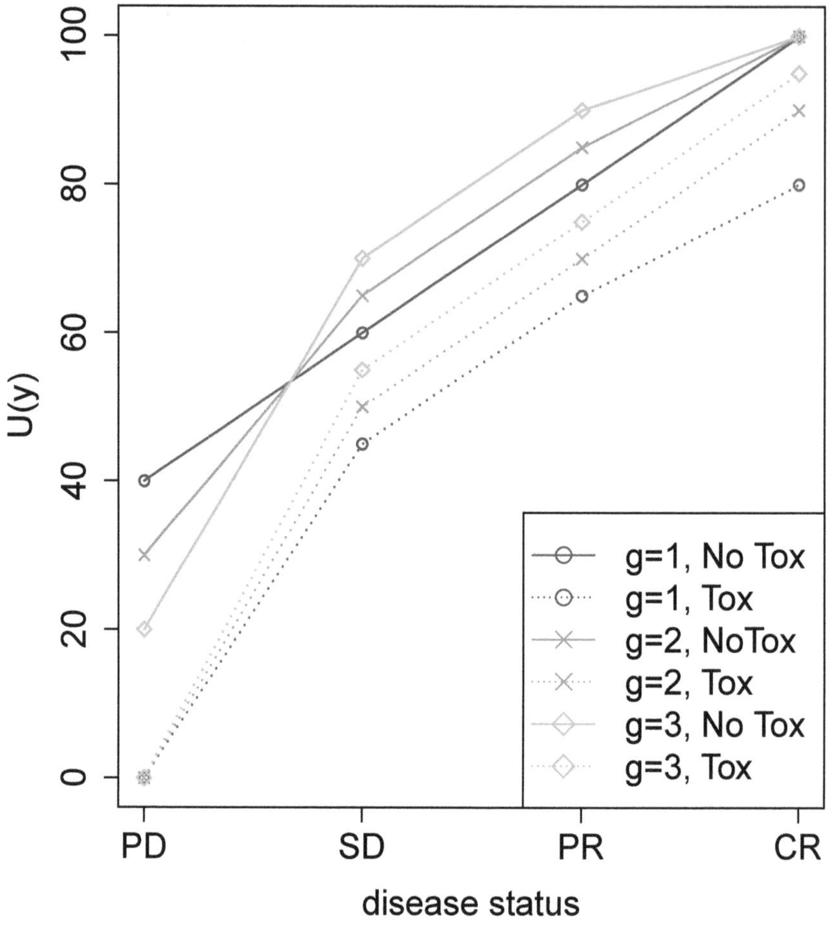

FIGURE 5.2
Illustration of elicited subgroup-specific utilities $U_g(\boldsymbol{Y})$ of bivariate outcome $\boldsymbol{Y} = (Y_T, Y_R)$, for subgroups $g = 1$ (Favorable IMDC score), $g = 2$ (Intermediate IMDC score, and $g = 3$ (Poor IMDC score). $Y_T = 1$ indicates the occurrence of severe toxicity, and $Y_R = 0, 1, 2,$ and 3 represent PD, SD, PR and CR.

The design includes rules to stop accruing patients to an E_k in any subgroup where the data show that E_k has an unacceptably high rate of either toxicity or PD compared to C. If all E_k's are found to be unacceptable for a subgroup, the design stops accrual and does not select any E_k in that subgroup. The design terminates a trial if the accrual is stopped for all subgroups. Recall that $(Y_T = 1)$ = Toxicity and $(Y_R = 0)$ = PD. Denote the probabilities of these events with treatment k in subgroup

g by $\lambda_T(k,g)$ and $\lambda_{\text{PD}}(k,g)$. Accrual to E_k is stopped in subgroup g if

$$P(\lambda_T(k,g) < \lambda_T(0,g) \mid \mathscr{D}_{n(t)}) < p_T^\star \quad \text{(Safety Criterion)} \tag{5.3}$$

or

$$P(\lambda_{\text{PD}}(k,g) < \lambda_{\text{PD}}(0,g) \mid \mathscr{D}_{n(t)}) < p_R^\star \quad \text{(Futility Criterion)}. \tag{5.4}$$

Since a high PD probability is equivalent to a low probability of SD, PR or CR, (5.4) may be considered a futility rule. Typically, small positive values between 0.01 and 0.20 are chosen for $p_T^\star$ and $p_R^\star$. The values $p_T^\star = p_R^\star = 0.10$ were used for the renal cancer trial design based on preliminary simulations. The rule in (5.3) says that, given the data $\mathscr{D}_{n(t)}$, it is unlikely that the probability of toxicity with E_k is lower than that with C for subgroup g. (5.4) says it is unlikely that the probability of PD with E_k is lower than that with C in subgroup g. At the end of the trial, in each subgroup, an optimal treatment is selected from the set of acceptable E_k's. In the simple case where Y_R is a binary response indicator with $\lambda_R(k,g) = \Pr(Y_R = 1 \mid k, g, \boldsymbol{\zeta}, \boldsymbol{\theta})$, the stopping rule given in terms of PD in (8.2) may be replaced by the more conventional futility stopping criterion, similar to that of Wathen and Thall (2017),

$$P(\lambda_R(k,g) > \lambda_R(0,g) + \delta_R \mid \mathscr{D}_{n(t)}) < p^\star, \tag{5.5}$$

where $\delta_R \geq 0$ is a fixed targeted improvement.

An important property of the design is that, because patients are randomized, the E_k-versus-C comparisons in (5.3), (5.4), and (5.5) are unbiased. Specifically, aside from shrinkage toward the prior, the posterior means of $\lambda_T(k,g) - \lambda_T(0,g)$, $\lambda_{\text{PD}}(k,g) - \lambda_{\text{PD}}(0,g)$, and $\lambda_R(k,g) - \lambda_R(0,g)$ are unbiased for each $k > 0$. This is in contrast with the inherently biased comparisons obtained if historical data on C are used to construct such rules, rather than including a randomized control. Because there are no closed forms for the posterior distribution of $(\boldsymbol{\theta}, \boldsymbol{\zeta})$, or for the posterior predictive distribution of $\boldsymbol{Y}$ for a future patient, numerical approximation is used to compute the posterior decision criteria. The posterior mean $u_g(k \mid \mathscr{D}_{n(t)})$ is approximated using a Monte Carlo sample of $(\boldsymbol{\zeta}, \boldsymbol{\theta})$ values simulated from the joint posterior $p(\boldsymbol{\zeta}, \boldsymbol{\theta} \mid \mathscr{D}_{n(t)}, \tilde{\boldsymbol{\theta}})$.

The design includes L interim analyses, done after successive cohorts of size $\lfloor \frac{1}{L+1} N_{\max} \rfloor$ each, with cumulative sample size $n_\ell = \lfloor \frac{\ell}{L+1} N_{\max} \rfloor$ at the ℓth analysis for $\ell = 1, \ldots, L$, with $n_{L+1} = N_{\max}$. Based on expressions (5.3) and (5.5) computed using the current data $\mathscr{D}_{n_\ell}$, for subgroup g let $\mathscr{A}_\ell(g)$ denote the set of acceptable E_k's and let $n_{k,g}(t)$ denote the number of patients in subgroup g given treatment k up to trial time t. For the motivating renal cancer trial, $L = 1$ with one interim analysis done based on data from $n_1 = \lfloor N_{\max}/2 \rfloor$ patients and the final selection performed at $n_2 = N_{\max}$.

In each subgroup g, the randomization for cohort $\ell + 1$ is restricted to the current set $\mathscr{A}_\ell(g)$ of acceptable treatments for g. If it is determined that g has no acceptable treatments, formally if $\mathscr{A}_\ell(g) = \emptyset$, then enrollment to g is terminated permanently, so $\mathscr{A}_{\ell'}(g) = \emptyset$ for all $\ell' > \ell$. At the end of the trial, for each g the final set $\mathscr{A}_{L+1}(g)$ is computed and the best acceptable E_k is selected, with no treatment selected for g

if it has no acceptable treatments, i.e. if $\mathscr{A}_{L+1}(g) = \emptyset$. Regardless of early treatment terminations, the planned overall maximum sample size $N_{\max}$ is maintained rather than being reduced. This has the advantage that, if some treatments are terminated interimly in a subgroup g, the sample sizes of the acceptable treatments in g are enriched, and consequently the final optimal treatment selection is more reliable. These design steps are given formally as follows:

Steps for Trial Conduct

1. Record the subgroup g_i of each patient enrolled at trial time e_i, and randomize fairly among $E_1, \cdots, E_K$ and C with probability proportional to $1/\{n_{k,g}(e_i)+1\}$.
2. At each interim analysis $\ell = 1, \ldots, L$,
   (a) if there is no acceptable E_k for subgroup g, i.e. $\mathscr{A}_\ell(g) = \emptyset$, then terminate accrual to g, with $\mathscr{A}_{\ell'}(g) = \emptyset$ for all $\ell' > \ell$ and no E_k selected in subgroup g;
   (b) if $\mathscr{A}_\ell(g) = \emptyset$ for all g, terminate the trial and do not select any E_k for any g;
   (c) for a patient in the $\ell + 1^{st}$ cohort with $g_i = g$, if $\mathscr{A}_\ell(g) \neq \emptyset$, assign the patient to treatment arm $k \in \{0\} \cup \mathscr{A}_\ell(g)$ with probability proportional to $1/\{n_{k,g}(e_i)+1\}$.
3. Select a final treatment for each subgroup g with $\mathscr{A}_{L+1}(g) \neq \emptyset$ based on all data $\mathscr{D}_{N_{\max}}$ subject to (5.3) and (5.4) using the criterion

$$X_{\text{sel}}(g) = \underset{k \in \mathscr{A}_{L+1}(g)}{argmax} \; u_g(k \mid \mathscr{D}_{N_{\max}}),$$

   with $X_{\text{sel}}(g) = 0$ denoting the case where no E_k is selected.

To evaluate the design's performance, Lee et al. (2023) simulated the renal cancer trial design under eight scenarios, six of which are given here. For all scenarios, three subgroups, three treatments including a control, and binary toxicity outcomes and four-level ordinal efficacy outcomes were assumed, so $G = 3$, $K = 2$, $M_T = 2$ and $M_R = 4$. Each g_i was simulated from a multinomial distribution with equal probabilities 1/3 for the prognostic subgroups. Each trial had $N_{\max} = 90$, so 10 patients were expected, on average, for each combination of (k, g) of E_k and subgroup. This maximum sample size is needed to evaluate three treatments reliably while accounting for three subgroup-specific effects. In terms of practicality, it is important to note that if the three subgroups have equal proportions of 1/3 then the expected subsample size in each subgroup is 30, which is close to what is used conventionally for phase II trials assuming patient homogeneity. The simulated design had one interim analysis at $n(t) = \lfloor N_{\max}/2 \rfloor$, so $L = 1$. Each scenario included the true clustering of the three subgroups, $\boldsymbol{\zeta}^{true} = (\zeta_1^{true}, \zeta_2^{true}, \zeta_3^{true})$, the variance $\sigma^{2,true}$ of the probit scores, the covariance matrix Ω^{true} for the frailty vectors, and $\eta_{j,k}^{true}$, for $j = T, R, k = 0, \ldots, K$, and $\alpha_{j,k,h}^{\star,true}$, $h = 2, \ldots, H^{true}$ with $\alpha_{j,k,1}^{\star,true} = 0$, $u_{R,m}^{k,true}$, $m = 2, 3$ while fixing $u_{j,0}^{k,true} = -\infty$,

$u_{j,1}^{k,\text{TR}} = 0$ and $u_{j,M_j}^{k,true} = \infty$. For each (k,g), given assumed true parameter values, the probabilities $\pi_{k,g}^{true}(\boldsymbol{y})$ of the $M_T \times M_R$ = 8 possible elementary outcomes were computed. The true parameter values were specified arbitrarily to examine robustness. The outcomes $\boldsymbol{Y}_i = (Y_{i,T}, Y_{i,R})$ were simulated with probabilities $\pi_{k,g}^{true}(\boldsymbol{y})$ conditional on $g_i = g$ and $X_i = k$.

The upper portion of Table 5.2 gives true values of the marginal probabilities of severe toxicity and PD,

$$\pi_T^{true}(k,g) = \sum_{Y_R=0}^{3} \pi_{k,g}^{true}(1, Y_R) \quad \text{and} \quad \pi_{\text{PD}}^{true}(k,g) = \sum_{Y_T=0}^{1} \pi_{k,g}^{true}(Y_T, 0)$$

and of the mean utility, $U_{k,g}^{true}$, $k > 0$. Truly unacceptable E_k's and truly optimal treatments are given in red and blue, respectively. With $G = 3$, four configurations of $\boldsymbol{\zeta}^{true}$ are possible due to subgroup ordinality. Scenario 1 has $\boldsymbol{\zeta}^{true} = (1,1,1)$, Scenarios 2-3 have $\boldsymbol{\zeta}^{true} = (1,1,2)$, Scenarios 4-5 have $\boldsymbol{\zeta}^{true} = (1,2,2)$, and Scenarios 5 and 6 have $\boldsymbol{\zeta}^{true} = (1,2,3)$. Due to subgroup-treatment interactions and the subgroup-specific utility function, the pattern of true mean utilities across treatments varies with subgroups in all scenarios. In Scenario 1, E_2 is optimal for all subgroups, but the pattern of $U_{k,g}^{true}$ varies with subgroups. In Scenarios 2 and 4-6, acceptability of the E_k's and truly optimal treatments vary by subgroups. In Scenario 2, E_1 is optimal for g = 1 and 2, but E_2 optimal for g = 3. In Scenario 4, no E_k's are acceptable for g = 1. Both E_1 and E_2 are acceptable for g = 2 and 3, and E_1 is optimal for those subgroups. In Scenario 4, no E_k is acceptable for any subgroup, and the optimal decision is to stop the trial early and not select either E_k.

The design is called "Sub" in the simulation table. Three conventional comparators are evaluated. "Sep", runs a separate trial for each subgroup, "Comb", ignores patient subgroups and makes decisions for all patients, and "Eff" selects the optimal E_k in each subgroup by maximizing the posterior predictive probability of PR or CR. While Comb and Sep are simpler than Sub, because they use a utility based on bivariate ordinal (Y_T, Y_R), both still are more sophisticated than most phase II screening designs used in practice, which typically are based on a single binary response variable. Comb and Sep are based on the same assumed model used for Sub, but with the key simplification that no subgroup effects are included, so $\mu_{j,k} = \tilde{\eta}_{j,k}$ with prior $\tilde{\eta}_{j,k} \sim N(\bar{\eta}'_j, w_j^2)$, where $\bar{\eta}'_j$ is specified using the elicited probabilities. Sep runs separate trials in the three subgroups, and no information is borrowed between trials. Since $N_{\max} = 90$ in Sub, to ensure a fair comparison each subgroup-specific trial in Sep has $N_{\max} = 30$ patients, with an interim analysis performed at $n_1 = 15$. Since Comb ignores subgroups, for this design $U_2(\boldsymbol{Y})$ waas sued as the common utility function. Under Comb, if an E_k is found to be unacceptable, no later patient is treated with E_k regardless of subgroup, if all E_k's are identified as unacceptable the trial is terminated, and a treatment is selected as optimal for all subgroups. Eff assumes the same model as Sub and selects the optimal treatments for subgroups, but uses the criterion

$$X_{\text{sel}}(g) = \underset{k \in \mathscr{A}_{L+1}(g)}{argmax} \, \text{P}(Y_R = \text{CR or PR} \mid k, g, \mathscr{D}_{N_{\max}})$$

TABLE 5.2

Simulation Results. $p^{\text{safe}}_{k,g}$ = P(declare E_k safe for subgroup g), $p^{\text{sel}}_{k,g}$ = P(select E_k as optimal for subgroup g), $k = 1, 2$, and $p^{\text{sel}}_{k,0}$ = P(do not choose any E_k as optimal in subgroup g). $n^{\text{trt}}_{k,g}$ = mean number of patients treated with k in subgroup g. Values for *truly unacceptable* and **true optimal** treatments are given in *red italics* and **blue bold**, respectively.

Treatment Arms			C	E_1	E_2	C	E_1	E_2
			Scenario 1, $\boldsymbol{\zeta}^{true} = (1,1,1)$			Scenario 2, $\boldsymbol{\zeta}^{true} = (1,1,2)$		
$\pi^{true}_T(k,g)$		$g=1$	0.20	0.15	0.15	0.20	0.10	0.15
		$g=2$	0.20	0.15	0.15	0.20	0.10	0.15
		$g=3$	0.20	0.15	0.15	0.46	0.44	0.25
$\pi^{true}_{\text{PD}}(k,g)$		$g=1$	0.25	0.20	0.10	0.20	0.05	0.15
		$g=2$	0.25	0.20	0.10	0.20	0.05	0.15
		$g=3$	0.25	0.20	0.10	0.46	0.44	0.34
$U^{true}_{k,g}$		$g=1$		65.65	**73.77**		**78.02**	69.62
		$g=2$		67.96	**77.59**		**81.86**	73.02
		$g=3$		70.02	**80.67**		46.40	**57.69**
$p^{\text{sel}}_{k,g}$	Sub	$g=1$	0.06	0.06	**0.88**	0.01	**0.92**	0.06
		$g=2$	0.05	0.06	**0.89**	0.01	**0.91**	0.08
		$g=3$	0.05	0.09	**0.86**	0.03	0.22	**0.74**
	Sep	$g=1$	0.21	0.24	**0.55**	0.09	**0.83**	0.08
		$g=2$	0.21	0.23	**0.56**	0.08	**0.85**	0.07
		$g=3$	0.20	0.25	**0.55**	0.10	0.18	**0.73**
	Comb	all g	0.05	0.05	**0.90**	0.02	**0.72**	**0.26**
	Eff	$g=1$	0.06	0.11	**0.83**	0.01	**0.86**	0.13
		$g=2$	0.05	0.10	**0.84**	0.01	**0.86**	0.13
		$g=3$	0.05	0.10	**0.85**	0.03	0.26	**0.71**
$p^{\text{safe}}_{k,g}$	Sub	$g=1$		0.89	0.91		0.98	0.95
		$g=2$		0.90	0.92		0.98	0.96
		$g=3$		0.90	0.92		0.86	0.94
	Sep	$g=1$		0.73	0.59		0.86	0.73
		$g=2$		0.73	0.59		0.87	0.73
		$g=3$		0.74	0.60		0.70	0.87
	Comb	all g		0.88	0.92		0.96	0.96
$n^{\text{trt}}_{k,g}$	Sub	$g=1$	10.14	9.51	9.93	10.41	9.97	9.80
		$g=2$	10.03	9.52	9.65	10.13	9.98	9.57
		$g=3$	9.91	9.58	9.70	10.09	9.38	9.82
	Sep	$g=1$	10.35	8.75	9.94	10.43	9.52	9.14
		$g=2$	10.36	8.78	9.90	10.47	9.55	9.20
		$g=3$	10.31	8.82	9.88	10.32	8.66	10.25
	Comb	$g=1$	8.63	8.55	9.36	8.58	8.83	9.45
		$g=2$	12.01	10.99	11.96	12.03	11.42	12.05
		$g=3$	9.02	8.88	8.53	9.05	9.15	8.54

Table 5.2 **Simulation Results** (continued).

treatment arms			C	E_1	E_2	C	E_1	E_2
			Scenario 3, $\boldsymbol{\zeta}^{true} = (1,1,2)$			Scenario 4, $\boldsymbol{\zeta}^{true} = (1,2,2)$		
$\pi_T^{true}(k,g)$		$g=1$	0.10	0.25	0.30	0.10	0.20	0.25
		$g=2$	0.10	0.25	0.30	0.49	0.28	0.58
		$g=3$	0.15	0.38	0.44	0.49	0.28	0.58
$\pi_{\text{PD}}^{true}(k,g)$		$g=1$	0.15	0.25	0.20	0.15	0.30	0.30
		$g=2$	0.15	0.25	0.20	0.54	0.39	0.44
		$g=3$	0.22	0.34	0.32	0.54	0.39	0.44
$U_{k,g}^{true}$		$g=1$		58.24	63.44		60.83	59.52
		$g=2$		59.90	65.85		54.77	46.66
		$g=3$		52.76	55.95		54.04	45.32
$p_{k,g}^{\text{sel}}$	Sub	$g=1$	0.74	0.11	0.16	0.29	0.54	0.17
		$g=2$	0.77	0.09	0.14	0.04	0.91	0.05
		$g=3$	0.78	0.12	0.10	0.04	0.85	0.11
	Sep	$g=1$	0.70	0.13	0.18	0.63	0.31	0.06
		$g=2$	0.73	0.10	0.16	0.11	0.77	0.12
		$g=3$	0.81	0.15	0.04	0.11	0.73	0.16
	Comb	all g	0.87	0.09	0.04	0.06	0.90	0.04
	Eff	$g=1$	0.74	0.10	0.16	0.29	0.48	0.23
		$g=2$	0.77	0.09	0.14	0.04	0.72	0.24
		$g=3$	0.78	0.10	0.12	0.04	0.72	0.24
$p_{k,g}^{\text{safe}}$	Sub	$g=1$		0.18	0.16		0.62	0.44
		$g=2$		0.15	0.14		0.94	0.69
		$g=3$		0.15	0.13		0.94	0.69
	Sep	$g=1$		0.20	0.20		0.34	0.16
		$g=2$		0.16	0.19		0.86	0.70
		$g=3$		0.16	0.05		0.87	0.70
	Comb	all g		0.11	0.05		0.92	0.68
$n_{k,g}^{\text{trt}}$	Sub	$g=1$	8.41	7.45	7.24	10.65	10.04	8.41
		$g=2$	8.16	7.15	6.72	10.38	10.26	8.69
		$g=3$	8.15	7.13	6.82	10.42	10.24	8.66
	Sep	$g=1$	8.70	7.47	7.56	8.66	7.54	7.29
		$g=2$	8.78	7.51	7.62	10.53	9.75	8.79
		$g=3$	7.83	6.78	7.02	10.55	9.77	8.79
	Comb	$g=1$	7.39	6.66	6.44	9.05	9.23	8.20
		$g=2$	9.17	7.96	7.24	12.55	12.12	10.15
		$g=3$	7.38	6.73	6.27	9.33	9.40	7.63

Table 5.2 **Simulation Results** (continued).

treatment arms			C	E_1	E_2	C	E_1	E_2
			Scenario 5, $\boldsymbol{\zeta}^{true}=(1,2,3)$			Scenario 6, $\boldsymbol{\zeta}^{true}=(1,2,3)$		
$\pi_T^{true}(k,g)$		$g=1$	0.10	0.20	0.20	0.15	0.15	0.10
		$g=2$	0.54	0.32	0.61	0.22	0.22	0.15
		$g=3$	0.80	0.90	0.70	0.22	0.22	0.15
$\pi_{\text{PD}}^{true}(k,g)$		$g=1$	0.10	0.30	0.30	0.20	0.15	0.20
		$g=2$	0.49	0.44	0.54	0.32	0.30	0.32
		$g=3$	0.80	0.84	0.68	0.51	0.49	0.51
$U_{k,g}^{true}$		$g=1$		63.82	62.84		66.70	**68.39**
		$g=2$		**53.76**	40.81		58.59	**61.27**
		$g=3$		12.42	**26.92**		46.19	**47.16**
$p_{k,g}^{\text{sel}}$	Sub	$g=1$	**0.49**	0.37	0.14	0.09	0.31	**0.60**
		$g=2$	0.12	**0.72**	0.15	0.08	0.34	**0.59**
		$g=3$	0.11	0.14	**0.74**	0.10	0.39	**0.51**
	Sep	$g=1$	**0.86**	0.10	0.03	0.16	0.44	**0.40**
		$g=2$	0.13	**0.75**	0.12	0.23	0.36	**0.41**
		$g=3$	0.15	0.09	**0.76**	0.29	0.32	**0.39**
	Comb	all g	**0.14**	**0.33**	**0.52**	0.13	0.19	**0.68**
	Eff	$g=1$	**0.49**	0.31	0.20	0.09	0.36	**0.56**
		$g=2$	0.12	**0.53**	0.35	0.08	0.32	**0.60**
		$g=3$	0.11	0.15	**0.73**	0.10	0.32	**0.58**
$p_{k,g}^{\text{safe}}$	Sub	$g=1$		*0.42*	*0.30*		0.82	0.86
		$g=2$		0.81	*0.67*		0.82	0.86
		$g=3$		*0.59*	0.85		0.78	0.81
	Sep	$g=1$		*0.12*	*0.05*		0.70	0.71
		$g=2$		0.83	*0.60*		0.64	0.63
		$g=3$		*0.51*	0.80		0.61	0.59
	Comb	all g		*0.75*	*0.77*		0.70	0.80
$n_{k,g}^{\text{trt}}$	Sub	$g=1$	9.77	*8.97*	*8.06*	10.29	9.21	9.61
		$g=2$	10.27	9.91	*8.78*	10.34	9.22	9.57
		$g=3$	10.44	*8.55*	9.87	10.22	9.22	9.29
	Sep	$g=1$	8.49	*7.06*	*7.46*	10.15	8.67	9.38
		$g=2$	10.62	9.64	*8.66*	9.97	8.54	9.33
		$g=3$	10.49	*7.59*	10.33	10.09	8.48	9.42
	Comb	$g=1$	9.04	8.46	*8.59*	8.82	8.43	8.94
		$g=2$	12.38	10.92	*10.86*	12.20	10.79	11.34
		$g=3$	9.13	*8.77*	8.07	9.10	8.67	8.30

Eff does not account for the risk of severe toxicity in its treatment selection.

The simulations evaluated each of the four designs using the following subgroup-specific criteria. In subgroup g,

$p_{k,g}^{\text{safe}}$ = probability of declaring E_k safe compared to C, for $k = 1, \ldots, K$.

$p_{k,g}^{\text{sel}}$ = probability of selecting E_k as optimal, for $k = 1, \ldots, K$.

$p_{0,g}^{\text{sel}}$ = probability of not selecting any E_k as optimal.

$n_{k,g}^{\text{trt}}$ = mean # patients in subgroup g treated with $k = 0, 1, \ldots, K$.

While $p_{k,g}^{\text{safe}}$ and $p_{k,g}^{\text{sel}}$ vary with g under Sub, Sep and Eff, they are the same for all g under Comb. For each simulated trial, $b = 1, \ldots, B$ under each design, let $X_{sel}^{(b)}(g) \in \{0, 1, \ldots, K\}$ denote the treatment selected for subgroup g, with $X_{sel}^{(b)}(g) = 0$ if no E_k is selected, and denote $w_k^{(b)}(g) = 1$ if treatment E_k is identified as safe for subgroup g and 0 if not, with the number of patients treated denoted by $N^{(b)}$. For each scenario and design, the simulation results are summarized by the following subgroup-specific sample proportions:

$$p_{k,g}^{\text{safe}} = \frac{1}{B} \sum_{b=1}^{B} w_k^{(b)}(g), \; k = 1, \ldots, K,$$

$$p_{k,g}^{\text{sel}} = \frac{1}{B} \sum_{b=1}^{B} \mathscr{I}(X_{sel}^{(b)}(g) = k), \; k = 0, \ldots, K,$$

$$n_{k,g}^{\text{trt}} = \frac{1}{B} \sum_{b=1}^{B} \sum_{i=1}^{N^{(b)}} \mathscr{I}(X_i^{(b)} = k \text{ and } x_i^{(b)} = g), \; k = 0, \ldots, K.$$

A total of 500 trials with $N_{\max} = 90$ were simulated under each scenario using each design. Simulation results are summarized in Table 5.2. Overall, for each subgroup, the Sub design reliably identifies E_k's that are either excessively toxic or have low efficacy compared to C, with small $p_{k,g}^{\text{safe}}$ obtained for truly unacceptable E_k's. When either of $\pi_T^{true}(k,g)$ or $\pi_{\text{PD}}^{true}(k,g)$ for E_k is clearly greater than the corresponding value for C, $p_{k,g}^{\text{safe}}$ is particularly small. When E_k is truly unacceptable for all subgroups, $p_{k,g}^{\text{safe}}$ is small for all subgroups, possibly because the model borrows information across subgroups through $\alpha_{j,k,\zeta_g}^{\star}$ and thus improves reliability. For example, under Scenario 3, where all E_k's are unacceptable for all subgroups, Sub yields at most 18% for $p_{k,g}^{\text{safe}}$ for all (E_k, g) and selects no E_k with probabilities .74, .77 and .78 for the three subgroups. In Scenarios 4 and 6, $p_{k,g}^{\text{safe}}$ is large for truly unacceptable E_k in $g = 1$ while the arm is truly acceptable for the other subgroups g = 2 and 3. In Scenario 4, E_1 and E_2 are truly unacceptable in subgroup 1 but acceptable for subgroups 2 and 3. Sub incorrectly identifies those arms as acceptable with probabilities .62 and .44 for subgroup 1, but the differences $\pi_T^{true}(1,1) - \pi_T^{true}(0,1) = .20 - .10 = .10$ and $\pi_T^{true}(3,1) - \pi_T^{true}(0,1) = .25 - .10 = .15$ are small, with differences of .15 for the corresponding PD probabilities. It is unrealistic to expect any screening rule with overall $N_{\max} = 90$ to reliably detect such small differences in a setting with three treatments and three subgroups. The values .62 and .44 for E_1 and E_2 for $g = 1$ obtained with $N_{\max} = 90$ drop to .39 and .20 for the larger sample size $N_{\max} = 180$. The reliability of the design's screening rules thus increases with sample size, even in cases where the true toxicity and PD probabilities of the E_k's are very different between subgroups.

In general, Sub selects truly optimal safe treatments with high probabilities. When an E_k is optimal in more than one subgroup, $p^{\text{sel}}_{k,g}$ is especially high. In Scenario 1, where E_2 is optimal for all subgroups, Sub selects E_2 as optimal with probabilities .88, .89 and .86 for subgroups g = 1-3. In Scenario 2, where E_1 is optimal for g = 1 and 2, E_1 is chosen as optimal with probabilities .92 and .91 for those subgroups. When a subgroup's optimal treatment is different from those of the other subgroups and the other subgroups have the same optimal treatment arm, $p^{\text{sel}}_{k,g}$ is smaller for the subgroup having a different optimal treatment arm. In scenario 5 where E_1 is truly optimal for $g = 1$, and E_2 is optimal for g = 2 and 3, Sub selects the true optimal E_k with probabilities .72 for $g = 1$, and .93 and .92 for g = 2 and 3. In Scenario 8, although $U^{true}_{k,g}$ is similar for all treatments in each subgroup, Sub selects E_2 as optimal with probabilities .60, .59 and .51 for g = 1-3. Mean sample sizes $n^{\text{trt}}_{k,g}$ show that the design assigns fewer patients to truly unacceptable E_k's, seen in Scenarios 3, 4, and 6. When any E_k is truly acceptable in those scenarios, more patients were treated at the truly safe E_k or C. Recall that the true values of the parameters in the simulation setup are arbitrarily specified and very different from the prior means. Thus, in terms of all criteria, Sub is robust in that it performs well in a variety of scenarios not matching any particular model.

Probabilities of identifying an E_k as safe and of treatment selection for Sep, Comb and Eff also are summarized in Table 5.2. Sub has greatly superior performance compared to these comparators, or very similar performance, in nearly all scenarios. Sep performs similarly to Sub, or slightly better in some scenarios, e.g. subgroup 1 in Scenario 4, subgroup 2 in Scenarios 6 and 7. In Scenario 7, where not choosing any E_k, choosing E_1 and choosing E_2 are optimal for subgroups 1-3, respectively, Sub chooses them as optimal with probabilities .49, .72 and .74, compared to .86, .75 and .76 with Sep. When more than one subgroup has the same true optimal treatment, Sub performs better, in many cases by a wide margin. In Scenario 1, where E_2 is truly optimal for all subgroups, $p^{\text{sel}}_{g,2}$ is .86 to .89 under Sub versus .55 to .56 under Sep. In Scenario 2, subgroups 1 and 2 have E_1 as truly optimal, while subgroup 3 has E_2 as truly optimal. $p^{\text{sel}}_{k,g}$ of the truly optimal treatments is .92, .91 and .74 under Sub versus .83, .85 and .73 under Sep, for subgroups 1-3. When the truly optimal E_k is not selected, Sub selects the second best E_k more often than Sep. Sep also is safe, as seen in Scenario 3, where Sep selects no E_k with probabilities .70, .73, and .81 for subgroups 1-3.

Not surprisingly, Comb behaves very poorly when optimal decisions should differ between subgroups. In Scenario 2, Comb selects E_1 as optimal with probabilities .72 for all subgroups, but the true optimal treatment for subgroup 3 is E_2. In Scenario 4, all E_k's are truly unacceptable for subgroup 1, but Comb selects E_1 for subgroup

1 with probability .90 since E_1 is truly optimal for subgroups 2 and 3. Moreover, E_1 and E_2 are identified as acceptable with probabilities .92 and .68 for all subgroups including subgroup 1, when they are not acceptable for subgroup 1. Again, ignoring subgroups when treatment effects are truly heterogeneous is a complete disaster.

Eff performs very similarly to Sub in most scenarios. However, because Eff ignores toxicity, when response or PD distributions over subgroups are similar for treatments but their toxicity distributions are very different, with toxicity levels high in some subgroups but not others, the comparative performance of Eff becomes very poor. In Scenario 4, the difference in $\pi_{\text{PD}}^{true}(k,g)$ between E_2 and E_3 is 0.05 for subgroups 2 and 3, but $\pi_T^{true}(k,g)$ is much greater for E_2 than for E_1, resulting in E_1 being truly optimal for both subgroups. Sub accounts for the difference in $\pi_T^{true}(k,g)$ through the utility function, with 54.77 versus vs 46.66 in subgroup 2, and 54.04 vs 45.32 in subgroup 3 for $U_{k,g}^{true}$ of E_1 and E_2, respectively. Sub thus selects E_1 as optimal with high probabilities, .91 and .85 for subgroups 2 and 3. Because Eff does not penalize E_2 for having a much greater toxicity probability, it incorrectly selects E_2 as optimal with a probability of .24 in each of subgroups 2 and 3.

To examine the effect of a larger sample size, if the sample size is doubled to $N_{\text{max}} = 180$, the performances of Sub and Sep are improved in all scenarios, and their difference in $p_{k,g}^{\text{sel}}$, $p_{k,g}^{\text{safe}}$ and $n_{k,g}^{\text{trt}}$ are smaller. Comb and Eff do not improve with larger sample size, but rather their comparatively poor performances are more pronounced, essentially because each of these designs is doing the wrong thing. This illustrates that, while both Comb and Eff are based on very commonly made assumptions, each is fundamentally flawed, and often unsafe, because they ignore key information about patient heterogeneity or treatment safety.

Computer Software A zip file of computer software for implementing the Sub screen and select design is available from Biometrics as Supporting Information for Lee et al. (2023).

6

Precision Randomized Phase III Designs

6.1 Nutritional Prehabilitation and Post-Operative Morbidity

Esophageal cancer patients who undergo chemoradiation and surgical resection are at risk of a large number of postoperative morbidities. These include anastomotic leak from the surgical connection, lymphatic fluid leaks into the space between the lung and chest wall, gastric dumping syndrome, pneumonia, and cardiac complications such as atrial fibrillation. These adverse events may be summarized by post operative morbidity (POM), a composite variable that is scored within 30 days following surgery. POM was defined by Clavien et al. (1992) and Dindo et al. (2004) as an ordinal categorical variable taking the six possible values in given in Table 6.1. It was hypothesized that nutritional prehabilitation (N) prior to surgery might reduce the risk of these morbidities. To address this, Murray et al. (2018) designed a randomized group sequential trial to compare N to the standard of care, C, using POM as the primary outcome.

Denoting POM by Y, if patient prognosis is ignored, then the inferential focus might be on the probabilities $\pi_{X,k} = \Pr(Y = k \mid X)$ for treatment $X = N$ or C and POM score $k = 0, \cdots, 5$. The effect of nutritional prehabilitation could be characterized by comparing the POM score probability vectors $\boldsymbol{\pi}_X = (\pi_{X,0}, \cdots, \pi_{X,5})$ for $X = N$ and C. The trial's entry criteria included both primary patients (P) who had not been treated previously, and salvage patients (S) who had suffered disease recurrence after previous therapy. The tests described below are based on elaborations of the above probabilities, to account for both treatment arm and prognostic subgroups, allowing different comparative conclusions within subgroups.

Because the differences in numerical utilities between the successive POM levels are far from equal, using the integer-valued scores 0, 1, 2, 3, 4, 5 is a poor way to quantify POM outcome. Consequently, the nutritional prehabilitation trial design was based on the elicited numerical utilities of the possible POM outcomes, given in the upper portion of Table 6.2. The numerical utilities were determined by first fixing the two extreme utilities to be $U(0) = 100$ and $U(5) = 0$, and then eliciting the four intermediate values from the trial's Principal Investigator. The elicited utilities show that, due to the steep drop from $U(2) = 65$ to $U(3) = 25$, the event $Y \leq 2$ is much more desirable than its complement, $Y \geq 3$. However, dichotomizing POM score in this way would throw away a great deal of information, since the numerical utility values within each of these two subgroups still vary substantially. The probabilities

DOI: 10.1201/9781003474258-6

TABLE 6.1
Definitions of Clavien-Dindo post operative morbidity scores.

Score	Definition
0	Normal recovery
1	Minor complication
2	Complication requiring pharmaceutical intervention
3	Complication requiring surgical, endoscopic, or radiological intervention
4	Life-threatening complication requiring ICU admission
5	Death

$\Pr(Y \geq 3 \mid g, X)$ for subgroup $g = P, S$ and treatment arm $X = N, C$ are used to evaluate the design's power, however.

It is useful to examine how the numerical utilities in Table 6.2 turn a vector $\boldsymbol{\pi} = (\pi_0, \pi_1, \cdots, \pi_5)$ of POM score probabilities into a single mean utility $\overline{U}$ that quantifies the desirability of $\boldsymbol{\pi}$. This is at the heart of how the utility-based design works. In the lower portion of Table 6.2, the probability vector $\boldsymbol{\pi}_1$ may be considered a null distribution and $\boldsymbol{\pi}_2$ a targeted alternative distribution for salvage patients. Going from $\boldsymbol{\pi}_1$ to $\boldsymbol{\pi}_2$ reduces the probability $\Pr(Y \geq 3)$ of an undesirably high POM score from .35 to .14, and increases $\overline{U}$ from 60.0 to 81.4. Similarly, $\boldsymbol{\pi}_3$ may be considered a null distribution and $\boldsymbol{\pi}_4$ a targeted alternative distribution for primary patients. Going from $\boldsymbol{\pi}_3$ to $\boldsymbol{\pi}_4$ reduces the probability $\Pr(Y \geq 3)$ of a high POM score from .20 to .08, and increases $\overline{U}$ from 75.5 to 89.0.

An important property of the design is that it accounts for patient heterogeneity by allowing possibly different conclusions within the two subgroups P and S when comparing treatment N and S. That is, it is a precision medicine design. It uses a group sequential testing procedure based on the posterior utilities for the POM score computed under the following Bayesian probability model. For Y = POM score, a

TABLE 6.2
Elicited utilities of Clavien-Dindo post operative morbidity (POM) scores, and four examples of POM probability distributions and mean utilities. Each $\boldsymbol{\pi}_j$ is vector of hypothetical POM score probabilities.

POM Score	0	1	2	3	4	5	
Elicited Utility	100	80	65	25	10	0	
	Mean Prior Probabilities						$\overline{U}$
$\boldsymbol{\pi}_1$	.30	.25	.10	.10	.10	.15	60.0
$\boldsymbol{\pi}_2$	.60	.14	.12	.08	.04	.02	81.4
$\boldsymbol{\pi}_3$	.50	.20	.10	.10	.05	.05	75.5
$\boldsymbol{\pi}_4$	.73	.14	.05	.05	.03	0	89.0

non-proportional odds (NPO) model is assumed in which the linear term is

$$\begin{aligned}\eta_{NPO}(y,g,X) &= \text{logit}\{\Pr(Y \le y \mid g,X)\} \\ &= \alpha_y + \gamma_{1,y} I_g + \gamma_{2,y} I_X + \gamma_{3,y} I_g I_X,\end{aligned}$$

for $y = 0,1,2,3,4$, where the subgroup covariate is defined as $I_g = -.5$ for P and $+.5$ for S, and the treatment variable is defined similarly as $I_X = -.5$ for C and $+.5$ for N. The subgroup and treatment variables are defined in this way numerically, rather than as more traditional 0/1 variables, in order to balance the variances under the Bayesian model. The probability $\Pr(Y \le y \mid g,X)$ of the ordinal POM variable is increasing in y due to a constraint in the prior of the parameters $(\boldsymbol{\alpha},\boldsymbol{\gamma})$. A hierarchical model is assumed, to borrow strength between subgroups while allowing shrinkage, with location parameters determined from elicited prior information on POM score. Details are given in Murray et al. (2018).

The NPO model differs from the usual proportional odds (PO) model of McCullagh (1980) in that all elements of the treatment and subgroup effect parameter vector $\boldsymbol{\gamma}_y = (\gamma_{1,y}, \gamma_{2,y}, \gamma_{3,y})$ vary with the level y of POM score. The corresponding PO model has the simpler linear term

$$\eta_{PO}(y,g,X) = \alpha_y + \gamma_1 I_g + \gamma_2 I_X + \gamma_3 I_g I_X. \tag{6.1}$$

Under the PO model, the intercept parameters $\boldsymbol{\alpha} = (\alpha_0,\alpha_1,\alpha_2,\alpha_3,\alpha_4)$ vary with the levels of Y, but the treatment effects do not vary with y. Thus, the NPO model is much more flexible, since it accounts for treatment and subgroup effects that vary with the level of POM score, in addition to including treatment-subgroup interactions.

Denoting

$$\pi_y(g,X \mid \boldsymbol{\alpha},\boldsymbol{\gamma}) = \Pr(Y = y \mid g,X,\boldsymbol{\alpha},\boldsymbol{\gamma}),$$

for $y = 0,1,2,3,4$, the mean utility of treatment X in subgroup g is defined as

$$\overline{U}(g,X,\boldsymbol{\alpha},\boldsymbol{\gamma}) = \sum_{y=0}^{5} U(y)\pi_y(g,X \mid \boldsymbol{\alpha},\boldsymbol{\gamma}). \tag{6.2}$$

The trial has a maximum of 200 patients, with interim and final tests in each subgroup $g = P$ and S performed at $n = 100$ and 200 patients, based on the posterior probability predictive probabilities

$$\Pr\{\overline{U}(g,X,\boldsymbol{\alpha},\boldsymbol{\gamma}) > \overline{U}(g,X',\boldsymbol{\alpha},\boldsymbol{\gamma}) \mid data\} > p_{cut}, \tag{6.3}$$

for treatments $(X,X') = (N,C)$ or (C,N). The numerical value of p_{cut} was determined to control the two subgroup-specific type I error probabilities to be $\le .025$. In subgroup g, a large value of (6.3) for $(X,X') = (N,C)$ corresponds to superiority of N over C. A large value of (6.3) for $(g,g') = (C,N)$ corresponds to superiority of C over N, equivalently inferiority of NuPrehab to standard.

The power figures of the tests were determined to detect the ratio

$$\frac{\Pr(Y \ge 3 \mid g,N)}{\Pr(Y \ge 3 \mid g,C)} = .60$$

for each subgroup $g = P$ and $g = S$. This translates into a reduction in $\Pr(Y \geq 3 \mid P)$ from 0.20 to 0.08 for primary patients, and a reduction in $\Pr(Y \geq 3 \mid S)$ from 0.35 to 0.14 for salvage patients. This, in turn, determines targeted mean utility increases with N compared to C from 75.5 with C to 89.0 with N for $g = P$ patients, and from 60.0 with C to 81.3 with N for $g = S$ patients.

As a comparator, one may consider a conventional PO-model-based design that ignores subgroups and makes "one size fits all" treatment comparisons. For this conventional design, the linear term in the PO model is the additive model

$$\eta_{PO,conv}(y, X) = \alpha_y + \gamma_1 I_g + \gamma_2 I_X$$

for $y = 0, 1, 2, 3, 4$, so γ_1 is the subgroup effect and γ_2 is the only treatment effect parameter. The posterior decision criterion under this model is simply $\Pr(\gamma_2 > 0 \mid data) \geq p_{cut,\gamma}$, where the cut-off $p_{cut,\gamma}$ is calibrated to ensure a type overall I error $\leq .05$. Both designs were simulated under each of the four scenarios given in Table 6.3. For example, under Scenario 3, Nuprehab achieves an improvement of 89.0 - 75.5 = 13.5 in POM score mean utility in the primary patients, but there is no difference between the two treatments in the salvage patients. Thus, there is a treatment-subgroup interaction in Scenario 3. With either design, to ensure balance and also comparability, patients are randomized using stratified block randomization with blocks of size four, so that for each block of four patients within each prognostic subgroup, two patients receive Nuprehab and two receive the control.

After the paper by Murray et al. (2018) describing the trial design was published, the trial's Principal Investigator recruited other medical centers to participate, so it then became feasible to double the maximum sample size, from 100 to 200. Consequently, the following numerical simulation results are different from those given by Murray et al. (2018), which were based on a maximum sample size of 100. Table 6.4 summarizes the simulation results for a 200 patient trial, showing that both the conventional "one size fits all" design that makes the same decision for the Primary and Salvage subgroups because it ignores them, and the design with subgroup-specific decisions, have overall type I error probabilities very close to .05, due to the way that they are calibrated. Scenarios 2 and 3 are treatment-subgroup interaction cases where one treatment is superior to the other in one subgroup but the two treatments are equivalent in the other. These scenarios show the great advantage of making subgroup-specific decisions. In Scenario 2, N is truly superior to C in Salvage patients, since N gives an increase in mean POM utility of 81.3 − 60.0 = 21.3, but the two treatments are equivalent in the Primary patient subgroup. For this scenario, the subgroup-specific design has a power of .808 in Salvage patients and Type I error probability of .029 in Primary patients.

Comparisons of the subgroup-specific design to the conventional design that assumes homogeneity in Scenarios 2 and 3 illustrate numerically how wrong a test's conclusions can be if patient heterogeneity is ignored. In Scenario 2, the conventional design that ignores subgroups has power .440, equivalently, Type II error probability 1 – .440 = .560, in Salvage patients, and Type I error probability .440 in Primary patients. The same pattern is seen in Scenario 3. These huge within-subgroup false positive and false negative probabilities with the conventional design in Scenarios 2 and 3, where there are treatment-subgroup interactions, suggest that one could do

TABLE 6.3
Simulation scenarios for evaluating the two nutritional prehabilitation trial designs, in terms of the true mean utility of the POM score for each treatment-subgroup combination.

Scenario	Treatment	Subgroup	Mean Utility
1 (Null/Null)	Control	Primary	75.5
	NuPrehab	Primary	75.5
	Control	Salvage	60.0
	NuPrehab	Salvage	60.0
2 (Null/Alt)	Control	Primary	75.5
	NuPrehab	Primary	75.5
	Control	Salvage	60.0
	NuPrehab	Salvage	81.3
3 (Alt/Null)	Control	Primary	75.5
	NuPrehab	Primary	89.0
	Control	Salvage	60.0
	NuPrehab	Salvage	60.0
4 (Alt/Alt)	Control	Primary	75.5
	NuPrehab	Primary	89.0
	Control	Salvage	60.0
	NuPrehab	Salvage	81.3

just as well by not running a trial at all, and instead deciding which treatment is better by flipping a coin. In Scenario 4, however, where N provides an improvement over C in both subgroups, the conventional design has substantially larger power than the subgroup-specific design. That is, if there is improvement and no treatment-subgroup interaction, assuming homogeneity happens to be correct. The loss of power in Scenario 4 appears to be the price that must be paid if one chooses to use a comparative design that allows different decisions to be made within subgroups. This should be considered along with the Type I and II error probabilities .440 and .560 using the conventional design in the mixed Scenarios 2 and 3. These simulation results suggest that the choice to use the conventional methodology, which may have within-subgroup Type I and Type II error probabilities in the range of .46 to .56, should reflect a strong belief that any N-versus-C effect will be the same in the two prognostic subgroups. However, a useful extension of the design would include a provision for adaptively combining the two prognostic subgroups, if the interim data show that they have very similar treatment effects, as done by Lin et al. (2021b) in the group sequential precision phase III BAGS design for survival time outcomes.

The properties of the different designs considered for the NuPrehab trial illustrate an important general point about treatment comparison in settings where there may be treatment-subgroup interactions. If, in fact, treatment effects are

TABLE 6.4
Simulation results for the two nutritional prehabilitation trial designs. Each cell gives the proportion of simulated trials that declared the NuPrehab superior or inferior compared to the control within a subgroup, Prim = Primary Treatment, Salv = Salvage Treatment. Each correct superiority decision probability is given in **boldface blue**. $\bar{n}$ = mean sample size.

Scenario	Prim/Salv	Prim	Salv	Prim	Salv	$\bar{n}$
		N Superior		*N* Inferior		
Precision Design Accounting for Subgroups						
1	Null/Null	.023	.025	.027	.031	199.2
2	Null/Alt	.029	**.808**	.022	.000	189.6
3	Alt/Null	**.777**	.036	.000	.023	187.0
4	Alt/Alt	**.823**	**.845**	.000	.000	172.4
Conventional Design Ignoring Subgroups						
1	Null/Null	.024	.024	.028	.028	199.4
2	Null/Alt	.440	**.440**	.000	.000	193.0
3	Alt/Null	**.562**	.562	.000	.000	189.6
4	Alt/Alt	**.977**	**.977**	.000	.000	145.1

substantively different between subgroups due to treatment-subgroup interactions, then assuming homogeneity and estimating one overall effect essentially averages the different subgroup-specific effects. As shown above, this easily can produce extremely large probabilities of false positive or false negative decisions within subgroups. An implication is that conventional clinical trials that make overall decisions based on the assumption that patients are homogeneous with regard to treatment effects may have been a scientific disaster spanning many decades in the medical research community. Designing a trial from the start to make reliable within-subgroup decisions certainly requires a larger sample size, but not doing so may produce incorrect and misleading conclusions.

Computer Software R programs for implementing the probability models and reproducing the simulations of the designs is available as *biom12842-sup-0002-SuppSimulation.R9.9* at the Biometrics website of Wiley Online Library.

6.2 Precision Confirmatory Survival Time Comparisons

Lin et al. (2021a) proposed a precision Bayesian adaptive GS (BAGS) design for a randomized phase III clinical trial to compare survival distributions of an experimental treatment, E, and a control, C. BAGS includes latent subgroup membership variables and decision rules that adaptively combine subgroups having similar estimated

E-versus-C effects on survival. This is similar to how several designs discussed earlier behave when dealing with subgroups, including the SubTITE design of Chapple and Thall (2018) for phase I dose-finding, the phase I-II design of Lee et al. (2021), and the randomized phase II design of Lee et al. (2023).

BAGS is illustrated by a phase III trial in patients with non-small-cell lung cancer, called OAK (Rittmeyer et al., 2017). In the OAK trial, patients were randomized to receive either docetaxel, which has been the standard of care for second-line or third-line treatment, or atezolizumab, an antibody that targets the humanized antiprogrammed death-ligand 1 (PD-L1) pathway. Docetaxel has efficacy against lung cancer, but it also carries a substantial risk of toxicity. As an immunotherapy, atezolizumab is safer and was shown to be promising in phase II studies. The objective of the OAK trial was to do a confirmatory comparison of survival between atezolizumab and docetaxel. Patients were stratified into four subgroups based on PD-L1 expression: subgroup 1 (TC0 or IC0) was defined as PD-L1 expression on less than 1% of tumor cells or tumor-infiltrating immune cells; subgroup 2 (TC1 or IC1) was defined as PD-L1 expression on 1% to 5% on these cells; subgroup 4 (TC3 or IC3) was defined as PD-L1 expression on 50% or more of tumor cells, or 10% or more tumor-infiltrating immune cells; and the remaining patients were stratified into subgroup 3 (TC2 and IC2). Between-subgroup heterogeneity observed in the OAK trial is illustrated in Figure 2 of Rittmeyer et al. (2017), which shows that, for patients treated with atezolizumab, the median survival time was 12.6 months in the TC0 or IC0 patients, and 20.5 months in the TC3 or IC3 subgroup. In contrast, for patients treated with docetaxel, both of these subgroups had a median survival 8.9 months. Thus, the estimated atezolizumab-versus-docetaxel differences in median survival were 12.6 - 8.9 = 3.7 months in subgroup 1, and 20.5 - 8.9 = 11.6 months in subgroup 4. See Figure 6.1. Additional *post hoc* analyses were reported by Rittmeyer et al. (2017) in which the data from different subgroups were combined iteratively. While these numerical results may seem compelling because they suggest a much larger atezolizumab-versus-docetaxel effect in the TC3 or IC3 subgroup, they are a typical example of *post hoc*, unplanned estimation or testing of comparative effects within subgroups. This common practice has been the subject of much debate for many years, and may be criticized as data dredging, since the probability of reaching a false positive conclusion in at least one subgroup increases with the number of analyses. Thus, a large estimated subgroup-specific effect detected in this way may be due to the play of chance. Bechhofer et al. (1995) provide a general developments of selection and multiple testing, and practical recommendations for subset analysis are given by Pocock et al. (2002) and Wang et al. (2007), among many others.

In precision medicine, a major goal is to identify subgroups where a targeted agent is most effective. For example, in oncology, due to heterogeneous tumor genomics, cancer cells may interact differentially with their surrounding microenvironment. This may cause patients in different biomarker subgroups to respond differently to a targeted treatment. Such biological heterogeneity brings many new challenges to clinical trial design and analysis (Simon, 2010; Garralda et al., 2019), including planning a trial where the subgroups are given *a priori* but heterogeneity

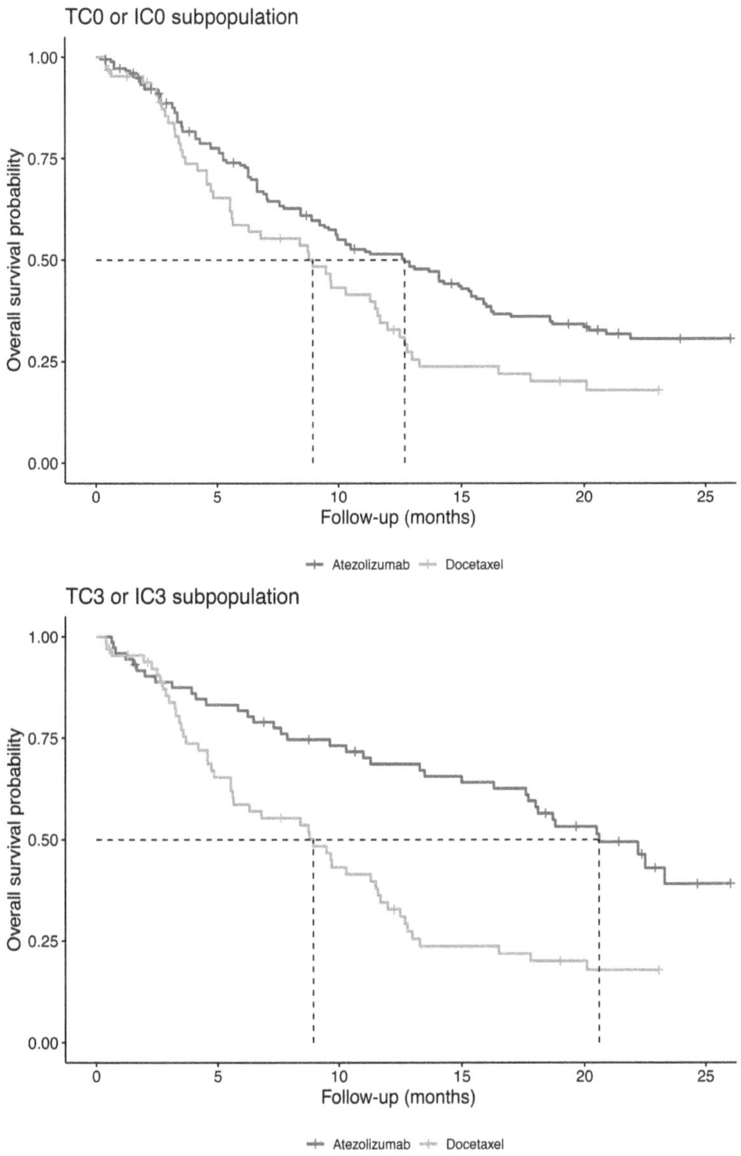

FIGURE 6.1
Treatment-subgroup interactions. Kaplan-Meier estimates of overall survival distributions for the Atezolizumab and Docetaxel arms in two selected subgroups in the OAK trial. The TC0 or IC0 subgroup was defined as PD-L1 expression on $< 1\%$ of tumor cells or tumor-infiltrating immune cells. The TC3 or IC3 subgroup was defined as PD-L1 expression on $> 50\%$ of tumor cells or 10% of tumor-infiltrating immune cells.

of treatment effects between subgroups is unknown (Thall et al., 2003; Wathen et al., 2008; Murray et al., 2018; Lee et al., 2019); addressing the multiple testing issue if heterogeneity is known (Rosenblum et al., 2016); and identifying a subgroup that may respond more fully to a new agent (Simon and Roychowdhury, 2013).

The BAGS design addresses the problem of comparing survival time distributions of E versus C in settings where, *a priori*, the patient population has been partitioned into subgroups. These often are defined biologically, and it is not known whether the E versus C effects will differ between the subgroups. Figure 6.2 illustrates a case where six subgroups are prespecified, but the data show that, in terms of the E effect on survival, they can be combined into the three induced subgroups, $Z_1 = \{1,2,3\}$, $Z_2 = \{4,6\}$, and $Z_3 = \{5\}$. Combining subgroups 1, 2, and 3 into Z_1 increases the power of the subgroup-specific tests. Subgroups 4 and 5 have different E versus C effects, so they are not combined. BAGS addresses two statistical goals. The first is to adaptively determine a partition of empirically induced subgroups for which E has similar effects within each induced subgroup, but different effects between induced subgroups, while controlling the misclassification rate. The second is to carry out GS comparisons within subgroups that control within-subgroup and overall false positive rates and provide high subgroup-specific power. Because BAGS addresses both goals prospectively, it avoids the problems that arise from doing *post hoc* subgroup analyses, as in the OAK trial.

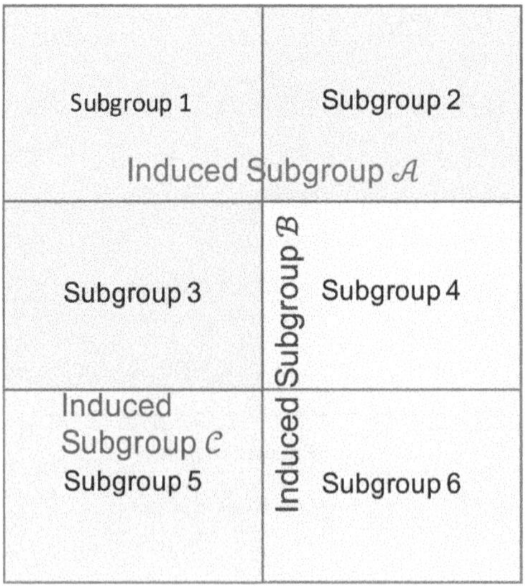

FIGURE 6.2
Illustration of a possible subgroup clustering. Starting with six subgroups, it is determined adaptively that subgroups 1–3 are homogeneous and hence are induced subgroup $\mathscr{A}$, subgroups 4 and 6 are induced subgroup $\mathscr{B}$, and induced subgroup $\mathscr{C}$ is subgroup 5.

Given subgroups $g = 1, \ldots, G$, denote the survival distributions for subgroup g in arms E and C by $S_{g,E}(t)$ and $S_{g,C}(t)$. For each g, the design's tests address the null hypothesis

$$H_{g,0}: \; S_{g,E}(t) = S_{g,C}(t), \quad \text{for all } \; t > 0, \tag{6.4}$$

and alternative hypothesis

$$H_{g,1}: S_{g,E}(t) \neq S_{g,C}(t) \quad \text{for some } \; t > 0. \tag{6.5}$$

Most approaches to this problem either test one global null hypothesis that ignores the subgroups, or apply a multiple testing procedure to do independent subgroup-specific tests, which leads to an inflated type I error rate and reduced power. Numerous approaches have been proposed for dealing with heterogeneity in survival analysis. Schumacher et al. (1987) discussed the effect of ignoring population heterogeneity when comparing survival time distributions for two heterogeneous populations. Aalen (1988) used mixture distributions to model the impact of patient heterogeneity. Previous methods focused on survival analysis with heterogeneity explained by prognostic variables under a proportional hazard assumption (Cox, 1972). In contrast, BAGS is based on a flexible survival model, and uses a GS design that includes adaptive subgroup combination, with possible subsequent re-splitting, and tests done within induced subgroups.

To provide a basis for adaptive subgroup combination that allows subgroups that have been combined to later be separated, BAGS takes a Bayesian hierarchical latent variable approach that facilitates subtype-specific survival comparisons. At each MCMC step of the posterior computation, a vector of latent variables is used to identify each subgroup's empirically induced subgroup. This allows the data from different subgroups to be combined or split adaptively, based on the observed survival data. For robustness, a piecewise constant hazard function for survival time in each treatment arm is assumed. Each survival hazard function is endowed with a three-level Bayesian hierarchical Markov-gamma process prior. This model accommodates changing latent subgroup classification during the trial. The interim decision rules are constructed to control the family-wise type I error rate and maximize subgroup-specific power.

In arm $j = E, C$, let $g_{i,j} \in \{1, \ldots, G\}$ denote the subgroup of patient $i = 1, \cdots, n_j$ where n_j is the number of patients. Denote the survival time of the ith patient in arm j by $T_{i,j}$, and assume that each patient is followed until death or administrative right-censoring. For current trial time t where a decision is made, denote $r_{i,j} = I(T_{i,j} < t)$, so $r_{i,j} = 1$ indicates that death was observed before time t and $r_{i,j} = 0$ denotes the event that $T_{i,j}$ was administratively censored at t. Thus, the observed time to death or censoring is $Y_{i,j} = \min\{T_{i,j}, t\}$. The design allows the possibility that a prognostic covariate or biomarker, $Z_{i,j}$, possibly associated with $T_{i,j}$, may be available at enrollment. Here, $Z_{i,j}$ is a patient characteristic other than those used for defining subgroups. In immunotherapy trials, $Z_{i,j}$ may be a binary or real-valued immune response variable. In addition to the regression of $T_{i,j}$ on $Z_{i,j}$, BAGS allows the possibilities that the distributions of $T_{i,j}$ and $Z_{i,j}$ each may differ between two or more subgroups, and there may be treatment-subgroup interactions. The efficiency and accuracy of the adaptive subgroup identification process may be improved by borrowing strength from $Z_{i,j}$. To focus on the main ideas of the BAGS design, a one-dimensional continuous $Z_{i,j}$ that can be observed quickly is assumed.

The Bayesian hierarchical model for the survival time distribution includes the latent subgroup membership variables, and the data from the C and E treatment groups are considered independent. For most oncology trials, a common assumption is that the standard of care under C induces response distributions that are homogeneous across subgroups. Following this, it is assumed that the distributions of $T_{i,C}$ and $Z_{i,C}$, $i = 1, \cdots, n_C$, are identical for all $g = 1, \cdots, G$. In contrast, potential heterogeneity of the distributions of T_{iE} and Z_{iE} between subgroups is assumed. The assumption of homogeneity under C is made here to simplify the development, but it can be relaxed without changing the decision rules of the design.

Temporarily suppress the treatment arm index $j = E$ to reduce notation. To account for patient heterogeneity in arm E, a latent subgroup variable $\zeta_g \in \{1, \ldots, G\}$ for each patient subgroup g is introduced, with indicator $\xi_g = I(\zeta_g = g)$. If $\zeta_g = \zeta_{g'} = g'$ for $g \neq g'$, then $\xi_g = 0$ and $\xi_{g'} = 1$, subgroups g and g' are homogeneous, and these two subgroups are combined into one induced subgroup g'. As a result, the distributions of (T_i, Z_i) and all parameters associated with subgroups g and g' are identical. This allows the data from these two subgroups to be combined as one induced subgroup in the likelihood function, and reduces the number of treatment-subgroup interaction parameters. This increases the power of the subgroup-specific E-versus-C survival comparison in the combined subgroup. If $\zeta_g = g$ for all $g \in \{1, \ldots, G\}$, then the G subgroups are fully heterogeneous. At the other extreme, $\zeta_1 = \cdots = \zeta_G$ corresponds to the completely homogeneous case. Let $\mathscr{S} = \{g : \zeta_g = g\}$ denote the set of induced subgroups, with $|\mathscr{S}|$ denoting its cardinality, so $|\mathscr{S}| \leq G$.

Mimicking Chapple and Thall (2018), the following prior distribution is assumed for (ξ_g, ζ_g), which is used at each MCMC posterior sampling step for adaptive subgroup combination,

$$\begin{aligned} \xi_g &\sim \text{Bernoulli}(p_g), \\ \zeta_g \mid \xi_g &\sim \xi_g \delta_g(\zeta_g) + (1 - \xi_g)\text{Unif}(\mathscr{S}), \end{aligned} \tag{6.6}$$

where $p_g = \Pr(\xi_g = 1) = \Pr(\zeta_g = g)$, $\delta_g(\cdot)$ is the Dirac distribution with point mass 1 on g, and Unif$(\mathscr{S})$ is a uniform discrete distribution on $\mathscr{S}$ with $Pr(\zeta_g = g') = 1/|\mathscr{S}|$ for all $g' \in \mathscr{S}$. If $\xi_g = 1$, then subgroup g is in its own induced subgroup. If $\xi_g = 0$, then $\zeta_g = g' \neq g$ for some $g' \in \mathscr{S} \backslash \{g\}$, and subgroups g and g' both are in the induced subgroup g'. According to the prior in (6.6), different subgroups are likely to be clustered into the same induced subgroup if the observed data indicate strong evidence that the distributions of T and Z are similar. This leads to the model dimension being changed adaptively. When the number of subgroups is relatively large, it may not be feasible to enumerate all possible models. To deal with the problem of repeatedly changing model dimension, Bayesian reversible jump MCMC is used to adaptively identify the latent subgroup indicators based on the observed data (Green, 1995). A possible subgroup clustering is illustrated in Figure 6.2.

To model the biomarker, assume that it is continuous and normally distributed, with

$$Z_i \mid g_i = g \sim \mathrm{N}(\mu_g, \sigma_Z^2), \tag{6.7}$$

where μ_g is the mean of Z for patients in subgroup g and σ_Z^2 is a common variance. While the same variance σ_Z^2 is assumed for all subgroups, if desired, the proposed model and method can be generalized easily to accommodate different variances across subgroups. Because the sample size of a randomized confirmatory study typically is large, any vague prior distribution for σ_Z^2 will work well. For example, one may assume an inverse gamma prior $\sigma_Z^2 \sim \mathrm{IG}(a_0, b_0)$, where (a_0, b_0) are fixed hyperparameters. To account for cases where some subgroups are homogeneous while others are heterogeneous with respect to Z_i, given the latent subgroup membership prior (6.6) on (ζ_g, ξ_g), the following spike-and-slab prior on μ_g is assumed,

$$\mu_g \mid \zeta_g, \xi_g \sim \xi_g \mathrm{N}(\mu_0, \sigma_{\mu_0}^2) + (1 - \xi_g)\delta_{\mu_{\zeta_g}}(\mu_g), \tag{6.8}$$

where $(\mu_0, \sigma_{\mu_0}^2)$ are prespecified hyperparameters. When there are some homogeneous subgroups, i.e., some $\xi_g = 0$, this prior facilitates adaptively borrowing information across subgroups that belong to the same induced subgroup. If Z_i is binary, the model (6.7) easily can be replaced by assuming, for example, that $\mathrm{logit}\{\Pr(Z_i = 1 \mid g_i = g)\} \sim \mathrm{N}(\mu_g, \sigma_Z^2)$.

Let $h_{g,i}(t)$ denote the hazard function of patient i in subgroup g at time t. A Cox proportional hazards model is assumed to quantify regression of survival time T_i on the marker Z_i, with hazard function

$$h_{g,i}(t) = \tilde{h}_g(t)\exp(\beta Z_i),$$

where $\tilde{h}_g(t)$ is the baseline hazard for subgroup g, and β is the log-hazard-ratio regression parameter. Any vague prior distribution can be placed on β, for example, $\beta \sim \mathrm{N}(\beta_0, \sigma_{\beta_0}^2)$ with hyperparameters $(\beta_0, \sigma_{\beta_0}^2)$, with suitably large $\sigma_{\beta_0}^2$. To construct flexible survival distributions, a piecewise constant hazard (Ibrahim et al., 2001) is obtained by partitioning the time scale $(0, \infty)$ into L intervals, $0 = s_0 < s_1 < \cdots < s_L = \infty$, and assuming a constant hazard $\lambda_{g,l}$ on interval $[s_{l-1}, s_l)$ for each subgroup $g = 1, \ldots, G$, and interval $l = 1, \ldots, L$. In subgroup g, denoting $\boldsymbol{\lambda}_g = (\lambda_{g,1}, \ldots, \lambda_{g,L})$, the baseline piecewise hazard function is

$$\tilde{h}_g(t \mid \boldsymbol{\lambda}_g) = \sum_{l=1}^{L} \lambda_{g,l} I(s_{l-1} \le t < s_l).$$

The baseline survival function is

$$S_{0,g}(t \mid \boldsymbol{\lambda}_g) = \exp\left\{ -\sum_{l=1}^{L} \lambda_{g,l} \Delta(t, s_{l-1}, s_l) \right\}, \quad t > 0,$$

where $\Delta(t, s_{l-1}, s_l) = \max\{0, \min(t, s_l) - s_{l-1}\}$, and the survival function for patient i in subgroup g is

$$S_g(t \mid Z_i, \boldsymbol{\lambda}_g) = \exp\left\{ -\sum_{l=1}^{L} \lambda_{g,l} \Delta(t, s_{l-1}, s_l) \exp(\beta Z_i) \right\}.$$

A robust, tractable survival model is obtained by assuming a spike-and-slab prior on the each subinterval's hazard,

$$\lambda_{g,l} \mid \zeta_g, \xi_g \sim \xi_g \pi(\lambda_{g,l}) + (1-\xi_g)\delta_{\lambda_{\zeta_{g,l}}}(\lambda_{g,l}), \tag{6.9}$$

where $\pi(\lambda_{g,l})$ denotes the following three-level hierarchical Markov gamma process (HMGP) (Nieto-Barajas and Walker, 2002; Murray et al., 2017) :

$$\begin{aligned} \lambda_{g,l} \mid \gamma_{g,l-1}, \eta_{g,l-1} &\sim \text{Gamma}(a_{g,l} + \gamma_{g,l-1}, b_{g,l} + \eta_{g,l-1}), \\ \gamma_{g,l} \mid \lambda_{g,l}, \eta_{g,l} &\sim \text{Poisson}(\eta_{g,l}\lambda_{g,l}), \\ \eta_{g,l} \mid w_g &\sim \text{Gamma}(1, w_g), \quad l = 1, \ldots, L \\ w_g &\sim \text{Gamma}(c_g, d_g), \quad g = 1, \ldots, G. \end{aligned} \tag{6.10}$$

For each g, $a_{g,l}, b_{g,l}, c_g$ and d_g are prespecified hyperparameters. The assumed priors (6.9) and (6.10) for the piecewise hazard facilitate borrowing information across homogeneous subgroups within the same induced subgroup, and also between adjacent subintervals in the partition of the survival time domain. Conditional on $\lambda_{g,l-1}$ and w_g, the prior mean of $\lambda_{g,l}$ under (6.10) is

$$E(\lambda_{g,l} \mid \lambda_{g,l-1}) = \frac{w_g}{w_g + 1/b_{g,l}} \frac{a_{g,l}}{b_{g,l}} + \left(1 - \frac{w_g}{w_g + 1/b_{g,l}}\right) \lambda_{g,l-1}.$$

The parameter w_g controls the smoothness of the estimated piecewise hazard. As $w_g \to 0$, the conditional prior mean of $\lambda_{g,l}$ converges to $\lambda_{g,l-1}$ and the prior variance converges to zero. As w_g increases, information borrowing between each pair of adjacent intervals decreases. As default values, one may use

$$a_{g,1} = \cdots = a_{g,L} = 1/L, \quad b_{g1} = \cdots = b_{gL} = a_{g,l}/\lambda_0$$

$c_g = .5$, and $d_g = c_g/\lambda_0$ with λ_0 being the prior mean for all $\lambda_{g,l}$'s. Based on this hyperparameter specification, the conditional prior mean of adjacent hazard parameters is

$$E(\lambda_{g,l} \mid \lambda_{g,l-1}) = \frac{1}{L+1}\lambda_0 + \frac{L}{L+1}\lambda_{g,l-1}.$$

This piecewise baseline hazard and hierarchical prior formulation provide reliable convergence of the MCMC computations, a robust survival model, and good performance of the BAGS design.

Let $\mathscr{D}_n = \{(Y_i, Z_i, g_i, r_i), i = 1, \ldots, n\}$ denote the observed data for the first n patients enrolled in the trial. The likelihood function takes the form

$$\begin{aligned} & L(\mu_1, \ldots, \mu_G, \boldsymbol{\lambda}_1, \ldots, \boldsymbol{\lambda}_G, \beta, \sigma_x \mid \mathscr{D}_n) \\ \propto & \prod_{i=1}^{n} \frac{1}{\sqrt{2\pi\sigma_x^2}} \exp\{-(Z_i - \mu_{g_i})^2/2\sigma_Z^2\} \\ & \times \prod_{l=1}^{L} \{\lambda_{g,l}\exp(\beta x_i)\}^{\delta_{i,l}} \exp\{-\lambda_{g,l}\exp(\beta Z_i)\Delta(Y_i, s_{l-1}, s_l)\}, \end{aligned}$$

where $\delta_{i,l} = 1$ if $r_i = 1$ and $Y_i \in [s_{l-1}, s_l)$ and $\delta_{i,l} = 0$ otherwise. Both the marker and survival outcomes contribute to identification of subgroups.

The task of identifying subgroups is turned into a Bayesian model selection problem, as follows. Denote $\boldsymbol{\zeta} = (\zeta_1, \ldots, \zeta_G)$, with posterior distribution $f(\boldsymbol{\zeta} \mid \mathscr{D}_n)$. Different combinations of $\boldsymbol{\zeta}$ create different induced subgroups, and thus different models. Enumerating all possible combinations of $\boldsymbol{\zeta}$ based on the posterior samples, one can assign the unique combinations of model indices, $M_1, M_2, \ldots, M_K$, with K denoting the total number of unique models sampled from the posterior. Denote the value of $\boldsymbol{\zeta}$ under M_k by $\boldsymbol{\zeta}^{(k)}$. Maximum a posteriori (MAP) estimation is used to select the model, M_{k^*}, corresponding to the most likely subgroup combination,

$$k^* = \underset{k=1,\ldots K}{\operatorname{argmax}}\ f(\boldsymbol{\zeta}^{(k)} \mid \mathscr{D}_n). \tag{6.11}$$

For each g with $\xi_g^{(k^*)} = 1$, the induced subgroup members are $\{g' : \zeta_{g'}^{(k^*)} = g\}$. The number of induced subgroups under model M_k equals

$$|M_{k^*}| = \sum_{g=1}^{G} I(\zeta_g^{(k^*)} = g) = \sum_{g=1}^{G} \xi_g^{(k^*)}.$$

Reintroducing the treatment index $j = E, C$, the model for C is as follows. Assume complete homogeneity across subgroups in C, so all subgroups in C have the same distribution, and there is no need to introduce latent subgroup variables. Formally,

$$\mu_{1,C} = \cdots = \mu_{G,C} = \mu_C \quad \text{and} \quad \boldsymbol{\lambda}_{1,C} = \cdots = \boldsymbol{\lambda}_{G,C} = \boldsymbol{\lambda}_C,$$

and the spike-and-slab priors (6.8) and (6.9) used for the subgroup-specific distributions under E are replaced with the simpler piecewise hazard distribution priors

$$\mu_C \sim \mathrm{N}(\mu_0, \sigma_{\mu_0}^2), \quad \lambda_{C,l} \sim \pi(\lambda_{C,l}), \quad l = 1, \ldots, L,$$

where $\pi(\lambda_{C,l})$ denotes the HMGP (6.10).

To conduct a trial using the BAGS design, the subgroup g_i and covariate Z_i of patient i are recorded at enrollment, followed by randomization between E and C, and follow-up for survival time. At each interim decision, BAGS has two steps. The first step identifies the induced subgroups $\mathscr{S}$ based on the current data, and the second step tests the null hypothesis for each induced subgroup $g \in \mathscr{S}$. Let N denote the maximum sample size, with interim sample sizes n_E, n_C, and $n = n_E + n_C$. To classify the G subgroups in E into induced subgroups $\mathscr{S}$ based on $\mathscr{D}_n$, posterior samples of $\boldsymbol{\zeta}$ are computed and each k^* is evaluated using the MAP formula (6.11). According to the prior distributions (6.7) and (6.9), if $\zeta_g^{(k^*)} = \zeta_{g'}^{(k^*)}$, then subgroups g and g' are in the same induced subgroup, and thus they share the same survival distribution, $S_g(t) = S_{g'}(t)$. The procedure is essentially testing $|M_{k^*}|$ hypotheses, for a total of $|M_{k^*}|$ different induced subgroups, where $|M_{k^*}|$ is random at each decision-making time.

In the second step, the average hazard ratio (AHR) (Kalbfleisch and Prentice, 1981) is used to sequentially test the survival difference between E and C for each induced subgroup $g \in \mathscr{S}$. Under non-proportional hazards, the AHR is a more valid measure of treatment effect than the standard hazard ratio, and the test provides greater power than a logrank test (Rauch et al., 2018). For arm $j = C,E$, let $h_{g,j}(t)$ denote the hazard function for subgroup g, and let $f(\boldsymbol{\lambda}_{g,j}, \mu_{g,j}, \beta_j \mid M_{k^*}, \mathscr{D}_n)$ denote the conditional posterior distribution of the parameters $(\boldsymbol{\lambda}_{g,j}, \mu_{g,j}, \beta_j)$ under model M_{k^*} given $\mathscr{D}_n$. Denoting

$$\bar{\lambda}_g = \frac{\lambda_{g,E,l} e^{\beta_E \mu_{g,E}} + \lambda_{g,C,l} e^{\beta_C \mu_{g,C}}}{2},$$

the average arm E *total hazard ratio* (Kalbfleisch and Prentice, 1981) is

$$\begin{aligned}
\theta_{g,E} &= -\int_0^\infty \frac{h_{g,E}(t)}{h_{g,E}(t) + h_C(t)} \mathrm{d}S_{g,E}^{1/2}(t) S_{g,C}^{1/2}(t) = -\int_0^\infty S_{g,C}^{1/2}(t) \mathrm{d}S_{g,E}^{1/2}(t) \\
&= \sum_{l=1}^{L} \frac{\lambda_{gEl} e^{\beta_E \mu_{g,E}}}{\lambda_{gEl} e^{\beta_E \mu_{g,E}} + \lambda_{gCl} e^{\beta_C \mu_{gC}}} \left\{ \exp(-\bar{\lambda}_g s_{l-1}) - \exp(-\bar{\lambda}_g s_l) \right\},
\end{aligned}$$

for each subgroup $g = 1, \ldots, G$. The average arm C total hazard ratio, $\theta_{g,C}$, is defined similarly. Using these definitions of AHR for E and C, if $S_{g,E}(t) = S_{g,C}(t)$ then $\theta_{g,E} = \theta_{g,C} = .5$. Ideally, $\theta_{g,E} + \theta_{g,C} = 1$, but since this equality does not always hold under the piecewise exponential structure, one may define standardized versions of the AHRs,

$$\tilde{\theta}_{g,E} = \frac{\theta_{g,E} + (1 - \theta_{g,C})}{2}, \quad \tilde{\theta}_{g,C} = \frac{\theta_{g,C} + (1 - \theta_{g,E})}{2}.$$

which guarantee that $\tilde{\theta}_{g,E} + \tilde{\theta}_{g,C} = 1$. If $\tilde{\theta}_{g,E} = 0.5$, then there is no survival difference between the two arms. If $\tilde{\theta}_{g,E} < 0.5$, then E is superior to C in terms of survival; and if $\tilde{\theta}_{g,E} > 0.5$, then E is inferior to C. The inequality $\tilde{\theta}_{g,E} < 0.5$ is equivalent to

$$\frac{\theta_{g,E} + (1 - \theta_{g,C})}{2} < 0.5 \quad or \quad \frac{\theta_{g,E}}{\theta_{g,C}} < 1,$$

where AHR $= \theta_{g,E}/\theta_{g,C}$.

The conditional posterior distribution $f(\tilde{\theta}_{g,j} \mid M_{k^*}, \mathscr{D}_n)$ is computed from $f(\boldsymbol{\lambda}_{g,j}, \mu_{g,j}, \beta_j \mid M_{k^*}, \mathscr{D}_n)$. When $|M_{k^*}| > 1$, the issue of multiple testing arises, and a multiplicity adjustment is needed to control the FWER. To do this in the BAGS design, at each interim analysis a Holm-like sequential testing procedure is used to gain more power, similarly to Ye et al. (2013) and Maurer and Bretz (2013). At each test for induced subgroup $g \in \mathscr{S}$, the specified Bayesian test statistic is computed. The probability cutoff for the test statistic must be a decreasing function of m = the number of active hypotheses, so that as m increases it becomes more difficult to reject the hypothesis. To implement this, each GS test uses a cutoff $c(n,m)$ dependent on both the current sample size n and number of active hypotheses, m. At each decision-making time, m is recalculated, and the procedure tests the hypotheses for each induced subgroup $g \in \mathscr{S}$ with $\xi_g = 1$, which includes homogeneous subgroups, i.e., $\mathscr{S}_g = \{g'; \zeta_{g'}^{(k^*)} = g\}$. The three possible subgroup-specific decisions are as follows:

1. **Superiority of** E **over** C: For each subgroup $g \in \mathscr{S}$, if

$$\Pr(\tilde{\theta}_{g,E} < .5 \mid \xi_g = 1, M_{k^*}, \mathscr{D}_n) \;>\; c(n,m),$$

then reject the composite null hypothesis $\cup_{g' \in \mathscr{S}_g} H_{g',0}$ and conclude that arm E is superior to arm C in the induced subgroup $\mathscr{S}_g$.

2. **Inferiority of** E: For each subgroup $g \in \mathscr{S}$, if

$$\Pr(\tilde{\theta}_{g,E} > .5 \mid \xi_g = 1, M_{k^*}, \mathscr{D}_n) \;>\; c(n,m),$$

then reject the composite null hypothesis $\cup_{g' \in \mathscr{S}_g} H_{g',0}$ and conclude that E is inferior to C in the induced subgroup $\mathscr{S}_g$.

3. **Inconclusive**: There is insufficient evidence in the current data to reject the union of null hypotheses.

If the composite null hypothesis is rejected for some induced subgroup $\mathscr{S}_g$, then the trial stops recruiting patients from this induced subgroup, and it also drops it from the union of null hypotheses $\cup_{g' \in \mathscr{S}_g} H_{g',0}$ for the remainder of the trial. The trial continues recruiting patients in the remaining induced subgroups until the maximum sample size N has been enrolled. If at some point there are no remaining induced subgroups, then the trial is terminated.

The probability cutoff $c(n,m)$ must be calibrated to ensure good operating characteristics in the multiple-testing framework. To facilitate the calibration, the flexible two-parameter function

$$c(n,m) = 1 - \frac{\kappa}{m}(n/N)^{\varepsilon} \,,$$

is assumed, where $\kappa > 0$ and $\varepsilon > 0$ are tuning parameters, recalling that m is the number of active hypotheses. The cutoff function $c(n,m)$ is similar to an α-spending function for a standard GS design (Lan and DeMets, 1983; Jennison and Turnbull, 1999) and the adaptive cut-off function used in Bayesian GS testing by Wathen and Thall (2008). It is monotonically decreasing in the interim sample size n. At the beginning of the trial, a more stringent stopping rule is imposed to control the risk of false discoveries due to sparseness of the early data. As more patients are accrued and longer follow-ups are observed, more survival time information is accumulated and there is less uncertainty, so the changing values of $c(n,m)$ can graduate promising subgroups more reliably. Since $c(n,m)$ depends on the number of active hypotheses, m, this also must be updated for each GS decision.

For example, suppose that there are $G = 4$ subgroups initially, and at an interim analysis with sample size n, the induced subgroups are $\{1,2\}$, $\{3\}$, and $\{4\}$. Because there are three induced subgroups, the number of active hypotheses is now $m = 3$, and the test cutoff is $c(n,3)$. If the hypothesis for induced subgroup $\{1,2\}$ is tested first and is rejected, then the number of active hypotheses is reduced to $m = 2$.

Consequently, provided that the induced subgroups $\{3\}$, and $\{4\}$ are not combined subsequently, the cutoff $c(n,2)$ is used to test the remaining two active composite null hypotheses. If the hypothesis for subgroup $\{4\}$ is rejected next, then $c(n,1)$ is used for testing the last null hypothesis for the final induced subgroup $\{2\}$. This sequential procedure is similar to the Holm multiple-testing procedure for frequentist GS designs (Ye et al., 2013; Maurer and Bretz, 2013), and is more powerful than the parallel testing procedure using the cutoff $c'(n,m) = 1 - \kappa(n/N)^{\varepsilon}$ that ignores m.

There is an important difference between the decision-making procedure of the BAGS design and a conventional GS procedure. For a conventional GS design with G hypotheses, the number of active hypotheses at an interim point in the trial equals G minus the number of previously rejected hypotheses. In contrast, since the BAGS procedure requires the induced subgroups to be identified before each test, the number of active hypotheses is bounded above by $|M_{k^*}|$. Consequently, for the BAGS design, m depends on both $|M_{k^*}|$ and the number of previously rejected hypotheses, so the BAGS design adds another layer of randomness to the number of active hypotheses.

There are several different types of errors associated with the BAGS design. For each $H_{g,0}$, $g = 1,\ldots,G$, the subgroup-specific type I error rate (SSER) is $\alpha_g = \Pr(\text{reject } H_{g,0} \mid H_{g,0})$. Because hypotheses $H_{g,0}$ are tested for multiple subgroups, there also is a family-wise type I error rate (FWER), defined as

$$\tilde{\alpha} = \Pr(\text{reject at least one } H_{g,0} \mid \cap_{g=1}^{G} H_{g,0}).$$

The FWER is very important for subgroup-specific comparisons when controlling false discoveries in testing multiple hypotheses. If $H_{g,1}$ is true for a subgroup g, then it is desirable to have large subgroup-specific power (SSP), defined as $1 - \beta_g = \Pr(\text{reject } H_{g,0} \mid H_{g,1})$. The SSP quantifies a design's ability to correctly identify a subgroup where E is promising. For BAGS, in addition to these error probabilities, there also is a misclassification rate (MCR), since BAGS adaptively determines induced subgroups as part of its GS procedure. The MCR is defined as

$$\alpha_c = Pr(\text{at least one subgroup g is misclassified}).$$

Because BAGS combines classification and testing in each GS step, the generalized family-wise power (GFWP) is defined as

$$1 - \tilde{\beta} = \Pr(\text{obtain correct induced subgroups \underline{and} make correct test decisions}).$$

Having a high GFWP is a very important property for the BAGS design, since correct induced subgroup selection is a key element of the decision-making process. Obtaining a good GFWP depends on well-calibrated design parameters and a sufficiently large sample size. The event in the definition of the GFWP is a subset of the event (make correct test decisions), so for a given N, it is more difficult to achieve a large GFWP than a large Pr(make correct test decisions).

The BAGS design requires the specification of fixed prior hyperparameters in the Bayesian models. In addition to the default priors discussed above, the remaining prior hyperparameters that must be specified are $(\mu_0, \sigma^2_{\mu_0}, a_0, b_0, \beta_0, \sigma^2_{\beta_0}, \lambda_0)$ for both treatment arms, and $\{p_g : g = 1, \ldots, G\}$ for E. Although independent modeling procedures are considered, to ensure fair comparisons the same prior distributions, including fixed hyperparameter values, should be used for the two arms. Since the trial is a large-sample comparative study, any vague prior is suitable. In the motivating example, the assumed values are $\mu_0 = \beta_0 = 0$, $\sigma^2_{\mu_0} = \sigma^2_{\beta_0} = 10^2$, $a_0 = b_0 = 0.1$, and $\lambda_0 = 0.1$. If desired, p_g, the prior probability that g is in its own induced subgroup, can be elicited from the clinicians planning the study, or estimated from historical data. The simulations will use $p_1 = \cdots = p_G = 0.8$, although different values of p_g have a small impact on the performance of BAGS.

Estimation of the Bayesian hierarchical piecewise model, and adaptive identification of the induced subgroups, both depend on the partition of the time scale in the survival time model's hazard function. There is a trade-off between subgroup identification accuracy and the flexibility of the estimated survival curves, in terms of the number of subintervals in the partition. The finer the partition, the more flexible and smooth the estimated survival curve will be. As the number of subintervals increases, the number of events contained in each subinterval decreases. Since this increases the variances of the estimated piecewise hazards, it might lead to inaccurate collapsing of different subgroups. In the simulations, the method works well for $L = 3$ to 9 subintervals. The BAGS model allows different partitions in the two treatment arm distributions, and it uses random partitions by assigning a probability distribution to the subinterval cutoff times $s_0, \ldots, s_L$. However, because there is a difficulty in interpreting random partitions, the illustrative trial uses $L = 5$ subintervals that span the time scale $[0, \infty)$ for each induced subgroup, and it uses the percentiles of the observed survival times as the interval cutoffs.

The maximum sample size N determines the design's power figures and induced subgroup identification accuracy, with larger N giving larger SSP and GFWP. Preliminary simulations may be used to calibrate N to obtain a desired GFWP. For the number of interim analyses, its determination depends on the specific trial requirements and available resources, although the first interim analysis should be performed at $n = N/2$ to prevent a high false discovery rate caused by sparse data, while still preserving efficiency. In the illustration, two interim analyses are performed when $n = N/2$ and $n = 3N/4$ patients have been accrued.

The design parameters κ, ε, and N can be determined by a grid search over all possible combinations, to optimize the BAGS design's performance. In the illustration, this was done for the three values $\varepsilon = 1, 2, 3$, six values $\kappa = .010, .015, .020, .025, .030, .035$, and five values $N = 600, 650, 700, 750, 800$. This gives a grid of $3 \times 6 \times 5 = 90$ combinations to study. One may choose (κ, ε, N) to maximize the GFWP of the BAGS design over the grid while controlling the FMER and SSP at prespecified levels. This can be done as follows.

Steps in the Grid Search for κ, ε, and N

Step 1: Ask the clinicians to specify desirable FMER and SSP values, and establish three sets of hypotheses: (1) Homogeneous null: Under $H_{g,0}$ for $g = 1, \ldots, G$, all subgroups are homogeneous and there is no treatment effect. (2) Heterogeneous null: Under $H'_{g,0}$ for $g = 1, \ldots, G$, the subgroups are heterogeneous and there is no treatment effect. (3) Alternative: Under $H_{g,1}$ for $g = 1,\ldots,G$, there are two subgroups, with one subgroup containing the responders, defined as patients who have a lower death rate with E, while the other subgroup contains non-responders. In addition, other trial parameters, such as patient accrual rate and follow-up time, also should be determined.

Step 2: Choose a sample size N, carry out simulations, and identify all combinations of (κ, ε) that control the FWER under both $H_{g,0}$ and $H'_{g,0}$. Given a set of (κ, ε), the FWER is unchanged regardless of the sample size N. Therefore, one only needs to consider one value of N in this step.

Step 3: Among the admissible set of (κ, ε) pairs identified in Step 2, carry out simulations under $H_{g,1}$ for different values of N, and select the combination (N, κ, ε) that yields the desired SSP while maximizing the GFWP as the optimal design parameters.

In Step 2, two sets of null hypotheses are considered, to guarantee that the BAGS design will maintain a specified FWER under a variety of different scenarios.

In the simulation study, the aim is to control the FWER $\leq .05$ and the SSP $\geq .90$. The data from the OAK trial (Rittmeyer et al., 2017) were used to construct the simulation scenarios. The trial treated 425 patients in the Docetaxel arm (C) and 425 patients in the Atezolizumab arm (E), and the 850 patients were classified into $G = 4$ subgroups based on PD-L1 expression. Patient accrual was assumed to follow a Poisson distribution with a rate of 15 patients per month. In the trial, each patient was followed until death due to any cause, with follow-up time after accrual completion up to 2.25 years. Let $\tilde{z}_g$ be the true subgroup for $g = 1,\ldots,4$. The simulations assumed identical prevalences for the four subgroups of .25 each.

Data for the Docetaxel patients were simulated with

$$Z_{i,C} \sim iid \; \mathrm{N}(\mu_C, 1), \qquad T_{i,C} \sim iid \; \mathrm{Exp}(\lambda_C \exp(\beta_C Z_{i,C}))$$

with $\beta_C = -0.25$. The data for the Atezolizumab patients were simulated with

$$(Z_{i,E} \mid g) \sim iid \; \mathrm{N}(\mu_{g_iE}, 1), \qquad T_{i,E} \mid g \sim iid \; \mathrm{Exp}(\lambda_{g_iE} \exp(\beta_E Z_{i,E}))$$

with $\beta_E = -0.25$. The parameters $\lambda_C, \mu_C, \lambda_{g,E}, \mu_{g,E}$ were determined based on the historical data from the OAK trial, which gave median survival times of 9.6 months for the Docetaxel arm and 13.8 months for the Atezolizumab arm. To mimic the published results, the parameters for the Atezolizumab patients were set to be $\mu_C = 0.5$ and $\lambda_C = \exp(-2.5)$. The homogeneous null case is $\mu_{g,E} = \mu_C$, $\lambda_{g,E} = \lambda_C$ for $g = 1,\ldots,4$, and thus $\zeta_1 = \ldots = \zeta_4 = 1$. The heterogeneous null requires that the four subgroups have different death rates for E, and for this, it was assumed that $\mu_{g,E} = -0.3 + 0.5g$, $\lambda_{g,E} = 2.7 - 0.2g$, and thus $\zeta_g = g$ for $g = 1,\ldots,4$. For an alternative hypothesis, $\mu_{1E} = \mu_{2E} = \mu_C$, $\mu_{3E} = \mu_{4E} = 1.22$, $\lambda_{1E} = \lambda_{2E} = \lambda_C$, $\lambda_{3E} = \lambda_{4E} = \exp(-2.7)$. Thus, two of the subgroups were responders who had a lower death rate with E, and two were non-responders. Two interim analyses were performed when $n = N/2$ and $n = 3N/4$ patients had been accrued, with a final analysis at the end of

TABLE 6.5
Simulation scenarios. For each subgroup $g = 1,2,3,4$, $\log(\lambda_{g,E})$ = log baseline hazard, $\mu_{g,E}$ = population mean, AHR$_g$ = average hazard ratio between arms E and C. In all scenarios, the log-hazard-ratio regression parameter $\beta_E = -0.25$.

Scenario	Subgroups	$-\log(\lambda_{g,E})$	$\mu_{g,E}$	AHR$_g$
1	$\{1,2,3,4\}$	(2.5, 2.5, 2.5, 2.5)	(.5, .5, .5, .5)	(1, 1, 1, 1)
2	$\{1\},\{2\},\{3\},\{4\}$	(2.5, 2.3, 2.1, 1.9)	(.5, 1.3, 2.1, 2.9)	(1, 1, 1, 1)
3	$\{1,2\},\{3,4\}$	(2.5, 2.5, 2.7, 2.7)	(.5, .5, 1.22, 1.22)	(1, 1, .7, .7)
4	$\{1,2,3\},\{4\}$	(2.5, 2.5, 2.5, 2.7)	(.5, .5, .5, 1.22)	(1, 1, 1, .7)
5	$\{1\},\{2,3,4\}$	(2.5, 2.7, 2.7, 2.7)	(.5, 1.22, 1.22, 1.22)	(1, .7, .7, .7)
6	$\{1\},\{2,3,4\}$	(2.2, 2.7, 2.7, 2.7)	(0, 1.22, 1.22, 1.22)	(1.5, .7, .7, .7)

follow-up. The design parameters were optimized using the specified homogeneous null, heterogeneous null, and alternative hypotheses. This resulted in $N = 700$, $\kappa = .02$, and $\varepsilon = 3$.

In the simulations, BAGS was compared to two alternative methods commonly used in multiple subgroup settings. The first, referred to as the *homo-logrank* design, assumes that all subgroups are homogeneous and implements a logrank test based on a GS design. The second approach, called the *hetero-logrank* design, does a separate logrank test in each subgroup and does not borrow information between subgroups. As benchmarks, two oracle designs also were included, *oracle-logrank-BAGS* and *oracle-BAGS*, both of which always use the correct subgroup classification in their inferences. For the logrank test-based GS design, O'Brien-Fleming boundaries (O'Brien and Fleming, 1979) were used with the Holm procedure to adjust for multiplicity (Ye et al., 2013). The oracle-BAGS design uses the same set of design parameters as the BAGS design.

Table 6.5 gives eight simulation scenarios, and Table 6.6 gives the operating characteristics of the design based on 5,000 simulated trials under each scenario. Scenario 1 is the homogeneous null case, where all subgroups have the same distributions and Atezolizumab gives no survival benefit over Docetaxel. All five designs control the probability of making one or more type I errors near 5%. The BAGS design has a very low misclassification rate which, together with the FWER, leads to a high GFWP. The hetero-logrank design assumes complete heterogeneity among the subgroups and uses a multiplicity adjustment to control the FWER. Consequently, the SSER of the hetero-logrank design is lower than that of the other designs. In scenario 2, where all four subgroups are different, it is desirable to not borrow information across subgroups. The BAGS design preserves the FWER at the prespecified level while maintaining a low MCR. Scenario 3 is the alternative hypothesis, where there is one induced subgroup, {3, 4}, with responders and another induced subgroup,

TABLE 6.6

Operating characteristics of BAGS and four comparators, with four subgroups. All values except $\bar{N}$ = mean sample size are percentages. $H_{g,0}$ is the subgroup-specific null hypothesis for $g = 1,2,3,4$, FWP = family-wise power, FWER = family-wise type I error rate, MCR = subgroup misclassification rate, and GFWP = generalized family-wise power. Homo-logrank = GS design based on the logrank test assuming homogeneity. Hetero-logrank = GS design based on the logrank test assuming heterogeneity. Oracle methods know the correct the subgroup classifications.

Design	% Reject $H_{g,0}$				FWP (FWER)	MCR	GFWP	$\bar{N}$
	1	2	3	4				
Scenario 1								
BAGS	5.3	5.4	5.3	5.4	5.6	0.4	94.2	696
Homo-logrank	6.2	6.2	6.2	6.2	6.2	0.0	93.8	695
Hetero-logrank	1.5	1.3	1.7	1.6	5.8	100.0	0.0	700
Oracle-logrank	6.2	6.2	6.2	6.2	6.2	0.0	93.8	695
Oracle-BAGS	5.2	5.2	5.2	5.2	5.2	0.0	94.8	696
Scenario 2								
BAGS	1.5	1.4	1.6	1.7	5.9	4.2	90.2	700
Homo-logrank	6.2	6.2	6.2	6.2	6.2	100.0	0.0	695
Hetero-logrank	1.5	1.3	1.7	1.6	5.8	0.0	94.2	700
Oracle-logrank	1.5	1.3	1.7	1.6	5.8	0.0	94.2	700
Oracle-BAGS	1.9	1.9	1.7	1.9	6.6	0.0	93.4	700
Scenario 3								
BAGS	8.2	8.2	92.3	92.5	93.6	9.2	80.6	693
Homo-logrank	66.5	66.5	66.5	66.5	66.5	100.0	0.0	647
Hetero-logrank	1.7	1.7	64.2	63.3	84.1	100.0	0.0	700
Oracle-logrank	4.4	4.4	92.6	92.6	92.8	0.0	90.3	699
Oracle-BAGS	5.7	5.7	94.4	94.4	94.6	0.0	88.9	698
Scenario 4								
BAGS	5.6	5.2	5.2	77.9	78.3	9.7	68.9	697
Homo-logrank	23.7	23.7	23.7	23.7	23.7	100.0	0.0	685
Hetero-logrank	1.5	1.3	1.7	62.1	64.4	100.0	0.0	700
Oracle-logrank	3.9	3.9	3.9	71.8	72.4	0.0	68.5	699
Oracle-BAGS	4.6	4.6	4.6	79.2	79.6	0.0	75.4	699
Scenario 5								
BAGS	21.1	96.0	95.1	95.1	97.2	23.2	69.1	675
Homo-logrank	93.7	93.7	93.7	93.7	93.7	100.0	0.0	583
Hetero-logrank	2.0	65.8	65.1	64.3	91.9	100.0	0.0	700
Oracle-logrank	4.7	97.6	97.6	97.6	97.7	0.0	95.2	699
Oracle-BAGS	5.9	97.9	97.9	97.9	98.0	0.0	92.6	697
Scenario 6								
BAGS	91.3	97.9	96.2	96.2	99.8	8.7	83.7	678
Homo-logrank	79.3	79.3	79.3	79.3	79.3	100.0	0.0	637
Hetero-logrank	83.3	68.4	67.7	66.9	93.9	100.0	0.0	699
Oracle-logrank	95.5	98.4	98.4	98.4	99.9	0.0	93.9	649
Oracle-BAGS	92.6	98.7	98.7	98.7	99.8	0.0	91.6	672

{1,2}, with non-responders. It is highly desirable to combine the data within each of the two induced subgroups to make the inferences more efficient. In Scenario 3, fully borrowing information across all subgroups, which is what the homo-logrank does, leads to a high subgroup-specific error rate, SSER, which in this case is % Reject $H_{g,0}$ for each g. Because the hetero-logrank does not borrow information, it has a low power for detecting promising subgroups. In contrast, because the BAGS design adaptively creates induced combined subgroups, it can exploit the borrowed information in its tests.

The BAGS design is as powerful as the oracle designs in identifying the promising subgroup {3,4}. However, since there is a nonzero chance of misclassification, around 9.8%, the BAGS design has a slightly higher SSER and lower GFWP than the oracle designs. In scenario 4, only patients in subgroup 4 have a higher response rate with Atezolizumab. This scenario is included to assess whether the decision for subgroup 4 based on BAGS would be contaminated by the majority of patients, which includes the non-responders. Surprisingly, the BAGS design still is able to detect subgroup 4 reliably and achieve subgroup-specific power similar to that of oracle-BAGS, and better than that of oracle-logrank. In addition, the SSERs for subgroups 1–3 are also close to the 5% nominal level. In scenario 5, the majority of subgroups are responders, which can be regarded as an opposite case of scenario 4. Compared to scenario 4, the MCR of BAGS in scenario 5 is higher, leading to a higher SSER for subgroup 1. BAGS still yields GFWP = .691 in scenario 5, compared to GFWP = 0 for the commonly used homo-logrank and hetero-logrank designs. Scenario 6 is a completely homogeneous case where patients in all subgroups respond to Atezolizumab. Compared to the hetero-logrank design, BAGS has much higher power and lower achieved sample size. In this case, the performance of BAGS is almost identical to that of the oracle designs. Scenario 7 is a difficult case because there are four heterogeneous subgroups, and the low sample size per subgroup tends to lead to a higher MCR, which in turn causes a higher SSER for subgroup 1. In scenario 8, subgroup 1 and subgroups 2–4 have opposite treatment effects, with Atezolizumab inferior to Docetaxel in subgroup 1, and Atezolizumab superior in subgroups 2–4. Despite this, the BAGS design still maintains a high GFWP.

The simulation study indicates that, when the MCR is low, the operating characteristics of the BAGS design are nearly identical to those of its oracle counterpart. There are some scenarios that cause a higher MCR for BAGS and thus a higher SSER. This is mainly due to the fact that sample sizes for misclassified subgroups typically are small. Nevertheless, the BAGS design shows greatly superior performance compared to the homo-logrank and hetero-logrank designs, which are by far the most commonly used designs in practice.

7

Enrichment Concepts and Methods

7.1 Disease Heterogeneity and Targeted Agents

In oncology, it is well known that the molecular biology of nearly all cancers is heterogeneous. Using new biotechnologies to characterize cancers at the cellular or molecular level, what once was considered to be a single disease often turns out to be a collection of biologically distinct but related diseases. Tumors may have several different types of heterogeneity:

1. between different patients who have what nominally is the same type of cancer (Dagogo-Jack and Shaw, 2018);
2. between different tumors in a patient who has multi-focal disease;
3. between tissue samples taken at the same time from different locations in a single solid tumor, (Prasetyant and Madema, 2017; Lee et al., 2016);
4. for a single patient's disease evaluated repeatedly over time, due to clonal evolution and metastasis (McGranahan and Swanton, 2017).

Characterizing specific sources and forms of such heterogeneities requires formulating and solving both technological and statistical problems. Using microarrays, cytometry, or DNA sequencing, heterogeneities can be identified and quantified by genomic, proteomic, or other biological variables. This has motivated the widespread development of *targeted agents*, which are molecules designed to disrupt functional pathways of particular types, or subtypes, of cancer cells based on their molecular biology. The clinical goals are to reduce and, ideally, to eradicate a patient's disease. Four strategies for biologically specific treatments are as follows:

1. Create an immunological response that attacks the disease.
2. Cause cancer cells to die by inducing programmed cell death (apoptosis).
3. Cut off a solid tumor's blood supply by inhibiting vascular endothelial growth factor (VEGF).
4. Infuse cells designed to recognize, attack, and kill specific types of cancer cells.

A therapeutic agent may be a designed molecule, or a specific type of cell obtained from a donor or umbilical cord blood, including NK cells, mesenchymal stem cells,

DOI: 10.1201/9781003474258-7

and regulatory T-cells. To use such cells as treatments, they are engineered biologically by first selecting and modifying the cells to have particular anti-disease functions, and then expanded *ex vivo* to a sufficient number that constitutes a therapeutic dose. See, for example, Brentjens et al. (2013), Grupp et al. (2013), Porter et al. (2011), Davila et al. (2014), or Liu et al. (2020). The wide range of possible biological functions of such molecular or cellular treatments has led to their broad use to treat many cancers and also diseases outside of oncology, including immune disorders, viral infections, and acute respiratory distress syndrome.

A general statistical paradigm for thinking about a biologically targeted experimental agent, E, is as follows. Suppose that E is designed to hit a biological target thought to be associated with the disease being treated, and that whether or how extensively the target is present in a patient can be represented by a function of a vector $\mathbf{Z} = (Z_1, \cdots, Z_p)$ of biomarkers. Two major statistical goals when studying and developing E are to (1) determine a function of $\mathbf{Z}$ that characterizes a subgroup of patients who are likely to respond to E, and (2) do a confirmatory comparison of E to a standard active control treatment, C, in the identified subgroup, and possibly also in the overall patient population.

It is useful first to consider how this might be done with one binary covariate, $Z = 1$ if a patient has the biomarker for the target, referred to as the patient being *biomarker positive* or E*-sensitive*, and $Z = 0$ if not. The primary clinical outcome of interest, Y, most often is either an early treatment response indicator or survival time. If the targeted agent works as it was designed, then patients who receive E will have a higher response probability, or live longer, on average, and the beneficial effect will be larger if $Z = 1$ than if $Z = 0$. That is, E-sensitive patients will be more likely to benefit from E. In practice, patients usually have more than one important clinical outcome, and Y may be a vector including two or more co-primary endpoints, such as a response indicator Y_{RES} of early anti-disease effect, an indicator Y_{TOX} of one or more severe adverse events (toxicity), and Y_S = survival time. In this case, $Y = (Y_{RES}, Y_{TOX}, Y_S)$. While designs using multidimensional Y potentially are useful, to focus on enrichment trials, for now I will discuss only designs with one primary outcome. If the effects of E on Y in the two subgroups defined by $Z = 1$ and $Z = 0$ are different, then Z is said to be a *predictive* covariate. Denote the treatment indicator $X = 1$ if a patient is treated with $E + C$ and $X = 0$ if treated with C.

As an illustration, suppose that the disease is newly diagnosed AML, C is the well established chemotherapy agent cytosine arabinoside (ara-C), and E is a new molecule designed to inhibit leukemia cell proliferation by targeting a mutation in the FLT3 tyrosine kinase enzyme. Suppose that a clinical trial randomizes patients fairly between C and $E + C$. The outcome, Y, may be either the indicator of complete remission (CR) within 42 days, or survival time. The trial's main goal is to determine whether E-sensitive patients treated with $E + C$ have a larger probability of CR, or longer mean survival time, than patients treated with C. In terms of treatment effects, at one extreme it may be the case that if $Z = 0$ then $E + C$ provides no benefit over C, and if $Z = 1$ then $E + C$ provides a substantial benefit over C. At the opposite extreme Z is irrelevant, and either the effect of adding E to C is the same for $Z = 0$ and $Z = 1$, or E has no effect whatsoever. A regression model for the distribution of Pr(CR) as a function of (X,Z) may take the form

$$\theta_{X,Z} = \Pr(Y = 1 \mid X,Z) = \text{logit}^{-1}(\eta_{X,Z}),$$

where $\text{logit}(p) = \log\{p/(1-p)\}$ for $0 < p < 1$, and a linear term $\eta_{X,Z}$ accounting for the possible effects of both treatment and biomarker may be

$$\eta_{X,Z} = \mu + \gamma X + \beta Z + \xi XZ.$$

In this model, the effect of E is $\gamma + \xi$ for $Z = 1$ and γ for $Z = 0$, so the treatment-biomarker interaction that quantifies the additional effect of E in biomarker positive patients is $(\gamma + \xi) - \gamma = \xi$. If $\xi = \gamma = 0$, then E has no anti-disease effect at all, regardless of whether $Z = 1$ or 0.

If, instead, Y is survival time, then η typically appears in the model for the logarithm of the hazard of death. For example, if one assumes a Weibull distribution, this has the flexible hazard (death rate) function $h(t) = \lambda \alpha t^{\alpha-1}$, where λ is a rate parameter, and α is a shape parameter. The mean survival time is $\theta = \Gamma(1+1/\alpha)\lambda^{-1/\alpha}$, where $\Gamma(\cdot)$ denotes the gamma function. One may assume that the rates are $\lambda_{X,Z} =$ $= \exp(\eta_{X,Z})$ to model the effects of treatment X and biomarker status Z on survival time, with the mean parameterized as

$$\theta_{X,Z} = \Gamma(1+\alpha^{-1}) \exp^{-\eta_{X,Z}/\alpha} = \frac{\Gamma(1+\alpha^{-1})}{\exp\{\alpha^{-1}(\mu + \gamma X + \beta Z + \xi XZ)\}}$$

In this case, the parameters μ, γ, β, and ξ have the same interpretation as given above for response probability, but now in terms of death rate.

7.2 Enrichment Using a Vector of Biomarkers

In many settings, a vector $\mathbf{Z}$ that includes multiple candidate biomarkers and possibly known prognostic covariates is available, rather than only a single binary Z. An important statistical problem is to construct a *discrimination function*, $f(\mathbf{Z})$, and a fixed cutoff f^*, such that $f(\mathbf{Z}) > f^*$ identifies a patient as being E-sensitive. Statistical methods for identifying such a discrimination function will be discussed below. If an E-sensitive patient subset $\{\mathbf{Z} : f(\mathbf{Z}) > f^*\}$ in which $E + C$ has a substantively larger anti-disease effect than C can be identified, then $\mathbf{Z}$ may be used by practicing physicians to guide precision medicine, wherein each physician uses each patient's vector $\mathbf{Z}$ to guide treatment decisions. The availability of modern biomarkers notwithstanding, physicians have been using patient covariates to guide their therapeutic decision-making for thousands of years. That is, precision medicine is not new. Rather, the technologies used to generate $\mathbf{Z}$ vectors are new, and the biomarkers often have complex biological interpretations that provide a causal basis for explaining their effects on clinical outcomes.

An adaptive enrichment design, probability model, and parameter estimation method must address two closely related statistical problems. These are identification of a subpopulation of *E-sensitive* patients, and estimation or comparative testing of the effects of E on Y in both E-sensitive (biomarker positive) and non-E-sensitive (biomarker negative) patients. One may think of statistical estimation as evaluation, and comparative statistical hypothesis testing as validation. As seen in several of the designs discussed earlier, such as BAGS, the decision of whether $E+C$ is superior to C within a given subgroup is more general than simply testing hypotheses. Doing identification, estimation, and evaluation reliably in the same clinical trial, or in a series of trials, can be a very challenging problem. Preclinical *in vitro* data on the molecular biology of a given disease and *in vivo* data on the effects of E in rodents xenografted with the disease may suggest which elements of $\mathbf{Z}$ are more likely to identify E-sensitive patients. Ultimately, however, data on humans treated with $E+C$ and C from a randomized clinical trial are required to make reliable inferences. Several strategies and designs for clinical identification, evaluation, and validation will be reviewed below.

It might seem counterintuitive to evaluate the possible effects of E in biomarker negative patients. In practice, however, in terms of clinical outcomes, a new targeted agent often does not affect a particular human disease as expected based on preclinical *in vitro* or *in vivo* data. Freidlin et al. (2014) argue that, rather than comparing E or $E+C$ to C overall and in an identified biomarker-positive subgroup, the comparison should be done in both the biomarker-positive and negative subgroups. They propose a *Marker Sequential Test* (MaST) design that does this while controlling false positive error rates in both subgroups. They show that the MaST design has higher power than a sequential subgroup-specific design if the E-versus-C effect is homogeneous across the biomarker subgroups, and also has high power if the benefit of E is limited to the biomarker-positive subgroup. Due to complexity of the biological process, E may have a direct effect on clinical outcome that is not mediated by Z, that is, there is some other biomarker, Z', for the effect of E that has not been identified. Additionally, any statistical rule for dichotomizing patients into E-sensitive and non-E-sensitive subgroups based on an available biomarker vector $\mathbf{Z}$ cannot be perfect, because it is based on data, and data vary randomly. For any of these reasons, E may turn out to have a large anti-disease effect in biomarker-negative patients. Numerical examples of this will be given below.

The two main goals of any clinical trial are to treat the patients in the trial, and to provide high-quality data for making statistical inferences that may benefit future patients. In most enrichment designs, decisions and actions include identifying an E-sensitive subgroup, increasing or possibly restricting enrollment to E-sensitive patients, and testing whether $E+C$ provides a substantive benefit over C either within the E-sensitive subgroup, or overall. A central statistical problem in any adaptive enrichment design is that its statistical decisions include some combination of selection of a subvector $\mathbf{Z}^*$ of elements of $\mathbf{Z}$, construction of a function $f(\mathbf{Z}^*)$ that is used to define an E-sensitive subgroup, or possibly multiple subgroups, and one or more comparative tests to assess the effect of E, conducted in sequence and possibly within different subgroups. Conventional Type I error and power of one such test viewed in isolation are incorrect and misleading, since the computation ignores all of the other statistical decisions made before performing the test. As an example, consider the probabilities of the following two actions:

Action 1. Determine $\mathbf{Z}^*$, $f(\mathbf{Z}^*)$, and a cutoff c such that $[f(\mathbf{Z}^*) > c]$ = [E-sensitive].

Action 2. Test whether E is superior to C in the E-sensitive patient subset determined in Action 1.

Let θ_{E+C} and θ_C denote the response probabilities, or mean survival times, for the two treatments. The probability of [Action 1 and Action 2] for a given value $\theta^*_{E+C} > \theta_C$ may be called the *generalized power* (GP) of the entire statistical procedure at θ^*_{E+C}. By definition, the GP must be smaller than the power of the test in Action 2 considered alone, as if the subset of E-sensitive patients were known and not determined from data. This is simply because the GP is the probability of a smaller event. Conventional power may be considered a hypothetical version of GP when an oracle provides the E-sensitive subset. A conventional power figure thus is misleading if the test for which the power is computed has been preceded by data-based decisions such as model selection, variable selection, subset selection, or treatment selection. The relevant quantity is the GP of the entire decision process, not the oracle-based power of the final tests considered alone. Additionally, to compute a GP in the above example of two actions, one must assume both a true subset of E-sensitive patients and a numerical value of the parameter θ_{E+C} in that subset. This problem has been well known for many decades, much more generally. In regression analysis, if a dataset first is used to determine a particular model, that model then is fit to the data, and the distributions of parameter estimates or predicted values obtained from the final fitted model are computed under the assumption that the model was prespecified and not obtained via preliminary data analyses to select a model. This process sometimes is referred to as "overfitting". The actual distribution of the parameter estimates obtained at the end of a model selection and model fitting process is a function of both the model selection algorithm and the final model. This problem arose due to the combination of adaptive subgroup clustering and hypothesis testing in the BAGS design.

A very common example of this issue may arise when using least squares to fit a linear model $Y_i = \boldsymbol{\beta}'\mathbf{Z}_i + \varepsilon_i$, for $i = 1, \cdots, n$, where $\varepsilon_i \sim$ *iid* $N(0, \sigma^2)$. If this model is known to be correct, then denoting the matrix of covariate vectors by $\mathbf{Z}$, the least squares estimator $\hat{\boldsymbol{\beta}} = (\mathbf{Z}'\mathbf{Z})^{-1}\mathbf{Z}'\boldsymbol{Y}$ has a p-dimensional normal distribution, $\hat{\boldsymbol{\beta}} \sim N_p(\boldsymbol{\beta}, \sigma^2(\mathbf{Z}'\mathbf{Z})^{-1})$. If, instead, a preliminary variable selection was done to determine the elements of $\mathbf{Z}$, from a larger vector $\mathbf{Z}^+$ of dimension $q > p$, then this distribution for $\hat{\boldsymbol{\beta}}$ is incorrect. By assuming that the model was known, provided by an oracle, one ignores the variable selection process. This problem arises quite generally in frequentist data analysis, and it implies that many reported p-values and 95% confidence intervals are incorrect. One possible solution is to bootstrap the entire variable selection and least squares estimation process, with final inferences pertaining to a q-vector of parameters for the initial $\mathbf{Z}^+$ vector.

GP also arises in settings where a comparative test is preceded by treatment selection. Suppose that, for simplicity, patients are homogeneous. Consider a multi-arm, randomized, select-and-test trial of experimental treatments E_1, E_2, E_3 and an active control C, that selects the treatment E_{j^*} having largest estimated E_j-versus-C effect based on data from a first stage. For such two-stage phase II-III designs, as given by Thall et al. (1988), if $\hat{\theta}_{E_{j^*}} - \hat{\theta}_C$ is sufficiently large in stage 1, then additional data are obtained in a second stage by randomizing patients between the selected E_{j^*} and C, thus enriching the E_{j^*} arm. A final test of $\theta_{E_j^*} = \theta_C$ is done, based on all of the data, to decide whether E_{j^*} is superior to C. In the global null case where all four treatments have identical means, $\theta_{E_1} = \theta_{E_2} = \theta_{E_3} = \theta_C$, the statistical estimate $\hat{\theta}_{j^*}$ will have an expected value larger than the true value. Again, this is because a selected maximum produces upward estimation bias. If a two-arm test comparing E_{j^*} to C is constructed while ignoring the preliminary selection, it will not have the nominal size and power figures, since it ignores the fact that j^* is a statistic that depends on the stage 1 data from all four treatment arms. The relevant quantity is not conventional power, but the GP, which in this setting is the probability of (1) correctly identifying an E_{j^*} that truly provides an improvement over C in stage 1 and (2) correctly concluding that $\theta_{E_j^*} > \theta_C$ in stage 2 for some assumed true value of $\theta_{E_j^*} = \theta_C + \delta$ for a meaningfully large $\delta > 0$. In general, since the GP is the probability of an event that is smaller than the testing event (2), in practice it is difficult to achieve a numerical GP value that is close to a conventional power figure, unless the overall sample size is larger. A similar two-stage phase II-III design based on Y = survival time that allows more than one E_j to be selected for stage 2 is given by Schaid et al. (1990). This design controls the pairwise type I error rate and pairwise power when testing $\theta_{E_j^*} = \theta_C$ for each selected E_j^*. A multistage version is given by Stallard and Todd (2003).

7.3 Outcome Adaptive Randomization

Outcome-adaptive randomization (OAR) may be considered an extreme form of enrichment, since it is done continuously during a clinical trial. Rather than randomizing patients among K treatments with equal probabilities $1/K$, OAR repeatedly unbalances the randomization probabilities for the treatment arms by using the current data to favor the arm, or arms, seen to have more favorable outcomes. This continuously enriches the treatment arms that have superior performance based on the interim data by giving them larger sample sizes. The main motivation of OAR is ethical, to enroll a greater proportion of patients to the treatment arm or arms that, during the trial, show higher response rates, so that more patients may benefit from superior treatments. This may seem wonderful but, counter-intuitively, OAR often is likely to do more harm than good.

Before examining OAR, it is useful to think about a simple play-the-winner (PTW) rule that assigns each new patient in a two-arm trial to the arm having the best current empirical success rate based on the current data. While the PTW rule may seem like a good idea, it actually has terrible properties. To see this, suppose for example that, in a trial to compare the response probabilities of treatments A and B, an initial burn-in using ER is done for the first 10 patients and, due to the play of chance, 6 patients have been treated with A and 4 with B, with 4/6 (67%) responses observed with A and 1/4 (25%) with B. Suppose that the true but unknown response probabilities are π_A^{true} = .25 and π_B^{true} = .50, so in fact B is greatly superior to A. Thus, the empirical response rates of 67% for A and 25% for B seen with the first 10 patients are the opposite of the actual rates. But the PTW rule does not know this, and thus it will treat at least the next 10 patients with A since, even if all 10 do not respond, the empirical response rate with A would still be 25%. At that point, there would be 4/16 (25%) responses with A, finally equal to the empirical rate seen from getting 1 response in 4 patients treated with B. At that point, one might use ER, that is, a coin flip, to choose between A and B for the 21st patient In this example, PTW would assign the inferior treatment, which has π_A^{true} = .25, to 16 patients and assign the superior treatment, which has π_B^{true} = .50, to 4 patients, and the overall number of responses would be 4+1 = 5 in 20 patients (25%). Since getting a run of 10 non-responses if π_A^{true} = .25 has probability $(1\text{-}.25)^{10}$ = .056, it is not unlikely that 10 more patients would be treated with the inferior treatment A before its empirical response rate got down to 25%. What went wrong with the PTW rule was that, early on, due to the play of chance, the inferior treatment happened to have a higher observed response rate. An even more extreme example of this phenomenon is a setting where the burn-in is done with three patients treated with each of A and B, and by chance A has 1, 2, or 3 responses and B has 0 responses, which would occur with a probability of .125. The PTW rule then would treat *all remaining patients* with the inferior treatment A, since the empirical response rate with the truly superior treatment B would remain stuck at 0. Since numerous more complex versions of these examples can occur during a trial conducted using PTW, the chance of assigning more patients to the inferior treatment arm may be non-trivial. This problem with PTW often is called "stickiness", and it is seen more generally with any sequentially adaptive decision rule that is myopic in that it always takes the "optimal" action defined by the largest current empirical success rate. As the above examples show, the PTW rule easily can lead to a disaster. See, for example, Sutton and Barto (1998).

OAR may be considered a compromise between PTW and ER. OAR first was proposed by Thompson (1933), who considered a hypothetical two-arm trial comparing treatments A and B with response probabilities π_A and π_B. Under a Bayesian model in which these probabilities are assumed to be random and follow independent identical beta priors, given current (treatment, response) data $\mathscr{D}_n$ from n patients, the $n+1^{st}$ patient is given B with probability $r_{B,n} = \Pr(\pi_A < \pi_B \mid \mathscr{D}_n)$ and A with probability $r_{A,n} = 1 - r_{B,n}$. Since $r_{B,n}$ tends to be quite variable, in practice the modified the value

$$\tilde{r}_{B,n,1/2} = \frac{(r_{B,n})^{1/2}}{(r_{B,n})^{1/2} + (r_{A,n})^{1/2}}$$

with $\tilde{r}_{A,n,1/2} = 1 - \tilde{r}_{B,n,1/2}$ often is used. For example, with beta(1/2, 1/2) priors for both π_A and π_B, if the first 10 patients show 2/5 (40%) responses with A

and 4/5 (80%) responses with B, then a posteriori $\pi_A \sim beta(2.5, 3.5)$ and $\pi_B \sim beta(4.5, 1.5)$, which implies that $r_{A,10} = .10$ and $r_{B,10} = .90$. The modified value mitigates such extreme imbalances in randomization probabilities, and here the probabilities would be $\tilde{r}_{A,n,1/2} = .25$ and $\tilde{r}_{B,n,1/2} = .75$, with the extreme values .10 and .90 shrunk toward .50. There are numerous other ways to do OAR, (Sverdlov, 2015), and some methods incorporate the patient's covariate vector $\mathbf{Z}$, as in Thall and Wathen (2005).

Despite its intuitive appeal, some very undesirable properties of trials run using OAR designs have been seen in practice. The targeted agent trial BATTLE, described by Kim et al. (2011), accounted for five predefined biomarker subgroups and studied four targeted agents, started with ER among the agents, and later followed this with OAR. Unexpectedly, the BATTLE trial resulted in a lower observed response rate for patients in the portion of the trial run using OAR compared to the response rate observed for patients treated during the initial ER period. This result was the opposite of what was expected, based on the putative ethical attraction of OAR that, on average, OAR should result in better patient outcomes compared to ER. An additional disappointing result of the BATTLE trial was that the five predefined biomarker subgroups used in BATTLE were less predictive than the individual biomarkers. This made the subgroups of little use for clinical decision-making. These results suggest that, in precision medicine trials, the use of either predetermined subgroups that are fixed or OAR should be avoided.

Simulation studies of OAR have shed some light on why the BATTLE trial, and other trials that used OAR, should be expected to produce undesirable results that actually harm patients. Several undesirable, counterintuitive effects of OAR, as seen in the BATTLE trial, were identified in the simulation studies reported by Thall and Wathen (2007) and Thall et al. (2015) for two-arm trials, and by Wathen and Thall (2017) for multi-arm trials. To summarize these results, first consider a randomized 200-patient trial to compare treatments A and B based on a binary response outcome with probabilities π_A and π_B. As shown in Table 6.4 of Thall (2020), compared to ER with group sequential (GS) comparisons at 50, 100, 150, and 200 patients,

- OAR is very likely to inflate Type I error probability substantially;
- OAR is very likely to reduce power substantially;
- If an OAR method is calibrated to control Type I error probability to be .05, then the power reduction compared to a GS design with ER and Type I error probability .05 is even larger, and this power reduction is likely to be $\geq 50\%$;
- OAR is very likely to double or triple estimation bias compared to a GS design;
- In some cases, OAR has a non-trivial probability of unbalancing the sample sizes in the wrong direction, so that many more patients receive the inferior treatment.

For a multi-arm randomized 250- or 500-patient comparative trial of four experimental treatments E_1, E_2, E_3, E_4 and a control treatment C, with a binary response outcome, compared to a GS design with ER,

• OAR methods are less likely to correctly stop an E_j for which $\pi_{E_j}^{true} = \pi_C^{true}$ for futility;

• When exactly one E_j is truly superior to C, with $\pi_{E_j}^{true} = \pi_C^{true} + .20$, and all other $\pi_{E_{j'}}^{true} = \pi_C^{true}$, most OAR methods have much lower probabilities of correctly selecting E_j compared to a GS design with ER;

• In a "staircase" scenario where the true response probabilities are .20 for C and .25, .30, .35, .40 for the four E_j's, OAR methods have much lower probabilities of selecting either the best or second best E_j.

These problems with OAR are rooted in two of its inherent properties. First, OAR introduces much larger variability into the treatment arm sample size distributions, compared to the variability introduced by ER. Second, OAR has built-in bias due to the fact that it repeatedly selects the empirically better treatment with higher probability, which defeats the main reason for randomizing, which is to eliminate bias. While, from a naive viewpoint, OAR may appear to be ethically more desirable than fair randomization, in fact OAR is very likely to harm patients by increasing the false positive rate, reducing the probability of identifying a superior treatment, and in some cases having a non-trivial probability of treating many more patients with a treatment that actually is inferior. Although the simulations cited above did not consider cases with heterogeneous patients, the poor performance of OAR compared to ER in the homogeneous case strongly suggests even worse comparative performance when OAR probabilities are computed to account for $\mathbf{Z}$ as well as (treatment, outcome) data, due to the much smaller sample sizes for many patient subgroups. A review summarizing the many problems with OAR, and strongly advising against its use, is given by Proschan and Evans (2020). At the very least, when treatment comparison is the goal, the ethics of using OAR may be considered highly questionable.

7.4 Treatment-Biomarker Interactions

When using biomarkers for enrichment during a clinical trial, there are many possible cases of how biomarkers may or may not interact with treatments. Which case the data indicate is true, or is most likely, determines how precision medicine may or should be done based on the trial's results. The following toy numerical examples are constructed to illustrate qualitatively different cases, in a setting where E is compared to an active control treatment C in a randomized trial, in terms of Y = survival time, where one binary biomarker, Z, is available. In the cases considered, C has substantive anti-disease effect with a mean survival time of 24 months. It is desired to determine whether E improves mean survival time compared to C, either overall or within two subgroups determined by the biomarker. Each patient is either biomarker positive, pos = $(Z = 1)$, if the biomarker is detected as being present, or biomarker negative, neg = $(Z = 0)$. Table 7.1 gives the mean survival time $\theta_{\tau,Z}$ for

TABLE 7.1
Response probabilities or mean survival times for each of the four possible combinations of treatment arm and biomarker status. The biomarker sets are $pos = \{Z = 1\}$ and $neg = \{Z = 0\}$. The parametric differences are the E effects in these two biomarker subgroups, $\Delta_{pos} = \theta_{E+C,pos} - \theta_{C,pos}$ and $\Delta_{neg} = \theta_{E+C,neg} - \theta_{C,neg}$.

Treatment Arm	Biomarker Status	
	Positive	*Negative*
$E+C$	$\theta_{E+C,pos}$	$\theta_{E+C,neg}$
C	$\theta_{C,pos}$	$\theta_{C,neg}$
Effect of E	Δ_{pos}	Δ_{neg}

each (τ, Z) = (treatment, biomarker) combination, and also the E-versus-C effect in each biomarker subgroup. If Y were a response indicator rather than survival time, then the $\theta_{\tau,Z}$'s would be response probabilities. Denote the proportion of patients that test positive by $p_{pos} = \Pr(Z = 1)$. If there is no testing error, then p_{pos} is the *prevalence* of E-sensitive patients in the population of patients with the disease. The possibilities that the test may incorrectly classify a *negative* patent as *positive*, or vice versa, can be quantified by the test's sensitivity, π_{sens} = Pr(test is positive | patient is positive), and its specificity, π_{spec} = Pr(test is negative | patient is negative). The effects of these probabilities on how the test behaves can be evaluated if the actual disease prevalence is known, or in a sensitivity analysis using several assumed values. In the following examples, in order to focus on treatment-biomarker interactions, I will assume that $\pi_{sens} = \pi_{spec} = 1$, that is, the test for Z is perfectly accurate, so $Z = 1$ implies that the patient must be *pos*, and $Z = 0$ implies that the patient must be *neg*.

The mean subgroup-specific effects of E-versus-C are $\Delta_{pos} = \theta_{E,pos} - \theta_{C,pos}$ in biomarker-positive patients and $\Delta_{neg} = \theta_{E,neg} - \theta_{C,neg}$ in biomarker-negative patients. The overall effect of E in the patient population is the subgroup prevalence weighted average

$$\Delta = \Delta_{pos}\, p_{pos} + \Delta_{neg}\,(1 - p_{pos}),$$

and the vector of all parameters is

$$\boldsymbol{\theta} = (\theta_{E,pos}, \theta_{E,neg}, \theta_{C,pos}, \theta_{C,neg}, p_{pos}).$$

Different numerical configurations of a statistical estimate $\hat{\boldsymbol{\theta}}$ of $\boldsymbol{\theta}$ may motivate different precision treatment decisions by a physician, depending on each patient's biomarker status. This underscores the importance of conducting a randomized trial of E versus C that includes both *pos* and *neg* patients, although as data are accumulated during an enrichment trial the sample sizes of the four (treatment, biomarker) subgroups may be changed by a design's adaptive decision rules. In general, like

TABLE 7.2
Numerical illustration of 10 possible cases involving one binary biomarker that is either positive (pos) or negative (neg), an experimental treatment, E, and an active control treatment, C. Each case is characterized by the biomarker subgroup-specific mean survival times, in months, for E and C, and prevalence p_{pos} of biomarker-positive patients. Effects of E within the biomarker-positive and negative subgroups are $\Delta_{pos} = \theta_{E,pos} - \theta_{C,pos}$ and $\Delta_{neg} = \theta_{E,neg} - \theta_{C,neg}$. The overall effect is the weighted average $\Delta = p_{pos}\Delta_{pos} + (1 - p_{pos})\Delta_{neg}$. In all cases, $\theta_{C,pos} = \theta_{C,neg} = 24$ months.

	Mean Survival Time					
Case	$\theta_{E,pos}$	$\theta_{E,neg}$	p_{pos}	Δ_{pos}	Δ_{neg}	Δ
1	24	24	–	0	0	0
2	36	36	–	12	12	12
3	36	24	.10	12	0	1.2
4	84	24	.01	60	0	0.6
5	36	24	.50	12	0	6
6	36	30	.50	12	6	9
7	25	24	.50	1	0	.5
8	84	24	.40	60	0	24
9	60	48	.50	36	24	30
10	60	48	.05	36	24	24.6

OAR, any adaptive decision rules may introduce bias in the estimates of $\boldsymbol{\theta}$. This will be discussed below, in section 7.5. Table 7.2, which is similar to Table 2 of Thall (2021), illustrates 10 possible cases, each defined by fixed numerical values of $\boldsymbol{\theta}$.

For all cases, the mean survival time with C is 24 months, regardless of biomarker status. In Case 1, which may be considered a global null case, E has no effect, either overall or within either subgroup, and Z and p_{pos} are irrelevant. In Case 2, E provides a 12 month (50%) improvement in mean survival time over C in both subgroups, from 24 to 36 months, but again Z and p_{pos} are irrelevant. In Case 3 there is a biomarker-specific effect on mean survival time, with E providing a 12 month improvement over C if $Z = 1$, but E has no effect if $Z = 0$. While Z is very important in Case 3, only the 10% of patients who are biomarker positive benefit from E. If this were known with certainty or seen with high probability, then it would not make sense to give E to biomarker-negative patients. For example, the BRAF mutation, which is associated with colorectal cancer (CRC), encodes a serine–threonine protein kinase that is a downstream effector of activated KRAS, and consequently, BRAF is a popular target of therapies for CRC. The BRAF mutation has a low prevalence, however. Hsieh et al. (2012) cited BRAF rates of 6% to 13% in Spanish CRC populations, and 7% in Chinese or Greek CRC populations. An important point illustrated by Case 3 is that the overall average effect, which is $\Delta = 1.2$, is extremely misleading if considered alone, since what matters for therapeutic decision making is that $\Delta_{pos} = 12$ and $\Delta_{neg} = 0$. That is, the ability to do precision medicine by accounting for whether the patient is E-positive or not is very important in Case 3. This implies that using an adaptive enrichment design to identify a biomarker-positive subgroup and increase its sample size during a trial, and decrease the number of *neg* patients who receive E, is desirable in settings like Case 3. In practice, such adaptive rules are triggered by interim estimates showing that $\hat{\Delta}_{pos}$ is much larger than $\hat{\Delta}_{neg}$ and moreover that $\hat{\pi}_{neg}$ is small. Case 4 is a more extreme version of Case 3, with $\Delta_{pos} = 60$ months but $p_{pos} = .01$. One might call E a "home run" since it extends mean survival from to one to five years, but unfortunately it only does this in a very small subpopulation of 1% of patients who are biomarker positive. As a numerical illustration of what actually might happen in Case 4, a sample of 1000 patients randomized fairly between E and C would yield a subsample with an expected size of only 10 biomarker-positive patients, and consequently the reliability of either an estimate of $\hat{\Delta}_{pos}$ or a statistical test of $\Delta_{pos} = 0$ would be extremely low. Case 4 illustrates the difficulty of reliably detecting a large subgroup-specific benefit when the subgroup where it occurs is a very small subpopulation. The point is that, when the prevalence p_{pos} is very small, it may be very difficult to make a reliable inference about Δ_{pos}, even if the overall sample size is large. This underscores why adaptive enrichment by increasing the sample size of an identified biomarker-positive subgroup may be useful, since increasing the sample size of *pos* patients improves the reliability of inferences about the benefit Δ_{pos}. Case 5 also is like Case 3, but $p_{pos} = .50$ rather than .10 or .01. This is a much easier case to deal with statistically. The potential overall benefit of using E for treating biomarker-positive patients is much larger because, on average, there are a lot more of them. In Case 5, subsample size enrichment of biomarker-positive patients is far less useful. In Case 6, E provides a substantial advantage over C in both subgroups, but the expected survival benefit is twice as large in biomarker-positive patients, $\Delta_{E,pos} = 12$ months versus $\Delta_{E,neg} = 6$ months. This case shows why, perhaps counter-intuitively, a randomized trial should include both biomarker-positive and biomarker-negative patients. Despite what may be believed based on preclinical studies and the way that E was engineered, for unanticipated reasons Δ_{neg} may turn out to be meaningfully large, so E is beneficial in biomarker-negative patients. In practice, this sort of desirable anti-disease effect in biomarker-negative patients may be caused by a designed molecule that has multiple biological effects against a disease, only one of which is accounted for by the nominal biomaker. If the other effects could be identified, then there would be more than one biomarker to identify E-sensitive patients. In any case, regardless of what is known or not known about the biological effects of E, including nominally "biomarker negative" patients in the clinical trial provides an empirical way to detect an unanticipated desirable E-versus-C effect in *neg* patients. Case 7 is like Case 5 in that E provides a benefit over C only in biomarker-positive patients, but the expected improvement is only $\Delta_{pos} = 1$ month, a trivial clinical benefit. This makes using E in place of C nearly useless, and a waste of resources, especially if E is expensive to produce or has severe adverse side effects.

In each of Cases 8 – 10, the effect of E is very large in one or both biomarker subgroups. Case 8 is qualitatively identical to Case 7, but the numbers are very different. In Case 8, there is a subgroup of biomarker-positive "super-responders" who have an 84-24 = 60 month (five year) expected survival benefit with E compared to

C, they comprise 40% of all patients, and there is no E effect at all in the 60% of patients who are biomarker negative, with $\Delta_{neg} = 0$. Statistically, this is a very easy case to discover. In Case 8, it is highly likely that any reasonable adaptive design that prospectively accounts for Z and makes subgroup-specific inferences will conclude that $\Delta_{neg} = 0$, that Δ_{pos} is large, and that after the trial all future patients having $Z = 1$ will be treated with E and those having $Z = 0$ will be treated with C. While Case 8 is easy to deal with statistically, one still is quite likely to go astray if a conventional design that does not account for Z is used. If Z is ignored in Case 8, then a large estimate of the overall mean improvement $\hat{\Delta} = 24$ months, which doubles the overall mean survival time with C from 24 to 48 months, is very misleading, as in Case 3. This is simply because adding E to C only benefits biomarker-positive patients. In Case 8, if reliable estimates of $\boldsymbol{\theta}$ are obtained based on a large enough sample, $\hat{\Delta}$ would be likely to be close to 24 months. If the *pos* and *neg* subgroups are ignored, this would grossly underestimate $\Delta_{pos} = 60$ and grossly overestimate $\Delta_{neg} = 0$. In this case, it would be a waste of resources to continue to treat biomarker negative patients with E rather than C. Case 8 illustrates the importance of reliably identifying an E-sensitive subgroup, if it exists, estimating its prevalence, and not mistakenly concluding that E is superior to C in all patients when an overall beneficial effect is due entirely to a biomarker-positive subgroup. In general, if it can be inferred, interimly, that Δ_{pos} is substantively larger than Δ_{neg}, then enrichment of the $E + C$ subgroup is beneficial both during the trial and after its conclusion.

In Cases 9 and 10, E provides large advantages over C in both biomarker positive and biomarker negative patients, of $\Delta_{pos} = 36$ and $\Delta_{neg} = 24$ months. However, 40% of patients are biomarker positive in Case 9 compared to only 5% in Case 10. Consequently, in Case 10 the greatest advantage of giving E is seen in biomarker-negative patients, simply because there are a lot more of them. This is contrary to what may be anticipated if one relies on biomarker status alone, defined in terms of the biomarker that has been identified, without accounting for either prevalence or the possibility that E may benefit biomarker-negative patients.

Given the fact that most targeted agents are very expensive to produce, the question of whether a pharmaceutical company may consider it feasible or desirable to pursue development of a given targeted E is quite important. In many settings, the costs of preclinical experimentation to develop E in terms of time, money, and human resources are quite substantial. In terms of clinical evaluation, if the prevalence p_{pos} of E-sensitive patients turns out to be very small, then even if E is highly effective in E-sensitive patients it may not be economically worthwhile to produce the agent, unless patients or insurance companies pay a very high price for treatment with E. Consequently, a more useful approach would include consideration of both the disease prevalence and the expected number of people affected. For example, if the disease has annual prevalence .001 in a population of 200 million people and $p_{pos} = .10$, then one may expect 200,000 people to have the disease each year and 20,000 of these to be sensitive to E. So the actual number of people who may benefit from E is quite large, despite the low disease prevalence and small value of p_{pos}. Since lower values of either the disease prevalence or p_{pos} will reduce the number of people who have the disease and thus may benefit from E, from either a scientific or an economic viewpoint, this underscores the importance of obtaining reliable estimates of both p_{pos} and Δ_{pos}. For example, since the annual prevalence of influenza globally is about .05 to .20, although a given case of the flu may be due to numerous different strains of this class of viral diseases, the potential benefits of a highly effective vaccine or treatment targeting one or more specific strains are immense.

A common sociological problem seen with basic scientists is that reason may be clouded by optimism motivated by promising preclinical data about the effects of E at the molecular or cellular level, or using E to treat xenografted mice. This may lead to a randomized clinical trial of E versus C being conducted in biomarker-positive patients only. This optimistic approach of presumptive enrichment is dysfunctional if, unexpectedly, E also provides benefit in biomarker-negative patients, as illustrated above in cases where the known biomarker did an inadequate job of identifying E-sensitive patients. As noted, this may be due to the fact that many targeted agents act in more ways than attacking the one target for which they were engineered. A much worse approach is to conduct a non-randomized, single-arm trial of E in what are assumed to be E-sensitive patients. This may be called *presumptive enrichment*, which may be very bad scientific practice, as shown above in Case 1, that may lead to serious inferential errors with undesirable consequences for patients. Whatever is observed in molecules, cells, or mice may provide a useful basis for designing a clinical trial, but it cannot guarantee that a desired effect of E seen preclinically also will be seen in humans, or that a beneficial effect will not be seen in biomarker negative patients. Presumptively excluding putatively non-E-sensitive patients uses preclinical data as a basis for assuming that the effect Δ_{neg} in humans is negligible or 0. It also is based on the assumption that the discrimination function for identifying biomarker-positive patients never makes a classification error. A trial that does presumptive enrichment by ignoring the subgroups (E,neg) and (C,neg) cannot provide a basis for inferences about either Δ_{pos} or Δ_{neg}. Such a trial provides no data for making key inferences in any of cases 1, 2, 6, 7, 9, or 10 where Z has little or no relationship to clinical outcome because $\Delta_{pos} - \Delta_{neg}$ is 0 or small, or if $\Delta_{pos} = \Delta_{neg} = \Delta > 0$ is clinically meaningful (Case 2), or if $\Delta_{pos} > \Delta_{neg} > 0$ and Δ_{neg} is clinically meaningful (Cases 6, 9, and 10). In these cases, biomarker-negative patients would be deprived of the benefit of E. This also raises the question of whether conducting such presumptive enrichment trials based on preclinical data alone is ethical.

These numerical examples show the importance of obtaining reliable statistical estimates of the relevant parameters in $\boldsymbol{\theta}$, and more specifically doing precision medicine during a trial. This requires randomizing patients, reliably identifying an E-sensitive subgroup if it exists, and estimating comparative subgroup-specific treatment effects in both E-positive and E-negative patients. In Cases 3, 5, 7, or 10, in Table 7.1, given interim data to establish subgroups, an adaptive enrichment rule to treat a larger proportion of *pos* patients, or possibly to restrict subsequent enrollment to *pos* patients, would provide a better estimate of Δ_{pos}. The possibility of Cases 1, 3, 4, 5, or 7 suggests that it is desirable to use adaptive subgroup-specific rules to terminate enrollment of *neg* patients, or if appropriate of all patients, due to futility. However, the risk from using subgroup-specific futility rules is that, in Cases 2 or 6, incorrectly terminating accrual of *neg* patients and concluding that E is not superior to C in those patients when the magnitude of $\Delta_{neg} > 0$ is clinically meaningful would deprive future Z-negative patients of the benefit of E. That is, subgroup-specific futility rules should be constructed to have low false negative rates.

One reason for conducting a clinical trial is to make inferences to find out which case is most likely to be true. Table 7.2 shows that, given a reliable estimate of $\boldsymbol{\theta}$ from a well-designed trial, in some cases, physicians would have a well-informed basis for making precision treatment decisions using each patient's Z. The therapeutic value of using Z to make personalized decisions would be high in Cases 3, 4, 5, and 8, where the best decisions would be to treat biomarker-positive patients with E and biomarker-negative patients with C, which would avoid the treatment costs of adding E unnecessarily when it is not beneficial. In Cases 6, 9, and 10, E would be the best choice for all patients, but a better outcome could be expected in biomarker-positive patients. Here, Z still would be useful for predicting patient survival times.

Because adaptive enrichment designs account prospectively for possible treatment-subgroup interactions, they avoid *post hoc* data analyses to identify subsets where E may provide substantive benefit. When subgroup analyses are not planned ahead of time, discovery of a large treatment effect in some subgroup may be difficult to defend. As explained above in the discussion of selection bias, with many subgroups the maximum estimated treatment-subgroup effect among the subgroups will be large even if, in fact, there is no effect at all. Unfortunately, the practice of selecting a maximum without understanding its potentially misleading consequences is very common in medical research. There are many methods to correct for selection bias in *post hoc* searches for subsets with large treatment effects or treatment-covariate interactions. Bayesian hierarchical model-based approaches have been proposed by Dixon and Simon (1991) and Dixon and Simon (1992).

The multi-arm randomized phase II platform trial design of Lee et al. (2023), described above in Chapter 5, addresses these issues in a setting where the goals are screening and selection. In a phase III setting, more refined subgroup-specific adaptive rules are given in the GS enrichment design of Park et al. (2022), that uses both early response and progression-free survival time, along with a vector of patient covariates that may include biomarkers, to identify and refine an E-sensitive subgroup and test whether E provides an improvement over C in that subgroup. This design is reviewed below, in Chapter 8.

7.5 Estimation Bias

This section is not about making patient-specific decisions *per se*, but rather how adaptive decisions made during a clinical trial affect parameter estimates based on the trial's final data. Adaptively learning about how to use biomarkers to guide application of targeted agents is a basic idea underlying many enrichment designs, which include a variety of different types of sequentially adaptive decisions. The decision rules are "adaptive" in that current data on treatment, outcome, and covariates, $(X,Y,\mathbf{Z})$, from previous patients, and each newly enrolled patient's $\mathbf{Z}$, are used to make interim decisions. Many adaptive enrichment designs are simpler in that they are based on (X,Y) only, and do not involve patient covariates. In general, an adaptive *futility* rule stops accrual to a treatment found to be ineffective, and an adaptive *safety* rule stops accrual to a treatment found to be unsafe. The most common forms of these rules ignore $\mathbf{Z}$, so if accrual to an arm is stopped this is done for all patients. The examples given above illustrate how badly this "one size fits all" approach can go wrong. What may be less obvious is that, in general, making adaptive decisions during a trial can cause inferential problems at the end.

An important example is a randomized phase II screening trial of three experimental agents, E_1, E_2, and E_3, and a control, C, that includes adaptive futility and safety rules. See, for example, the design of Lee et al. (2023), discussed in Chapter 5. If interim data show that, for example, the E_1-versus-C efficacy effect is negligible, or that the E_1-versus-C toxicity difference is unacceptably high, then an enrichment rule will drop E_1 and subsequently increase the sample sizes of the remaining arms, E_2 and E_3. This then is followed by the confirmatory comparison of E_2 and E_3 to C. This is the approach taken in the STAMPEDE trial, described by James et al. (2009), which studied treatments for men with advanced prostate cancer, a very large randomized trial including many experimental treatments. In STAMPEDE, after an initial stage for evaluating safety, interim treatment comparisons are based on time to failure, defined as disease progression or death, and the trial design includes futility rules for termination of experimental treatment arms showing poor performance compared to the control arm. Final comparisons to the control are based on survival time data.

Any outcome adaptive rules applied during a clinical trial produce bias in parameter estimates computed at the end of the trial. This is the case for conventional phase II trials with futility stopping rules, and for GS phase III trials. Consequently, accounting for estimation bias is a major issue after a trial that includes adaptive rules for futility or safety monitoring, treatment selection, or subgroup selection, and it is an inherent problem in enrichment trials. Temporarily ignoring subgroups for simplicity, consider a simple conventional group sequential (GS) design for a randomized comparative two-arm trial of E versus C, as described generally in Jennison and Turnbull (2007). If a futility rule is included in the GS decision-making scheme, then the fact that the estimated difference $\hat{\Delta} = \hat{\theta}_E - \hat{\theta}_C$ between the response probabilities or mean survival times must be sufficiently large at each interim test to continue the trial causes the final statistic $\hat{\Delta}$ to over-estimate the true Δ. To see why this is so, first consider a simplified version of the trial conducted without any adaptive interim rules to stop the trial early if an interim estimate $\hat{\Delta}$ is too small. The fact that patients were randomized between E and C would ensure, by the statistical arguments given in Chapter 2, that the distribution of the estimator $\hat{\Delta}$ based on the final data will have a mean equal to the true Δ, that is, $\mathrm{E}(\hat{\Delta}) = \Delta$. See, for example, Fisher (1925), Rubin (1978), Rosenberger and Lachin (2004), or Chapter 6 of Thall (2020). Next, suppose that, instead, the trial has two stages with equal sample sizes of $N/2$ each, and that a futility stopping rule is applied after stage 1. The futility rule stops the trial and accepts the null hypothesis that $\theta_E = \theta_C$ if the interim estimator $\hat{\Delta_1} < c_1$ for some fixed cutoff c_1. Otherwise, if $\hat{\Delta}_1 \geq c_1$, the trial randomizes an additional $N/2$ patients between E and C and performs a final test that concludes $\theta_E = \theta_C$ is not true if the final estimator $\hat{\Delta}$ is sufficiently large, say using the rule $\hat{\Delta} > c_2$ or $\hat{\Delta} < -c_2$. If the futility rule does not stop the trial early, then only interim stage 1 data that give values of $\hat{\Delta}_1 \geq c_1$, large enough to continue the trial, are possible. This implies that, if the trial is continued, then $\mathrm{E}(\hat{\Delta}_1) > \Delta$ for true parameter Δ, since the expected value is really the conditional expectation $\mathrm{E}\{\hat{\Delta}_1 \mid \hat{\Delta}_1 > c_1\}$, and is not $\mathrm{E}\{\hat{\Delta}_1\}$. Since the conditional mean is bounded below by c_1, it must be larger than the unconditional mean $\mathrm{E}\{\hat{\Delta}_1\}$. If the estimator of Δ based on the stage 2 data is denoted by $\hat{\Delta}_2$, then the final estimator is $\hat{\Delta} = .50\hat{\Delta}_1 + .50\hat{\Delta}_2$. This implies that, regardless of what the final test of hypothesis concludes, if the trial is not stopped early by the futility rule then the upward bias of $\hat{\Delta}_1$ causes the final $\hat{\Delta}$ to have an expected value larger than Δ.

A similar problem arises in a two-arm enrichment trial comparing E to C where patient subgroups $\{S_1, \cdots, S_m\}$ have been identified before the start of the trial, possibly based on a biomarker vector $\mathbf{Z}$. Denote the true E-versus-C effects in the subsets by $\Delta_1, \cdots, \Delta_m$. If the final data from the trial are used to select the best subset, S_{j^*}, defined as that having the largest estimated E-versus-C effect, $\hat{\Delta}_{j^*}$, then the estimate of Δ in that subset will be upwardly biased. It is very important to bear in mind that the index j^* actually is a statistic, since it depends on the data through the subset-specific estimators $\hat{\Delta}_1, \cdots, \hat{\Delta}_m$.

Suppose that, in fact, the subsets actually do not matter in that they have identical E-versus-C effects, $\Delta_1 = \cdots = \Delta_m = \Delta$. In this case, $\hat{\Delta}_1, \cdots, \hat{\Delta}_m$ have identical distributions, all with mean Δ. But the fact that, by definition, the maximum $\hat{\Delta}_{j^*}$ must be larger than all of the other $\hat{\Delta}_j$'s implies that $\mathrm{E}(\hat{\Delta}_{j^*}) > \Delta$. For example, if 10 independent random variables are uniformly distributed between 0 and 1, so that each has a mean of .50, then the maximum of the 10 has an expected value of .91, not .50. Even if there are no treatment-subset interactions, a nominally "best" subset always can be identified, by choosing the j^* corresponding to the maximum among $\hat{\Delta}_1, \cdots, \hat{\Delta}_m$. Ignoring this elementary fact can lead to serious errors when making inferences about a selected nominally "best" E_{j^*}. These include estimation bias, miscalculation of a test's power, and the incorrect conclusion that E_{j^*} provides a treatment advance over C for patients in S_{j^*} when in fact it does not. This explains why the common practice of selecting a best subset based on *post hoc* data analyses, sometimes called "cherry picking", or "data dredging", is one of the major reasons why such results often cannot be replicated in later studies. In the context of adaptive enrichment trials, methods to correct for bias due to using the same dataset for both developing a classifier $\phi(\mathbf{Z}, X)$ and doing parameter estimation are given, for example, by Bai et al. (2017) and Zhang et al. (2017).

8

Adaptive Enrichment Designs

8.1 Adaptive Signature Designs

Freidlin and Simon (2005) proposed an adaptive signature design (ASD) for randomized clinical trials of molecularly targeted agents. The ASD is a two-stage design, with the following three goals:

Goal 1. Determine a subset of E-sensitive patients based on the stage 1 data.

Goal 2. Compare $E+C$ to C in the entire population using a test based on the data from all patients.

Goal 3. Compare $E+C$ to C in the E-sensitive subset that was determined in stage 1, based on the stage 2 data only.

Important requirements are that both comparative tests must control their overall Type I error probabilities to be $\leq .05$, and that they each must achieve specified power figures. There have been numerous extensions of the ASD, including the family of biomarker-based enrichment designs, which will be described later in this chapter.

Freidlin and Simon (2005) formulated the ASD model and decision structure for a two-stage design, as follows. For the ith patient, $i = 1, \cdots, N$, let $X_{i,E}$ denote the treatment E indicator, θ_i = Pr(Response), and $\mathbf{Z}_i = (Z_{i,1}, \cdots, Z_{i,K})$ a candidate biomarker vector. While they describe $\mathbf{Z}_i$ as a gene expression vector, the elements of $\mathbf{Z}_i$ can be quite general. The logistic regression model

$$\text{logit}(\theta_i) = \mu + X_{i,E}\left(\lambda + \sum_{k=1}^{K} \gamma_k Z_{i,k}\right)$$

is assumed. In this model, λ is the main E effect and γ_k is the additional E effect due to interaction between E and Z_k. The design accrues N_s patients in stage $s = 1$ and 2, uses the stage 1 data to develop a classifier, and applies it at the end of stage 2 to identify a subset of E-sensitive patients. For the two tests noted above, (1) overall comparison of $E+C$ to C is done by a test having type I error probability α_1, using the data from all $N_1 + N_2$ patients, and (2) comparison of $E+C$ to C in the selected E-sensitive subset is done by a test having type I error probability α_2, using only data from the subset of E-sensitive patients accrued in stage 2. If either test is significant, the trial is considered positive. The overall type I error, α, is bounded above by

DOI: 10.1201/9781003474258-8

$\alpha_1 + \alpha_2$. For example, one may use $\alpha_1 = .04$ and $\alpha_2 = .01$ to ensure an overall type I error rate $\leq .05$.

Freidlin and Simon (2005) proposed the following two-step algorithm for determining a set of E-sensitive patients.

Step 1. Using the fitted logistic regression model, for each $k = 1, \cdots, K$, declare biomarker Z_k to be "significant" if $\hat{\gamma}_k > c_1$, where c_1 is a fixed decision cutoff.

Step 2. For the stage 2 test, classify a patient as E-sensitive or not using the set of significant biomarkers selected in Step 1.

Step 2 is quite general, and a variety of different statistical rules may be used to define the set of biomarkers that characterize nominally "E-sensitive" patients. For example, one may declare the ith patient to be E-sensitive in Step 2 if $\hat{\gamma}_k Z_{i,k} > c_2$ for at least G of the significant biomarkers, where integer-valued G and real-valued c_2 are design parameters. This design has parameters $(N_1, N_2, \alpha_1, \alpha_2, c_1, c_2, G)$, which may be determined using various criteria, including achieving given power figures for the two tests, or deciding how large N_1 should be for given overall $N = N_1 + N_2$.

Subsequently, Freidlin et al. (2010) proposed a cross-validated ASD (CV-ASD), which generalizes the ASD and provides improved reliability. This generalization was motivated by three closely related problems. First, reliably determining a signature for E-sensitive patients based on a large dimensional $\mathbf{Z}$ requires a large sample. Second, if the proportion of E-sensitive patients is low, then for the test in the E-sensitive subset to have reasonably high power, the sample must be large. Finally, using the same dataset to determine an E-sensitive subset based on $\mathbf{Z}$ and test whether E provides an improvement may lead to inferential problems, including the distinction between the power of a final test assuming that the E-sensitive subset is known and the GP of the entire procedure.

The CV-ASD uses K-fold cross-validation (James et al., 2017) by first randomly partitioning the sample into K *validation cohorts*, $V_1, \cdots, V_K$, each consisting of $M = N/K$ patients. For each k, a signature based on $\mathbf{Z}$ is determined using the data in the complement of V_k, the *development cohort* $D_k = \{\cup_{r=1}^K V_r\} - V_k$. The signature obtained from D_k is applied to identify the subset, S_k, of E-sensitive patients in V_k. Since the sample consists of $\cup_{k=1}^K V_k$, each patient is classified as E-sensitive or not, and $S = \cup_{k=1}^K S_k$ is the set of all sensitive patients. A permutation test comparing $E + C$ to C then is carried out in S.

Freidlin et al. (2010) showed that the CV-ASD has much larger overall power than the ASD to detect very large treatment effects in cases where 10% of patients are sensitive, and the response probability is $\theta_{E,neg} = .25$ with E in the non-sensitive patients and $\theta_C = .25$. In this case, the CV-ASD using $K = 10$ validation cohorts has overall power (1) .71 compared to .35 with the ASD if the response probability $\theta_{E,pos} = .80$ in E-sensitive patients, and (2) overall power .91 compared to .60 with the ASD if $\theta_{E,pos} = .90$ in E-sensitive patients. Thus, the CV-ASD is useful when the prevalence of sensitive patients is small but the E effect is much larger in sensitive patients compared to nonsensitive patients.

8.2 Enrichment Designs with One Biomarker

Jiang et al. (2007) proposed a Biomarker Adaptive Threshold Design (BATD), with the aim to adapt the ASD of Freidlin and Simon (2005) to settings with one prespecified continuous biomarker, Z. The BATD is a phase III design that combines the ASD's test for an overall treatment effect with identification, validation, and estimation of a cutpoint for a single prespecified biomarker, Z, used to define an E-sensitive subpopulation.

Denote the hazard of death at time t with C by $h_C(t)$, let c_0 be a fixed cut-off for Z, and denote the treatment indicator X = 1 for E and 0 for C. The assumed model for the hazard of death with E for a patient with covariate Z is the step function

$$\log\{h_E(t \mid Z,X,c_0)\} = \log\{h_C(t)\} + \gamma X I(Z > c_0).$$

This reflects the strong assumption that E provides a benefit over C only in patients with $Z > c_0$. The log hazard ratio is $r(t) = \gamma I(Z > c_0)$, so $r(t) = 0$ if $Z < c_0$ and $r(t) = \gamma$ if $Z > c_0$. Since c_0 plays a central role in the BATD, to be useful it is essential that the assumed step-hazard model provides a reasonable representation of how Z actually works, and if so a reliable estimate of c_0 is needed. The BATD has three objectives, to determine whether

1. E is better than C for all patients,
2. E is better than C for an E-sensitive subset of patients determined adaptively by $(Z > c_0)$, or
3. E is not better than C, either overall or in a subset of E-sensitive patients.

Objective 2 requires establishing a value for the cut-off c_0, so that what is meant by Z being "sufficiently large" is made specific. Jiang et al. (2007) define two different but similar designs, procedures A and B, that combine two tests for the effect of E in the defined E-sensitive subgroup and overall. In procedure A, a test of $H_0 : r(t) = 0$ is conducted in all patients, with Type I error probability α_1. If this test rejects H_0, then the procedure is stopped. Otherwise, a second test of H_0 is conducted in the subset of E-sensitive patients defined by $Z > c_0$ with Type I error probability $\alpha_2 = \alpha - \alpha_1$, where α is an upper bound on the desired overall Type I error for both tests. In procedure B, which generalizes procedure A, the overall and subset test statistics are combined in the second stage to achieve greater efficiency by accounting for the correlation between the two test statistics.

The tests are constructed as follows. Assuming for simplicity that $0 < Z < 1$, for each of several candidate biomarker cutoff values $c_0 \in (0,1)$, the step-hazard model is fit to data from the subset of patients with $Z > c_0$ and a log likelihood ratio statistic $S(c)$ is calculated for the second test of $H_0 : \gamma = 0$, based on data from the patients in that subset. If the first test of Procedure B rejects $H_0 : r(t) = 0$, then an optimal cut-off c_0 is estimated as follows. Given a specified set of candidate cut-offs $0 < c_{0,1} < \cdots < c_{0,m}$, a test statistic is obtained by using the cut-off $c_{0,j}$ that maximizes the estimated E-versus-C effect. The test is fine-tuned by first adding a constant $R > 0$ to the stage 1 test statistic, which is $S(0)$, to obtain $S(0) + R$. This gives greater weight to the overall test statistic in stage 2, which is $T = \max\{[S(0)+R], \max_{0<c<1} S(c)\}$. In contrast, for procedure A the stage 2 test statistic is $T = \max_{0<c<1} S(c)$. Because standard distribution theory does not apply, Jiang et al. (2007) recommend that a permutation test be performed.

To evaluate procedures A and B of the BATD, Jiang et al. (2007) carried out a simulation study of a 200 patient trial with patients randomized fairly between E and C, Z distributed uniformly on [0, 1], outcomes generated from exponential distributions, administrative censoring from staggered entry between 10% and 20%, and using the candidate Z cut-off grid set $\{.1, .2, \ldots, .9\}$ for c_0. For procedure A, the values $\alpha_1 = .04$ and $\alpha_2 = .01$ were used. The second-stage biomarker-defined subset effect test was based on the permutation distribution of the maximized log likelihood ratio statistic with cutoff value restricted to (0.5, 1). This test focuses on treatment effects limited to a subpopulation that is unlikely to be detected by the test for an overall effect. For procedure B, R = 2.2 was used. The simulations evaluated scenarios with true cut-off values $c_0 = .25, .50, .75$, or .90, so the underlying model for the hazard function was assumed. To assess robustness, departures from the assumed model were evaluated, including a linear trend model in which $\log\{r(t)\}$ for benefit of E over C increases linearly as Z goes from 0 to 1; and a delayed trend model where $\log\{r(t)\} = 0$ for $Z < .5$ with $\log\{r(t)\}$ increasing linearly as Z goes from .5 to 1. Table 8.1 summarizes selected simulation results from Table 1 of Jiang et al. (2007) As expected, the overall test is best in Scenario 1 where there is no Z-specific subset effect, although both threshold-adaptive designs, procedures A and B, have similar power. In scenario 3 and especially scenario 5, the standard overall test has much smaller power than the threshold-adaptive designs. In scenario 6, where the HR increases linearly with Z, the three designs have similar power, with a slight advantage for the threshold-adaptive designs.

Simon and Simon (2013) proposed a general adaptive enrichment framework for both developing a classifier and using it to restrict enrollment when doing treatment comparisons. Their main focus was to preserve the overall Type I error rate. For each patient $i = 1, \cdots, N$, as above denote a single biomarker by Z_i, treatment indicator $X_i = 1$ for $E + C$ and $X_i = 0$ for C, and binary response indicator Y_i. Let $\theta_X(Z_i) = \Pr(Y_i = 1 \mid X, Z_i)$ denote the probability of response for a patient with biomarker Z_i treated with X. A discrimination function is defined as

$$f(Z_i) = I[\theta_C(Z_i) < \theta_{E+C}(Z_i)],$$

so $f(Z_i) = 1$ if the response probability for a patient with biomarker Z_i is larger with $E + C$ than with C, and $f(Z_i) = 0$ otherwise. Denoting an estimator based on

TABLE 8.1
Simulation results comparing the power figures of the biomarker adaptive threshold design procedures A and B and a conventional design with one overall test that ignores the biomarker. HR = E-to-C hazard ratio.

			Power		
				Procedure	
Scenario	True Model	HR	Overall Test	A	B
1	All patients	.80	.33	.30	.31
	benefit from E	.67	.78	.75	.73
		.57	.96	96	.94
3	Patients with $Z > .5$	.57	.50	.56	.61
	benefit from E	.40	.89	.93	.95
5	Patients with $Z > .9$	.40	.10	.24	.27
	benefit from E	.31	.16	.40	.41
		.21	.24	.63	.62
6	The HR increases	.57	.50	.50	.54
	linearly in Z	.40	.89	.89	.91
		.31	.97	.98	.99

the interim data from m patients who have been treated and evaluated by $\hat{f}_m(\mathbf{Z})$, the proposed design is as follows:

1. Randomize m_0 patients fairly between $E+C$ and C, and compute $\hat{f}_{m_0}$ from their data.
2. For $m > m_0$, compute the updated estimate $\hat{f}_m$ based on all accumulated data $\{(\mathbf{Z}_i, X_i, Y_i), i = 1, \cdots, m\}$.
3. Restrict trial entry to patients for whom $\hat{f}_m(\mathbf{Z}_i) = 1$, continuing this until a prespecified number N patients have been enrolled.
4. Perform a final test.

Defining the global null hypothesis $H_0 : \theta_{E+C}(Z) = \theta_C(Z)$ for all Z, they propose the test statistic

$$T = \sum_{i=1}^{N} [X_i Y_i + (1 - X_i)(1 - Y_i)],$$

which is the number of successes with $E+C$ plus the number of failures with C. Since T follows a binomial distribution with parameters $(N, .5)$ under H_0, by using appropriate cutoffs from this null binomial distribution to perform a test, the Type I error probability is controlled, regardless of how the discrimination function for adaptively changing the enrollment based on Z is defined.

To do adaptive threshold enrichment with a single biomarker Z, Simon and Simon (2013) suggested a practical approach for modeling f, since there are infinitely many possibilities for a "true" f. They started by assuming that the E effect $\Delta(Z) = \theta_{E+C}(Z) - \theta_C(Z)$ equals either 0 or δ, is monotone non-decreasing in Z, and jumps from 0 to δ at one of a set of candidate cutpoints, $c_1 < c_2 < \cdots < c_K$. At each interim decision, the cutpoint c_{j*} maximizing the likelihood of the current observed data is used as an estimate of the true cutpoint, and the rule $f(Z) = 1$ if $Z \geq c_{j*}$ and $f(Z) = 0$ if $Z < c_{j*}$ is used to restrict enrollment. As a practical matter, if the optimal cutpoint is not selected, but a nearby cutpoint is selected that still provides good discrimination between patients for whom the true $\Delta(Z)$ is large versus those for whom it is small, then the methodology has succeeded.

Simon and Simon (2013) provided simulations of this enrichment design for clinical trials with 200 patients, the biomarker uniformly distributed between 0 and 1, candidate cutpoints 1/(K+1), 2/(K+1), $\cdots$, K/(K+1), clinical outcome a binary response, and the adaptive enrichment rule applied once interimly at 100 patients. Accrual for the last 100 patients is restricted to "E-sensitive" patients having biomarkers $Z_i > c_{j^*}$. Their simulations show that, compared to a design that does not restrict enrollment using the biomarker, if 75% of patients are more likely to benefit from E, then the adaptive enrichment design has much larger power to detect differences $\Delta = \theta_{E+C} - \theta_C = .25$ or .30, in most cases considered. The price for restricting enrollment to E-sensitive patients defined in this way is that, for the sample size of 200, the trial duration will be increased, and this increase in duration will be larger if the proportion p_{pos} of E-sensitive patients is smaller. Simon and Simon (2013) also discussed extensions of their design to a group sequential structure with more than one interim decision, and settings with a continuous outcome rather than a binary response indicator.

8.3 A Hybrid Utility-Based Enrichment Design

Ondra et al. (2019) used Bayesian utility-based criteria to optimize frequentist single-stage and adaptive two-stage designs for randomized trials comparing a targeted therapy E to an active control C. They considered settings where a patient subgroup S of E-sensitive patients has been defined, presumably by one or more biomarkers. The complement consisting of non-E-sensitive patients, is denoted by S^c. The utility functions included costs of trial design, overall maximum sample size and, for both S and the overall population, (1) assumed true treatment effects, (2) fixed lower thresholds for effect sizes, and (3) indicators of testing procedures based on frequentist test statistics. The design was a hybrid that used frequentist test statistics and employed Bayesian machinery to choose sample sizes to maximize the expected value of a utility function. They considered one-sided tests of $H_F : \delta_F \leq 0$ for the full sample and $H_S : \delta_S \leq 0$ for the subpopulation of E-sensitive patients.

They defined two alternative utility functions for use as Bayesian decision criteria. The first utility reflected a pharmaceutical company whose goal is to maximize profit, while the second utility reflected a societal perspective with the goal to maximize cost-adjusted expected health benefits for a specified patient population. Each design included partial enrichment, defined as a prespecified proportion of patients in the trial recruited from the E-sensitive subgroup. Partial enrichment designs were derived by optimizing each of the two utilities, and compared to similar designs that either enforced full enrichment by restricting accrual to S, or did not do any enrichment. The "partial" and "full" enrichment designs are examples of presumptive enrichment, since they are based on the assumption that the subset S of E-sensitive patients that has been provided is correct.

For single-stage frequentist designs with partial enrichment, denote the subgroup sample sizes by n_S and n_{S^c}, with $n = n_S + n_{S^c}$. Thus, $\gamma = n_S/n$ is the empirical prevalence of the E-sensitive subgroup observed in the trial. Index the two patient subgroups by $j = S,\ S^c$, and the full population by $j = F = S \cup S^c$, and denote the E-versus-C effect estimates by $\hat{\delta}_j$ with standard errors $\sqrt{\nu}_j$. Each $W_j = \hat{\delta}_j/\sqrt{\nu}_j$ is approximately normally distributed, and may be used as a subgroup-specific test statistic. Denote the population prevalence of E-sensitive patients by λ. Denoting

$$\xi = \frac{\lambda^2}{\gamma} + \frac{(1-\lambda)^2}{1-\gamma},$$

the re-weighted test statistic for the overall population is

$$W_F = \frac{\frac{\lambda}{\sqrt{\gamma}} W_S + \frac{1-\lambda}{\sqrt{(1-\gamma)}} W_{S^c}}{\xi^{1/2}}.$$

For $j = S$ and S^c, denote $\nu_j = 2\sigma^2/n_j$, and $\nu_F = 2\sigma^2\xi/n$. The test statistic W_F is used to account for the possibility that standard Z-score statistics computed from the pooled sample may have a mean that does not equal 0, even if $\delta_F = 0$. However, in general (W_S, W_F) is asymptotically bivariate normal with mean $(\delta_S/\sqrt{\nu}_S,\ \delta_F/\sqrt{\nu}_F)$, variances 1, and covariances $\lambda/(\gamma\xi)^{1/2}$.

Ondra et al. (2019) used a conservative Bonferroni test to control type I error rate, and proposed the two tests

$$\psi_S = I(W_S \geq b_{\alpha/2}), \quad \text{and} \quad \psi_F = I(W_F \geq b_{\alpha/2},\ W_S \geq b_\eta,\ W_{S^c} \geq b_\eta)$$

The rule ψ_F rejects H_F only if there is a significant treatment effect in F at the Bonferroni adjusted level $\alpha/2$ and moreover, in each subgroup, there is a significant treatment effect at level η. Given these decision rules, a partially enriched single-stage design d is fully specified by the sample sizes (n_S, n_{S^c}) in the two subgroups. To obtain a fully enriched single-stage design, only biomarker-positive patients are enrolled and the decision functions are

$$\psi^f = (\psi_S^f, \psi_F^f) = I(W_F \geq b_\alpha, 0).$$

Similar two-stage group sequential rules are given, accommodating the need to split the sample sizes between the stages, account for GS testing, and for tests in S, S^c, and F.

For the two utilities, Ondra et al. (2019) first specified true treatment effects (δ_S, δ_F), $C(d)$ for the costs of a trial with design d, N the expected number of future patients treated, and μ_S, μ_F for minimally clinically relevant efficacy effect sizes, given the treatment costs and possible adverse effects. The societal benefit utility, $U_{societal}$, quantifies public health benefit,

$$
\begin{aligned}
U_{societal}(d) &= -C(d) + rN(\delta_F - \mu_F) \quad if\ \psi_F = 1 \\
&\quad -C(d) + r\lambda N(\delta_S - \mu_S) \quad if\ \psi_F = 0, \psi_S = 1
\end{aligned}
$$

with $U_{societal} = 0$ otherwise. The pharmaceutical company profit based utility was

$$
\begin{aligned}
U_{pharma}(d) &= -C(d) + rN(\hat{\delta}_F - \mu_F)^+ \quad if\ \psi_F = 1 \\
&\quad -C(d) + r\lambda N(\hat{\delta}_S - \mu_S)^+ \quad if\ \psi_F = 0, \psi_S = 1.
\end{aligned}
$$

Thus, the pharma company's utility does not depend on the true treatment effects. Instead, it depends on the treatment effect estimates and test decisions. The utility function U_{pharma} is defined to reflect the idea that, if the estimated effect is below the smallest clinically relevant value, then the drug will not be marketed, and consequently the utility is 0.

In the simulation study, attention was limited to the four pairs of effect sizes, $\boldsymbol{\delta} = (\delta_S, \delta_{S^c})$, in Table 8.2, where each pair corresponds to the effect sizes in the biomarker positive patient subset S and the biomarker negative subset S^c. These four pairs were assumed to be the support of the prior on $\boldsymbol{\delta}$, and two different priors on these four parameter pairs were assumed. One was nominally "weak" in that it assumed large effects and no effects were equally likely. The other prior was nominally "strong" in that it assumed a large prior probability of .60 on the parameter pair (δ_S, δ_{S^c}) = (.3, 0) reflecting an experimental treatment providing substantial benefit to biomarker-positive patients, and weak benefit to biomarker-negative patients. Denoting b = social or pharma, $\boldsymbol{\delta} = (\delta_F, \delta_S)$, with $\pi_0(\boldsymbol{\delta})$ the prior on the effect sizes, the mean utility for single-stage fully enriched design d is

$$
V_{\pi_0}^f = E_{\pi_0}\{E_{\boldsymbol{\delta}}[U_b(d)]\} = \sum_{\boldsymbol{\delta}} \int_{\mathbf{w}} U_b(d)\ f_d(\mathbf{w} \mid \boldsymbol{\delta}) d\mathbf{w}\ \pi_0(\boldsymbol{\delta})
$$

where $f_d(\mathbf{w} \mid \boldsymbol{\delta})$ is the distribution of the test statistics (W_S, W_{S^c}) given the effect sizes $\boldsymbol{\delta}$ and design $d = (n_S, n_{S^c})$. These definitions are suitably elaborated for a two-stage group sequential design. Given this structure, a design that is optimal under either $U_{societal}(d)$ or $U_{pharma}(d)$ may be obtained by maximizing the expected utility.

Ondra et al. (2019) illustrate the methodology assuming either a weak or a strong prior $\pi_0(\boldsymbol{\delta})$, and specifying clinically relevant thresholds $\mu_S = \mu_F = .10$. This was implemented by first assuming that all prior probability mass is on the four pairs

TABLE 8.2
Weak and strong priors $\pi_0(\delta_S, \delta_S^c)$ on the biomarker effect sizes.

Biomarker positive δ_S	0	.3	.3	.3
Biomarker negative δ_{S^c}	0	0	.15	.3
Weak Prior	.2	.2	.3	.3
Strong Prior	.2	.6	.1	.1

(δ_F, δ_S) = (0,0), (.3, 0), (.3, .15), and (.3, .3), and then assigning a prior probability mass to each. These priors are given in Table 8.2. They optimized the AEDs over stagewise sample sizes, each no smaller than 25, with a maximum overall sample size 265. They considered two scenarios. Assuming that the cost of a running a trial using a given design d takes the form

$$\kappa_{Total}(d) = \kappa_{setup} + \kappa_{biomarkerdevelop} + 2\kappa_{patient}(n^{(1)} + n^{(2)}) + \kappa_{screening}(d),$$

setting $n^{(2)} = 0$ for the single-stage design. Scenario 1 was characterized by biomarker screening costs of \$5000 and biomarker development costs of \$10,000,000. Scenario 2 set both of these values to 0. In both scenarios, for all designs considered, all expected utilities increased steeply and linearly in the population prevalence λ of biomarker sensitive patients.

The simulation results under Scenario 1 are summarized in Table 8.3. In all cases, mean per subgroup sample sizes in the subgroup are larger under $U_{societal}(d)$ compared to $U_{pharma}(d)$. A correct decision is to conclude there is efficacy in S only if there is no effect in S^c, and conclude there is an effect in the full population if there are effects in both S and S^c. Under the weak biomarker prior, the probability of a correct decision is larger under $U_{societal}(d)$ compared to $U_{pharma}(d)$. Under the strong biomarker prior, the optimal design under $U_{societal}(d)$ has a higher probability of a correct decision compared to $U_{pharma}(d)$, only in cases where the treatment effect is confined to the subgroup.

The subgroup-specific power figures in Table 8.3 are small in many cases. This indicates that, for these design settings and cases considered, the maximum overall sample size of 265 is inadequate. This illustrates the general fact that, when it is desired to make subgroup-specific decisions reliably, a larger sample size is required compared to what would be needed if homogeneity is assumed. Another important issue is that, when using Bayesian methods with a utility function to design a clinical trial, for the actual trial one may assume one utility and one utility only. Similarly, one may assume one prior and one prior only. This is in contrast with Bayesian data analysis, where sensitivity analyses may be done by carrying out the analyses repeatedly assuming different utilities to reflect different opinions about risk-benefits trade-offs, or different priors to reflect different prior beliefs. The simulation results strongly suggest that the societal benefit prior should be preferred, by both patients and regulatory agencies.

TABLE 8.3
Operating characteristics of adaptive two-stage designs of Ondra et al. (2019), optimized to maximize expected utility under either the weak or strong biomarker prior for the parameters. Computations reflect Scenario 1, with population prevalence λ = 0.5 for biomarker-positive patients.

	Societal Utility				Pharmaceutical Utility			
(δ_S, δ_F)	(0,0)	(.3,0)	(.3,.15)	(.3,.3)	(0,0)	(.3,0)	(.3,.15)	(.3,.3)
Weak Prior								
P(Futility Stop)	.58	.04	.02	.01	.43	.05	.03	.02
P(Full Enrich)	.12	.28	.08	.01	.10	.16	.06	.02
P(Part Enrich)	.30	.68	.90	.97	.47	.79	.90	.96
$\bar{n}_S$	164	212	199	179	162	174	174	172
$\bar{n}_{S^c}$	82	98	112	113	57	69	70	66
Power H_F	.01	.23	.63	.91	.01	.17	.45	.74
Power H_S Only	.01	.63	.27	.05	.01	.57	.34	.13
Strong Prior								
P(Futility Stop)	.64	.03	.02	.02	.50	.04	.03	.02
P(Full Enrich)	.25	.76	.59	.39	.16	.28	.14	.04
P(Part Enrich)	.11	.21	.38	.59	.34	.68	.83	.93
$\bar{n}_S$	188	237	231	221	164	187	187	184
$\bar{n}_{S^c}$	30	32	37	43	37	46	49	52
Power H_F	.01	.10	.26	.51	.01	.16	.38	.66
Power H_S Only	.01	.80	.64	.42	.01	.64	.44	.22

8.4 A Phase II-III Enrichment Design

Magnusson and Turnbull (2013) proposed an adaptive phase II-III group sequential enrichment design, called GSED, that compares an experimental treatment, E, to a control, C, within patient subgroups. Like many precision medicine designs, the GSED was motivated by the desire to avoid *post hoc* tests for treatment effects within subgroups. The use of such tests may produce an inflated overall false positive probability due to performing tests in multiple subgroups, and the small sample sizes within the subgroups make the tests unreliable and can lead to incorrect conclusions. The GSED includes both subgroup selection and testing, while controlling the familywise type I error rate (FWER), which is the probability of making at least one false conclusion in a set of multiple tests of hypotheses. The fact that it includes both selection and testing is why the GSED is nominally a phase II-III design. It uses upper and lower boundary spending functions to define group sequential test boundaries. Because the test statistics are defined using efficient scores, the design can accommodate settings with normal, binary, or time-to-event outcomes. The GSED requires a pre-specified partition that classifies patients into disjoint subgroups, denoted by $\{\Omega_1, \cdots, \Omega_K\}$, with the effect of E potentially differing between the subgroups. The subgroups may be determined from preclinical data, results of previous clinical trials, or from a biomarker vector. An important feature of the GSED is that, in its first stage, it eliminates subgroups where the data show that E has no effect. Simulations, given below, will show that this rule may produce false negative conclusions if a subgroup where E has an effect is eliminated in error in stage 1.

An important limitation of the GSED, that will arise later when evaluating several other precision medicine designs with the GSED as a comparator, is that it does not allow the possibility that the specified subgroups may be wrong. It thus does not include any provision for adaptively combining subgroups that interim data show have similar treatment-outcome parameters. This is not just an incidental feature of the AED, since it limits the inferences that may be made.

Denote the subgroup indices by $\mathscr{P} = \{1, \cdots, K\}$ and the prevalence of subgroup Ω_k by p_k, so $p_1 + \cdots + p_K = 1$. Any subpopulation of subgroups, $\mathscr{S} \subset \mathscr{P}$, is characterized by its subgroup indices, e.g. $\mathscr{S} = \{1,4,5\}$ is a subpopulation of $\mathscr{P} = \{1,2,3,4,5\}$. In general $\Omega_{\mathscr{S}} = \cup_{k \in \mathscr{S}} \Omega_k$ and $\mathscr{S}$ has prevalence $\sum_{k \in \mathscr{S}} p_k$. For each Ω_j, denote the response probabilities of E in the subgroups by $\theta_1, \cdots, \theta_K$, and the differences compared to C with response probability θ_0 by $\Delta_k = \theta_k - \theta_0$, for $k = 1, \cdots, K$. The effect in $\Omega_{\mathscr{S}}$ is the weighted average $\Delta_{\mathscr{S}} = \sum_{k \in \mathscr{S}} p_k \Delta_k$. In the above example, with $S = \{1, 4, 5\}$, $\Delta_S = p_1 \Delta_1 + p_4 \Delta_4 + p_5 \Delta_5$. For each Ω_k, the null hypothesis is $\Delta_k = 0$. For a given subpopulation $\mathscr{S}$, the composite null hypothesis $H_{0,S} : \Delta_{\mathscr{S}} = 0$ is tested against the one-sided alternative $H_{a,S} : \Delta_{\mathscr{S}} > 0$. The underlying idea is that, given the partition, the design identifies a possibly sensitive subpopulation $\mathscr{S}$ and then decides whether an an experimental treatment is effective in that subpopulation by performing a one-sided test.

The strategy for implementing the GSED is as follows. Stage 1 of the trial is conducted with no restriction of accrual, using equal randomization. A subpopulation, $\mathscr{S}^* \subset \mathscr{P}$, of E-sensitive patients then is determined adaptively using the stage 1 data. This is done using the standardized stage 1 efficient score statistics, denoted by $W_{1,1}, \cdots, W_{K,1}$ of the K subgroups. These are compared to a fixed stage 1 decision cut-off, ℓ_1, with subgroup k included in $\mathscr{S}^*$ if $W_{k,1} \geq \ell_1$ and excluded otherwise. Thereafter, in all subsequent group sequential stages, enrollment is restricted to $\mathscr{S}^*$, and the hypotheses $H_{0,\mathscr{S}^*} : \Delta_{\mathscr{S}^*} = 0$ versus $H_{a,\mathscr{S}^*} : \Delta_{S^*} > 0$ are tested using the updated efficient score statistics.

A key feature of the GSED is that the hypothesis to be tested isn't known at the start but rather is determined adaptively by identifying $\mathscr{S}^*$. This is the case with all adaptive designs that first use the early data to identify a subgroup or subgroups where treatment effects will later be tested, so the hypotheses are statistics because they are determined by the early data. An important limitation of the GSED is that, once $\mathscr{S}^*$ is determined in stage 1, this set is not refined in any subsequent stage by either dropping or adding subgroups adaptively. The GSED allows $H_{0,\mathscr{S}^*}$ to be rejected early, after only one stage, with the conclusion that $\Delta_{S^*} > 0$, that E provides an improvement in the identified subpopulation $\mathscr{S}^*$. If not, then stage 2 proceeds with enrollment restricted to $\mathscr{S}^*$, excluding future patients not in $\mathscr{S}^*$ who do not appear likely to benefit from E. The subsequent comparative tests of the group sequential design use all available data. Overall family wise error rate (FWER) is defined as the maximum probability of incorrectly rejecting at least one $H_{0,\mathscr{S}}$, when $\Delta_j = 0$ for all $j \in S$. The GSD controls the FWER by using a bootstrap algorithm to obtain point and interval estimates of treatment effect parameters, which then are adjusted for selection bias.

8.5 Biomarker Subgroups as Random Partitions

Xu et al. (2016b) proposed a Bayesian subgroup adaptive enrichment design, SUBA, for choosing each patient's treatment from a set of targeted therapies, $\mathscr{T} = \{1, \cdots, T\}$ based on a vector, $\mathbf{Z} = (Z_1, \cdots, Z_K)$, of numerical valued biomarkers available for each patient at enrollment. The idea underlying SUBA is that there may exist subgroups of patients, determined by $\mathbf{Z}$, who respond differently to each of the T treatments. Thus, choosing an optimal treatment should be done in a patient-specific way based on $\mathbf{Z}$, rather than choosing one nominally optimal "one size fits all" treatment that is likely to be suboptimal for many values of $\mathbf{Z}$. SUBA identifies subgroups having different response rates adaptively using a random partition of $\mathbf{Z}$ and assigns an optimal personalized treatment from $\mathscr{T}$ to each new patient based on their $\mathbf{Z}$ vector. The partition is used to define response probability as a function of the combination (biomarker subgroup, treatment). When a new patient is enrolled and their biomarker vector $\mathbf{Z}$ is recorded, based on the posterior of the current partition from the most recent data, the best treatment in $\mathscr{T}$ for the patient is identified by maximizing the posterior predictive response probability computed for all treatments $X \in \mathscr{T}$. As treatments are assigned adaptively on this basis and the patients' binary response variables are observed during the trial, SUBA uses this information to repeatedly refine the random partition of possible $\mathbf{Z}$ vectors. SUBA uses the well-established idea of a Bayesian regression tree (Chipman et al., 1998; Denison et al., 1998) to construct subgroups characterized as a partition of the $\mathbf{Z}$ vectors. The structure of the Bayesian partition model can be traced back to Müller et al. (2011). SUBA's approach of identifying subgroups adaptively during the trial using $\mathbf{Z}$ contrasts with designs, like the GSED, that begin with a predefined set of subgroups.

Denote the treatment response indicator by Y, with $\theta_X(\mathbf{Z}) = \Pr(Y = 1 \mid X, \mathbf{Z})$ the probability of response for a patient with biomarker vector $\mathbf{Z}$ treated with $X \in \mathscr{T}$. The design's goal is to choose an optimal treatment for each patient adaptively by accounting for the patient's biomarker vector $\mathbf{Z}$ when they are enrolled in the trial, and for all future patients after the trial's completion. SUBA does not suffer from the problems that arise with OAR, essentially because SUBA does sequential covariate-adaptive treatment optimization, rather than randomization to obtain unbiased treatment comparisons. To avoid confusing treatment index and patient index, denote the treatment given to the ith patient by $X_{[i]} \in \mathscr{T}$. Denote the data from the first n patients in the trial by

$$\mathscr{D}_n = \{(X_{[i]}, \mathbf{Z}_i, Y_i) : i = 1, \cdots, n\},$$

and denote the posterior predictive value of the response probability $\theta_X(\mathbf{Z})$ by

$$\hat{\theta}_X(\mathbf{Z}, \mathscr{D}_\mathbf{n}) = \mathbf{E}\{\theta_\mathbf{X}(\mathbf{Z}) \mid \mathscr{D}_\mathbf{n}\}.$$

The $n+1^{st}$ patient's treatment is chosen adaptively to be optimal for their covariate vector $\mathbf{Z}_{n+1}$, given formally as

$$X^{opt}(\mathbf{Z}_{n+1}) = \underset{x \in \mathscr{T}}{argmax}\ \hat{\theta}_x(\mathbf{Z}_{n+1}, \mathscr{D}_n).$$

Rather than using predefined fixed subgroups as done, for example, by the BATTLE and GSED designs, SUBA derives and repeatedly refines patient subgroups adaptively during the trial as new data are obtained. Let

$$\mathscr{P} = \{A_1, \cdots, A_M\}$$

be a random partition of the K-dimensional set of possible $\mathbf{Z}$ vectors. Each A_m is a subset of $\mathscr{R}^K$, the K-dimensional real numbers of possible biomarker vector values. Thus, the A_m's are mutually disjoint and $\cup_{m=1}^M A_m = \mathscr{R}^K$, or possibly a K-dimensional subset of $\mathscr{R}^K$ if, for example, all $Z_k > 0$. To compute $X^{opt}(\mathbf{Z}_{n+1})$, the posterior distribution of $\mathscr{P}$ is updated repeatedly as $\mathscr{D}_n$ is expanded during the trial. Each newly refined partition of $\mathbf{Z}$ corresponds to a set of prognostic subgroups. This, in turn, is the basis for assigning each new patient to the subgroup-specific treatment that is best based on $\mathscr{D}_n$.

The partition is exploited as follows. The probability of response is defined conditionally, not using the patient's biomarker vector $\mathbf{Z}_i$ directly, but rather in terms of the subset that contains $\mathbf{Z}_i$ in the current biomarker vector partition $\mathscr{P}$. Formally,

$$\theta_X(m) = Pr(Y_i = 1 \mid \mathbf{Z}_i \in A_m,\ X_{[i]} = X,\ \mathscr{P}). \tag{8.1}$$

Thus, the dependence of each $\theta_X(m)$ on the partition $\mathscr{P}$ plays a central role in SUBA. Denote the outcome, biomarker, and treatment vectors by

$$\mathbf{Y}^{(n)} = (Y_1, \cdots, Y_n),\quad \mathbf{Z}^{(n)} = (\mathbf{Z}_1, \cdots, \mathbf{Z}_n),\quad \mathbf{X}^{(n)} = (X_{[1]}, \cdots, X_{[n]})$$

and the response probability vector by

$$\boldsymbol{\theta} = \{\theta_x(m) : x = 1, \cdots, T,\ m = 1, \cdots, M\}.$$

During the trial, SUBA derives and repeatedly refines subsets in the random partition $\mathscr{P}$ of $\mathbf{Z}$. At each stage, the partition $\mathscr{P}$ is refined as a function of the current data $\mathscr{D}_n$. Given this, conditioning on $\mathscr{P}$ in the definition of $\theta_t(m)$ in (8.1) does away with the need to specify a functional form for the dependence of $\theta_t(\mathbf{Z}_i)$ on the vector $\mathbf{Z}_i$, as in a usual regression model. Instead, the response probability $\theta_t(m)$ is identified with the index m of the subset $A_m \in \mathscr{P}$ to which $\mathbf{Z}_i$ belongs. The dimension of $\boldsymbol{\theta}$

thus is TM, where T = number of treatments is fixed, and M = number of sets in $\mathscr{P}$, which may change adaptively during the trial as data accumulate and $\mathscr{P}$ is modified.

Denote the response (Y =1) and non-response (Y = 0) counts for all combinations of partition subsets and treatments (m,X) by

$$N_{m,X,y} = \sum_{i=1} I(\mathbf{Z}_i \in A_m, X_{[i]} = X, Y_i = y), \quad m = 1, \cdots, M, \ X = 1, \cdots, T,$$

so the total number of patients in subset m treated with t is $N_{m,X} = N_{m,X,1} + N_{m,tX,0}$. Given the partition, the conditional likelihood function can be written as the product

$$p(\mathbf{Y}^{(n)} \mid \mathbf{Z}^{(n)}, \mathbf{X}^{(n)}, \boldsymbol{\theta}, \mathscr{P}) = \prod_{X=1}^{T} \prod_{m=1}^{M} \{\theta_X(m)\}^{N_{m,X,1}} \{1 - \theta_X(m)\}^{N_{m,X,0}}.$$

Denoting the prior on $\boldsymbol{\theta}$ given the partition by $p(\boldsymbol{\theta} \mid \mathscr{P})$ and the hyperprior on the partition by $p(\mathscr{P})$, the full Bayesian hierarchical probability model is

$$p(\mathbf{Y}^{(n)}, \boldsymbol{\theta}, \mathscr{P} \mid \mathbf{X}^{(n)}, \mathbf{Z}^{(n)}) \propto p(\mathbf{Y}^{(n)} \mid \mathbf{Z}^{(n)}, \mathbf{X}^{(n)}, \boldsymbol{\theta}, \mathscr{P}) p(\boldsymbol{\theta} \mid \mathscr{P}) p(\mathscr{P}).$$

For the level 1 prior, it is assumed that $\theta_X(m) \sim$ iid Beta(a,b), with values such as $a = b = 1$ or $a = b = 1/2$ to ensure that it is suitably non-informative. The partition $\mathscr{P}$ is obtained by performing a sequence of recursive binary splits of the covariates $\mathbf{Z}$ to form a regression tree. Each node of the tree is a subset of $\mathscr{R}^K$, and it is either (1) a final leaf (bottom node) which defines one of the partition subsets A_m, or (2) it is split into two descendants. In case (2), the two descendants are determined by first randomly selecting a biomarker Z_k, and then splitting the current subset into two components by comparing each $Z_{i,k}$ to the median of the $Z_{i,k}$'s from the data points in A_m. Repeating this process, a sequence of splits generates a partition $\mathscr{P}$ of $\mathscr{R}^K$.

The hyperprior on $\mathscr{P}$ is determined by the probabilities $\boldsymbol{\nu} = (\nu_0, \nu_1, \cdots, \nu_K)$ that define the algorithm in the tree's random splitting rules, with $\nu_0 + \nu_1 + \cdots + \nu_K = 1$. At each stage of the adaptive splitting process, given the current subset partition $\mathscr{P}$, each set A_m either (1) is not split further with probability ν_0, or (2) with probability ν_k the biomarker Z_k is used to split A_m into two subsets. In practice, the values $\nu_k = 1/(K+1)$ can be used. For each m, let median$_k(A_m)$ denote the median of the $Z_{i,k}$'s computed from the set all data points in A_m, defined conditionally given the current partition. If, for some m, A_m is split using the kth biomarker, Z_k, then the partition is refined to contain two new subsets that replace A_m, defined as $A_{m,1} = \{i : Z_{i,k} < \text{median}_k(A_m)\}$ and $A_{m,2} = \{i : Z_{i,k} \geq \text{median}_k(A_m)\}$.

As a toy illustration of the splitting process that generates successive partitions, suppose that $\mathbf{Z} = (Z_1, Z_2)$, and three successive splitting steps are done, as follows, based on the prior probability vector $\boldsymbol{\nu} = (\nu_0, \nu_1, \nu_2)$. Let A_0 denote the set of all possible values of $\mathbf{Z}$.

Step 1. Z_1 is chosen randomly by $\boldsymbol{\nu}$ to do a first split, resulting in a partition with 2 subsets, $\mathscr{P}_1 = \{A_1, A_2\}$, where $A_1 = \{i : \ Z_{i,1} \leq \text{median}_1(A_0)\}$ and $A_2 = \{i : \ Z_{i,1} > \text{median}_1(A_0)\}$. Thus, M = 2 after step 1.

Step 2. Under $\boldsymbol{v}$, it is decided randomly to do two splits, with A_1 split into $\{A_{1,1}, A_{1,2}\}$, and A_2 split into $\{A_{2,1}, A_{2,2}\}$, producing the new partition $\mathscr{P}_2 = \{A_{1,1}, A_{1,2}, A_{2,1}, A_{2,2}\}$, so $M = 4$ after step 2.

Step 3. There is one final split, of $A_{2,2}$ into $\{A_{2,2,1}, A_{2,2,2}\}$ so $\mathscr{P}_3 = \{A_{1,1}, A_{1,2}, A_{2,1}, A_{2,2,1}, A_{2,2,2}\}$, and $M = 5$.

This final partition has probability

$$p(\mathscr{P}_3) \propto v_1 \times v_1 v_2 \times v_0 v_0 v_0 v_1 = v_0^3 v_1^3 v_2.$$

This construction may be elaborated slightly by multiplying the partition probability by ϕ^K for $0 < \phi < 1$, to favor partitions that re-split on the same biomarker, so the modified version of the above probability would be

$$p(\mathscr{P}_3) \propto v_0^3 v_1^3 v_2 \phi^2.$$

As the data in the trial accumulate, at each step the posterior of $\mathscr{P}$ is updated based on the current data $\mathscr{D}_n = \{(t_{[i]}, Y_i, \mathbf{Z}_i) : i = 1, \cdots, n\}$ from all previous patients. The posterior predictive probability of response with treatment X for a future $n+1^{st}$ patient with biomarker profile $\mathbf{Z}_{n+1}$ given treatment $X_{[n+1]} = X$ is

$$q(X, \mathbf{Z}_{n+1}) = \Pr(Y_{n+1} = 1 \mid \mathbf{Z}_{n+1}, X_{[n+1]} = X, \mathscr{D}_n).$$

This is computed by averaging over the posterior of $\mathscr{P}$, as follows. Denoting the set of all possible partitions by Π,

$$q(X, \mathbf{Z}_{n+1}) = \sum_{\mathscr{P} \in \Pi} p(Y_{n+1} = 1 \mid \mathbf{Z}_{n+1}, X_{[n+1]} = X, \mathscr{D}_n, \mathscr{P}) p(\mathscr{P} \mid \mathscr{D}_n),$$

where the posterior distribution of the partition is

$$p(\mathscr{P} \mid \mathscr{D}_n) \propto p(\mathscr{P}) \prod_{m=1}^{M} \prod_{x=1}^{T} \int_{\theta_x(m)} p(Y_i \mid X_{[i]} = x, \theta_x(m)) \, dp(\theta_X(m)),$$

bearing in mind that each $\theta_x(m)$ is a function of $\mathscr{P}$. Denoting the beta function $B(a,b) = \Gamma(a)\Gamma(b)/\Gamma(a+b)$ where $\Gamma(\cdot)$ is the gamma function, the above posterior of the partition can be written in the simple conjugate form

$$p(\mathscr{P} \mid \mathscr{D}_n) \propto p(\mathscr{P}) \prod_{m=1}^{M} \prod_{X=1}^{T} \frac{B(a + N_{m,X,1}, b + N_{m,X,0})}{B(a,b)}.$$

A further simplification is obtained by the computation

$$p(Y_{n+1} = 1 \mid \mathbf{Z}_{n+1}, X_{[n+1]} = X, \mathscr{P}, \mathscr{D}_n) = \sum_{m=1}^{M} I(\mathbf{Z}_{n+1} \in A_m) \frac{a + N_{m,X,1}}{a + b + N_{m,X}}.$$

Xu et al. (2016b) applied SUBA to design a trial with $T = 3$ treatments for breast cancer patients who previously received neoadjuvant systemic therapy (NST) and surgery. A vector $\mathbf{Z} = (Z_1, Z_2, Z_3)$ of quantitative protein biomarker levels were measured from biopsy samples taken after NST but before surgery, using reverse phase protein arrays. The three targeted treatments were ($X = 1$) a polymerase inhibitor designed to affect DNA repair and cell death programming (apoptosis), ($X = 2$) a PI3K pathway inhibitor to affects cell growth, proliferation, differentiation, and survival, and ($X = 3$) a cell cycle inhibitor targeting the cell cycle pathway. For this application, the partition $\mathscr{P}$ was constrained to have $M \leq 8$ biomarker subgroups, in order to avoid subsets with very small numbers of patients. Thus, there were only three rounds of random splits.

For trial conduct, SUBA begins with a burn-in with N_1 patients randomized fairly among the T treatments, to obtain data at the start of the trial. This is followed by adaptive decision-making using each patient's $\mathbf{Z}$ vector, including both treatment assignment and application of a futility rule to drop treatments that, based on the interim data, are found to be uniformly not promising. The futility rule is constructed as follows. Denote the observed values of the kth biomarker for the n patients in $\mathscr{D}_n$ by $\mathbf{Z}_k^{(n)} = (Z_{1,k}, \cdots, Z_{n,k})$. First, an equally spaced grid of size H_0 between the smallest and largest values of $\mathbf{Z}_k^{(n)}$ is constructed. The Cartesian product of these K grids is a K-dimensional grid $\tilde{\mathbf{Z}}^{(n)}$ of size H_0^K. Denote the list of all gridpoints by $\tilde{\mathbf{Z}}_h^{(n)}$, $h = 1, \cdots, H$. After the initial burn-in, the posterior predictive response probability $q(t, \tilde{\mathbf{z}}_h^{(n)})$ is computed for each $\tilde{\mathbf{z}}_h^{(n)}$. If a treatment t^* is predicted to be uniformly inferior to all other treatments, using the criterion

$$q(X^*, \tilde{\mathbf{z}}_h^{(n)}) < q(X, \tilde{\mathbf{z}}_h^{(n)}) \text{ for all } h = 1, \cdots, H \text{ and all } X \neq X^*, \tag{8.2}$$

then X^* is considered to have *uniformly unacceptably low efficacy* and is dropped from $\mathscr{T}$. If, at some interim analysis, only one acceptable treatment remains, then at that point the trial is stopped early.

An optimal treatment for the $n+1^{st}$ patient is chosen by maximizing $q(X, \mathbf{Z}_{n+1})$ over $X \in \mathscr{T}$. Thus, SUBA does sequentially adaptive treatment discovery, futility monitoring, and biomarker-specific treatment assignment.

Steps for designing and conducting a trial using SUBA are as follows:

Step 1. Specify the elements of the biomarker vector $\mathbf{Z}$, the T treatments, the beta prior hyperparameters (a, b), partition probabilities $\boldsymbol{v}$ that determine the hyperprior on $\mathscr{P}$, maximum sample size N, and burn-in subsample size $N_1 < N$.

Step 2. For the initial burn-in, randomize N_1 patients fairly among the T treatments, ignoring $\mathbf{Z}$.

Step 3. Discard any treatment X that has uniformly inferior response rate $q(X, \mathbf{Z})$ for $\mathbf{Z}$ in all subgroups according to the futility rule (8.2), and enrich the remaining treatments.

Step 4. For each new $n+1^{st}$ patient, update $\mathscr{P}$ based on $\mathscr{D}_n$, and choose the patient's treatment by maximizing $q(X, \mathbf{Z}_{n+1})$ over $X \in \mathscr{T}$.

TABLE 8.4
Simulation results comparing the ER, OAR, Regression, and SUBA designs in terms of mean number of patients assigned to treatment $X = 1, 2, 3$ within each subset. The number of patients given their truly optimal treatment in each case is given in **blue**.

	ER			OAR			Regression			SUBA		
t	1	2	3	1	2	3	1	2	3	1	2	3
	Scenario 1 ($X = 1$ optimal for all patients)											
–	67	67	67	83	65	52	120	70	10	177	19	4
	Scenario 2 ($X = 1$ optimal in A_1^{true} and $X = 3$ optimal in A_2^{true})											
A_1^{true}	33	33	33	33	33	33	35	33	32	73	18	9
A_2^{true}	33	33	33	33	33	33	35	33	32	8	18	74
	Scenario 3 (X optimal in A_t^{true} for all $X = 1, 2, 3$)											
A_1^{true}	19	19	19	22	17	18	19	16	23	41	9	8
A_2^{true}	25	25	25	21	27	28	24	22	30	14	36	26
A_3^{true}	22	22	22	25	21	21	21	20	26	11	11	44

Step 5. At the end of the trial, provide the final biomarker vector partition $\mathscr{P} = \{A_1, \cdots, A_M\}$ and the optimal treatment allocation rule to guide future personalized treatment selection. This rule treats a patient who has biomarker vector $\mathbf{Z}$ with the $X^{opt}(\mathbf{Z})$ that maximizes $q(X, \mathbf{Z})$ over all elements of $\mathscr{T}$ in the final acceptable treatment set.

Xu et al. (2016b) reported a simulation study showing that SUBA compares very favorably to designs using ER or OAR, and a design based on probit regression of Y on $\mathbf{Z}$ and X. The simulation study was designed as follows. A maximum of N= 300 patients were assigned to $T = 3$ treatments using $K = 4$ biomarkers (Z_1, Z_2, Z_3, Z_4), with each biomarker simulated as a random variable uniformly distributed between -1 and +1. All $\theta_X(m)$ were assumed to follow a Beta(1,1) prior, and the hyperprior probabilities for $\mathscr{P}$ were $v_k = 1/5$ for each $k = 0, 1, 2, 3, 4$, with $\phi = .5$. The trial began with equal randomization of $N_1 = 100$ patients among the three treatments. For the grid $\tilde{\mathfrak{z}}$, $H_0 = 10$ equally spaced points in the domain of each biomarker subspace were used, producing $H = 10{,}000$ grid points in the 4-dimensional grid of $\tilde{\mathfrak{z}}^{(n)}$ points. Of six scenarios studied by Xu et al. (2016b), results from the first three are summarized here in Table 8.4. In scenarios 1 and 2, biomarkers $Z_{i,1}$ and $Z_{i,2}$ affect $\theta_X(\mathbf{Z}_i)$ but $Z_{i,3}$ and $Z_{i,4}$ do not. Denoting the normal cdf with mean μ and standard deviation σ by $\Phi_{\mu,\sigma}(\cdot)$, the assumed true response probabilities were

$$\begin{aligned}
\theta_1(\mathbf{Z}_i) &= \Phi_{0,1.5}(Z_{i,1} + 1.5Z_{i,2}) \\
\theta_2(\mathbf{Z}_i) &= \Phi_{0,1.5}(Z_{i,1}) \\
\theta_3(\mathbf{Z}_i) &= \Phi_{0,1.5}(Z_{i,1} - 1.5Z_{i,2}).
\end{aligned}$$

These assumptions imply that $X = 3$ is always optimal when $Z_{i,2} < 0$, the three treatments have equal response probabilities when $Z_{i,2} = 0$, and $X = 1$ is superior when $Z_{i,2} > 0$. In scenario 1, $Z_{i,2} \equiv .8$ for all i, and in scenario 2 the $Z_{i,2}$'s are randomly generated as described above. For scenario 3, more complex interaction models were assumed, with

$$\begin{aligned}
\theta_1(\mathbf{Z}_i) &= \Phi_{0,1.5}(Z_{i,1} + 1.5Z_{i,2} - .5Z_{i,3} + 2Z_{i,1}Z_{i,3}) \\
\theta_2(\mathbf{Z}_i) &= \Phi_{0,1.5}(-Z_{i,1} - 2Z_{i,3}) \\
\theta_3(\mathbf{Z}_i) &= \Phi_{0,1.5}(Z_{i,1} - 1.5Z_{i,2} - 2Z_{i,1}Z_{i,2}).
\end{aligned}$$

As comparators, Xu et al. (2016b) studied designs with ER and OAR, and a design based on a probit regression model, all using the same burn-in of $N_1 = 100$ patients with ER. The OAR design used three predefined fixed biomarker subgroups, similarly to the BATTLE trial, given by $\{Z_{i,1} < -.5\}$, $\{-.5 \leq Z_{i,1} \leq +.5\}$, and $\{Z_{i,1} > +.5\}$. Denoting the true response probabilities in these fixed subgroups, indexed by $b = 1,2,3$, for the OAR design by $p_t(b)$, it was assumed that

$$Y_i \mid \mathbf{Z}_i, X \sim binomial(n_X(b), p_X(b)), \text{ and } p_X(b) \sim iid\ beta(1,1).$$

Denote $N_{X,b,1}$ = number of responses, $N_{X,b,0}$ = number of non-responses and $N_{X,b} = N_{X,b,1} + N_{X,b,0}$. The OAR probability of a patient in subgroup b for each X =1, 2, 3 was defined to be

$$\frac{\hat{p}_X(b)}{\hat{p}_1(b) + \hat{p}_2(b) + \hat{p}_3(b)}$$

where $\hat{p}_X(b) = (N_{X,b,1} + 1)/(N_{X,b} + 2)$, the posterior mean. For the Reg design, response probabilities were modeled as

$$p(Y_i = 1 \mid \mathbf{Z}_i, X) = \Phi(\beta_0 X_{[i]} + \boldsymbol{\beta}_1 \mathbf{Z}_i),$$

using maximum likelihood estimates of β_0 and $\boldsymbol{\beta}_1$. After the burn-in, a treatment t is chosen to maximize each new response probability estimate $\Phi(\hat{\beta}_0 t + \hat{\boldsymbol{\beta}}_1 \mathbf{Z}_i)$.

To evaluate the designs, denote the number of patients assigned to X in the dth simulation iteration, $d = 1, \cdots, 1000$, after the run-in phase, by

$$NP_X^d = \sum_{i=101}^{1000} I(X_i^* = X)$$

for $X = 1,2,3$. The average number of patients assigned to X is

$$\overline{NP}_X = \frac{1}{1000} \sum_{d=1}^{1000} NP_X^d,$$

given in Table 8.4. The results show that, in terms of the number of patients given the best treatment, SUBA is greatly superior to ER, OAR, and the regression-based method. In scenario 1, where $X = 1$ is uniformly most effective for all subgroups, most patients are allocated to $X = 1$ by SUBA. In scenario 2, the subsets are A_1^{true}

$= \{i : Z_{i,2} > 0\}$ and $A_2^{true} = \{i : Z_{i,2} < 0\}$, so $\{A_1^{true}, A_2^{true}\}$ is a partition of the simulation truth. In this case, SUBA assigns about twice as many patients to the best treatments as the three comparators. In scenario 3, the partition of the simulation truth is $\{A_1^{true}, A_2^{true}, A_3^{true}\}$ defined in a conceptually similar but more complex way that reflects the true probabilities. In this more difficult case, SUBA is again greatly superior to all the other methods.

Guo et al. (2017) proposed a design structured very similarly to SUBA, including a set $\mathscr{T}$ of targeted treatments, a biomarker vector **Z**, with the goals to define and adaptively refine a random subgroup partition in terms of **Z**, and obtain an optimal treatment for each subgroup. The design is called SCUBA, for Subgroup Cluster-based Bayesian Adaptive design. The main difference is that SCUBA uses random hyperplanes to partition the **Z** space, rather than SUBA's successive splits of subsets using randomly chosen covariates in a tree structure.

8.6 Enrichment Using Two Regressions on Biomarkers

Modern precision medicine often is based on the idea that heterogeneity of patient response to treatment can be explained by biological covariates, $\mathbf{Z} = (Z_1, \cdots, Z_p)$, that modify treatment effects at the cellular or molecular level. This is the basis for SUBA, which addresses this problem in the multi-treatment setting. Considering a single targeted experimental treatment, E, if its anti-disease effects involve drug metabolizing enzymes, signaling pathways, or cell surface markers, then it may be reasonable to assume that only a subset of "E-sensitive" patients, defined by **Z**, may be respond favorably to E. The design described in this section is based on the idea of determining this subset adaptively using **Z** in a GS randomized trial of E versus a control treatment C. Administration of E is restricted to the E-sensitive patient subset, that is, the identified subset is enriched. E-versus-C comparisons then are done prospectively in the identified subset, and this process is repeated group sequentially.

For example, 70% to 90% of hypertension patients respond to ACE inhibitors. Beta 2-agonists for asthma are effective for 30% to 60% of patients (Abrahams and Silver, 2009). Epidermal growth factor receptor is overexpressed in more than 60% of tumors from metastatic non-small-cell lung cancer patients, and is associated with poor prognosis (Sharma et al., 2007). There are numerous other examples. In such settings, traditional clinical trial designs that assume patients are homogeneous may be dysfunctional because they do not account for the empirically established fact that some patients may be more likely to respond to E than others. As shown earlier, a conventional E-versus-C effect estimate may show little or no benefit with E because the estimate is an average of high rate efficacy outcomes in an E-sensitive subpopulation and low rate efficacy outcomes in non-E-sensitive patients.

As a toy numerical illustration, suppose that mean survival time with a standard control therapy C is 18 months, 30% of patients are E-sensitive with E providing a 30-month mean survival time, and the remaining 70% of patients are not E-sensitive and have mean survival time 10 months with E. Consequently, if one ignores or is unaware of E-sensitivity and uses E to treat all patients, the overall mean survival time is the average over E-sensitive and non-E-sensitive patients, $.30\times30 + .70\times10 = 16$ months, which is below the mean of 20 months with C. A conventional comparative test ignoring subgroups would be very likely to conclude that E is no better than C, and miss the 12-month improvement in mean survival time over C in the E-sensitive patients. If the E-sensitive subgroup were identified after completion of a randomized trial, an estimated mean survival time of 30 months in this subgroup might be dismissed as being due to *post hoc* data dredging. This illustrates the problem that assuming homogeneity in a randomized trial, as done conventionally, quite easily may lead to the incorrect false negative inference that a new drug is ineffective for all patients, when in fact it is highly effective in a subgroup of E-sensitive patients.

These considerations imply that, in general, a more efficient and more realistic approach to evaluating a new targeted agent is a clinical trial with an enrichment design that identifies and focuses on E-sensitive patients adaptively during the trial. While solving this problem may seem straightforward, there are numerous ways that a clinical trial design that does enrichment may go astray. This is because at least two problems must be addressed. The first problem is reliably identifying an E-sensitive subgroup, if it exists. Given this, the second problem is to correctly conclude that E provides an improvement in that subgroup, if E actually provides such an improvement. As illustrated earlier, there are many other possibilities that must be taken into account. Because targeted agents do not always work as designed, a new agent may be either effective in non-E-sensitive patients, or not effective at all.

Many enrichment designs have been proposed that are based on the strong assumption that an E-sensitive subgroup is known. This assumption usually relies on preclinical studies in mice or rats, or historical data from earlier trials. Under this assumption, a simple trial design may be used that enrolls patients using predetermined eligibility criteria for the putative E-sensitive subgroup. Examples include the designs proposed by Brannath et al. (2009), Jenkins et al. (2011), Mehta et al. (2014), Uozumi and Hamada (2017), Kimani et al. (2015), and Rosenblum et al. (2016). Essentially, guessing an E-sensitive subgroup based on pre-clinical data regards treatment development as an engineering problem, because it ignores uncertainty. This also often ignores the fact that rats are very different from humans. In practice, there is a great deal of uncertainty about actual effects of a new targeted agent E in humans, and consequently assumptions about anti-disease activity or safety often turn out to be wrong. Using a pre-selected subgroup that has not been validated with human data easily may turn out to be a very poor way to dichotomize patients into an E-sensitive subgroup who are likely to respond and non-E-sensitive subgroup who are unlikely to respond to E. A more extreme version of this sort of behavior is to skip doing a clinical trial entirely, and do putatively "personalized therapy" based on informal assessments of preclinical data and *post hoc* laboratory based analyses of biomarkers from historical patients.

Park et al. (2022) proposed a group sequential adaptive enrichment design (AED) that addresses the dual problems of identifying an E-sensitive subgroup and testing whether E provides an improvement over C in that subgroup. The AED addresses the practical issue that, while extending survival or PFS time is the main goal of most phase III randomized trials, the data on time-to-event outcomes may be limited early in a GS trial. The AED considers settings where an early response indicator, R, is available, and it exploits possible relationships between R, Y = survival time, treatments E (X=1) and C (X=0), and biomarkers $\mathbf{Z}$, to improve reliability. This is particularly useful for earlier interim decisions. The AED addresses the following statistical goals:

Goal 1. Use $\mathbf{Z}$ to determine whether an E-sensitive subgroup exists, and if so identify it.

Goal 2. Test whether E provides an improvement in mean survival time over a standard control therapy, C, in the identified E-sensitive subgroup.

Goal 3. Evaluate and exploit relationships between an early outcome R, survival time, treatment, and $\mathbf{Z}$ to improve the reliability of inferences in Goals 1 and 2.

Several combinations of goals 1 and 2 have been addressed, under various sets of assumptions and taking various approaches, by the designs given earlier in this chapter. However, formally addressing goal 3, to exploit data from an early outcome that may be related to Y, is relatively new.

The AED proposed by Park et al. (2022) exploits the ideas that $\mathbf{Z}$ may be related to the effects of E on the distributions of Y and R, and that R also may help to predict Y. The AED addresses all three goals given above by repeatedly taking the following actions during each stage of the GS decision process:

Biomarker Selection Variable selection in $\mathbf{Z}$, based on regression models for both $p(Y \mid R,\mathbf{Z},X)$ and $p(R \mid \mathbf{Z},X)$, to identify an E-sensitive subpopulation.

Enrichment. Restrict enrollment to the current identified E-sensitive subpopulation.

Treatment Comparison. Compare the survival distributions of E and C in the identified subgroup of E-sensitive patients.

The AED uses a Bayesian model in which the distribution of Y is a mixture over responders ($R=1$) and non-responders ($R=0$). At each interim decision, regression sub-models for $p(Y \mid \mathbf{Z},X,R)$ and $p(R \mid \mathbf{Z},X)$ are fit to the current data, variable selection is done in each submodel to identify subvectors $\mathbf{Z}_R$ and $\mathbf{Z}_Y$ of $\mathbf{Z}$ that together are used to characterize E-sensitive patients, and comparative superiority and futility decisions for survival are made. Identification of an E-sensitive subset is done using a personalized benefit index (PBI), defined as the predictive probability that a patient with a given $\mathbf{Z}$ will benefit more from E than from C. The PBI is used to define the adaptive enrichment rule that restricts enrollment to the most recently identified E-sensitive patients. The GS procedure accounts for the sequentially adaptive variable selection, the adaptive enrichment, and multiple testing. If an E-sensitive subpopulation defined by $\mathbf{Z}$ actually exists, by reliably identifying it and determining whether E is superior to C in E-sensitive patients, results of a trial conducted using the design provide a basis for practicing precision medicine.

The AED was motivated by a trial to investigate effects of a PI3K pathway inhibitor combined with olaparib for treating high-grade serous or BRCA-mutant ovarian cancer, where the PI3K pathway is involved in cellular proliferation and likely to be upregulated. Olaparib is a potent inhibitor of poly(ADP-ribose) polymerase (PARP), an enzyme involved in repair of single-strand DNA breaks. Treatment with olaparib can lead to tumor regression by a process known as synthetic lethality, which results from accumulation of unrepaired DNA double-strand breaks and the resulting increase of genomic instability in the cancer cells. Combining PARP and PI3K pathway inhibition with olaparib may generate a synergistic treatment effect. A biomarker vector $\mathbf{Z}$ of 10 mutations related to the PARP and PI3K pathways are used to identify a genomic signature characterizing an E-sensitive subgroup, and E + olaparib is compared to the standard chemotherapy combination, cisplatin + paclitaxel, C, in the identified sensitive subgroup in terms of Y = PFS time.

For right-censoring of Y at a follow-up time when the data are evaluated for interim decision-making, the observed event time is Y^o, with the indicator $\delta = I(Y^o = Y)$. The relationship between Y and R is exploited by the design to allow adaptive decision-making to begin based on observed R values early in the trial, when there often is little information on the distribution of Y because few events have occurred. In treatment arm X, denote the response probability by

$$\pi(\mathbf{Z},X,\boldsymbol{\theta}_R) = \Pr(R=1 \mid \mathbf{Z},X,\boldsymbol{\theta}_R),$$

and let $h_X(y|\mathbf{Z},R,\boldsymbol{\theta}_Y)$ denote the hazard function of Y at time y for a patient with covariates $\mathbf{Z}$ and response indicator R treated with X, where $\boldsymbol{\theta}_X$ and $\boldsymbol{\theta}_Y$ are submodel parameter vectors. The subvector $\mathbf{Z}_R$ is identified by doing variable selection in the regression model $\pi(\mathbf{Z},X,\boldsymbol{\theta}_R)$, based on the difference between response probabilities,

$$\Delta_R(\mathbf{Z},\boldsymbol{\theta}_R) = \pi(\mathbf{Z},E,\boldsymbol{\theta}_R) - \pi(\mathbf{Z},C,\boldsymbol{\theta}_R).$$

This is a refinement of the indicator function $I[\pi(\mathbf{Z},E) > \pi(\mathbf{Z},C)]$ used by Simon and Simon (2013) to define an enrichment subset. The subvector $\mathbf{Z}_Y$ is identified by doing variable selection in the regression model for $p(Y \mid \mathbf{Z},X)$ based on the hazard ratio

$$\Delta_Y(\mathbf{Z},\boldsymbol{\theta}_Y) = \frac{h_E(y|\mathbf{Z},R,\boldsymbol{\theta}_Y)}{h_C(y|\mathbf{Z},R,\boldsymbol{\theta}_Y)}, \quad y > 0.$$

The subvectors $\mathbf{Z}_R$ and $\mathbf{Z}_Y$ are not necessarily identical, but they may share common terms when covariates predictive of a higher tumor response probability also are likely to be predictive of longer survival. To account for association, selection of $\mathbf{Z}_R$ and $\mathbf{Z}_Y$ are not done independently, but rather are based on correlated vectors of latent variable selection indicators, defined below.

The AED enrolls a maximum of N patients sequentially in cohorts of sizes $c_1,\ldots,c_K$, with $\sum_{k=1}^{K} c_k = N$. The trial initially enrolls patients under broad eligibility criteria for the first c_1 patients. When the first cohort's outcomes have been evaluated, $\mathbf{Z}_R$ and $\mathbf{Z}_Y$ are chosen and used to compute a PBI that defines the subgroup of E-sensitive patients, and comparative tests are done in this subgroup. Each test may terminate the trial due to either superiority or futility. If the trial is not stopped early then only E-sensitive patients are enrolled in the next cohort. The process of identifying $(\mathbf{Z}_R, \mathbf{Z}_Y)$, computing the PBI, defining the set of E-sensitive patients, and performing the tests is repeated group sequentially until the end of the trial. If the maximum sample size N is reached, a final analysis is done when the last patient's follow-up is completed.

A joint probability model for $p(Y,R \mid \mathbf{Z},X)$ is defined as a mixture of the conditional distribution $p(Y \mid \mathbf{Z},X,R)$, weighted by the marginal distribution $p(R \mid \mathbf{Z},X)$, assuming for simplicity that R always is observed before Y. If Y may be observed before R can be evaluated, which may arise with rapidly fatal diseases, then the necessary model elaboration is provided by Park et al. (2022). The AED assumes a probit model for the marginal response distribution, given by

$$\pi(\mathbf{Z}_i, X_i, \boldsymbol{\theta}_R) = \Phi\big(\mathbf{Z}_i\boldsymbol{\beta}_R + X_i\mathbf{Z}_i\boldsymbol{\gamma}_R\big), \quad i = 1,\ldots,n,$$

where $\Phi(\cdot)$ denotes the standard normal cdf, and $\boldsymbol{\theta}_R = \big(\boldsymbol{\beta}_R, \boldsymbol{\gamma}_R\big)$. Thus, $\boldsymbol{\beta}_R$ is the main covariate effect vector and $\boldsymbol{\gamma}_R = (\gamma_{R,0}, \gamma_{R,1},\ldots,\gamma_{R,p})$ is a vector of additional E-versus-C treatment-covariate interactions for R. Denote the vectors

$$\boldsymbol{R}_n = (R_1,\ldots,R_n), \ \boldsymbol{X}_n = (X_1,\ldots,X_n), \ \ \mathbf{Z}_n = (\mathbf{Z}_1,\ldots,\mathbf{Z}_n).$$

The marginal likelihood of the response vector $\boldsymbol{R}_n$ for the first n patients is

$$\mathscr{L}_n(\boldsymbol{R}_n, \boldsymbol{X}_n, \mathbf{Z}_n, \boldsymbol{\theta}_R) = \prod_{i=1}^{n} \Phi\big(\mathbf{Z}_i\boldsymbol{\beta}_R + X_i\mathbf{Z}_i\boldsymbol{\gamma}_R\big)^{R_i} \Big\{1 - \Phi\big(\mathbf{Z}_i\boldsymbol{\beta}_R + X_i\tilde{\mathbf{Z}}_i'\boldsymbol{\gamma}_R\big)\Big\}^{1-R_i}.$$

For the conditional distribution of $(Y \mid \mathbf{Z},X,R)$, a piecewise exponential hazard model is assumed (Ibrahim et al., 2005; Kim et al., 2007; McKeague and Tighiouart, 2000; Sinha et al., 1999). The model is defined by partitioning the time axis into M sub-intervals $\mathrm{I}_m = (u_{m-1}, u_m]$ for $m = 1,\ldots,M$, with fixed time grid $0 = u_0 < u_1 < \cdots < u_M < \infty$. The assumed piecewise exponential hazard function is

$$h(y \mid \mathbf{Z}, X, R, \boldsymbol{\theta}_Y) = \phi_m \exp\big(\mathbf{Z}\boldsymbol{\beta}_Y + X\mathbf{Z}\boldsymbol{\gamma}_Y + \alpha_Y R\big) I(y \in \mathrm{I}_m),$$

for $m = 1,\ldots,M$, with $\boldsymbol{\theta}_Y = \big(\boldsymbol{\phi}, \boldsymbol{\beta}_Y, \boldsymbol{\gamma}_Y, \alpha_Y\big)$, where $\phi_m > 0$ is the baseline hazard on the mth subinterval, $\boldsymbol{\phi} = (\phi_1,\ldots,\phi_M)$, and α_Y is the effect of response on the hazard of Y. For each $m = 1,\ldots,M$, $y_m = u_{m-1}$ if $y \le u_{m-1}$, $y_m = y$ if $u_{m-1} < y \le u_m$ and $y_m = u_m$ if $y > u_m$. Given this definition of y_m, the piecewise exponential cdf is

$$F(y \mid \mathbf{Z}, X, R, \boldsymbol{\theta}_Y) = 1 - \exp\Big\{ - \sum_{m=1}^{M} \phi_m (y_m - u_{m-1}) \exp\big(\mathbf{Z}\boldsymbol{\beta}_Y + X\mathbf{Z}\boldsymbol{\gamma}_Y + \alpha_Y R\big)\Big\}$$

for $y > 0$.

Denoting $\boldsymbol{Y}_n^o = (Y_1^o, \ldots, Y_n^o)$ and $\boldsymbol{\delta}_n = (\delta_1, \ldots, \delta_n)$, the joint likelihood function of the response and survival time outcomes is

$$\begin{aligned}
&\mathscr{L}(\boldsymbol{R}_n, \boldsymbol{Y}_n^o, \boldsymbol{\delta}_n, \boldsymbol{X}_n, \boldsymbol{Z}_n, \boldsymbol{\theta}_R, \boldsymbol{\theta}_Y) \\
&\quad = \mathscr{L}(\boldsymbol{R}_n, \boldsymbol{X}_n, \boldsymbol{Z}_n, \boldsymbol{\theta}_R) \\
&\quad \times \prod_{i=1}^{n} f(Y_i^o \mid R_i, X_i, \boldsymbol{Z}_i, \boldsymbol{\theta}_Y)^{\delta_i} \{1 - F(Y_i^o \mid R_i, X_i, \boldsymbol{Z}_i, \boldsymbol{\theta}_Y)\}^{1-\delta_i},
\end{aligned} \tag{8.3}$$

where $f(Y \mid R, X, \boldsymbol{Z}, \boldsymbol{\theta}_Y)$ is the conditional pdf of Y. The marginal likelihood of Y is obtained by averaging the joint likelihood of (Y, X) over R.

Bayesian posterior computation for $\boldsymbol{\theta}_R$ is done using data augmentation, as in Albert and Chib (1993), based on the iid latent real-valued variables $W_1, \ldots, W_n$, where observed $R_i = 1$ if and only if latent $W_i > 0$, assuming that

$$W_i \mid X_i, \boldsymbol{Z}_i, \boldsymbol{\theta}_R \sim \mathrm{N}(\boldsymbol{Z}_i \boldsymbol{\beta}_R + X_i \boldsymbol{Z}_i \boldsymbol{\gamma}_R,\ 1)$$

with prior

$$\boldsymbol{\theta}_R = (\boldsymbol{\beta}_R, \boldsymbol{\gamma}_R) \sim \mathrm{N}(\boldsymbol{\mu}_R, \boldsymbol{\Sigma}_R),$$

where $\boldsymbol{\mu}_R$ and $\boldsymbol{\Sigma}_R$ are fixed hyperparameters, and the normal variance is set to 1 to ensure identifiability. Vague normal priors with zero mean vector and diagonal covariance matrix with large diagonal element 10^6 are recommended generally in practice, and these are assumed in the simulations given below. For $\boldsymbol{\theta}_Y$, a normal prior is assumed on both $(\boldsymbol{\beta}_Y, \boldsymbol{\gamma}_Y)$ and α_Y and independent gamma distributions on $\boldsymbol{\phi}$ given by

$$(\boldsymbol{\beta}_Y, \boldsymbol{\gamma}_Y) \sim \mathrm{N}(\boldsymbol{\mu}_Y, \boldsymbol{\Sigma}_Y), \quad \alpha_Y \sim \mathrm{N}(a, \sigma_a^2), \quad \phi_m \sim \mathrm{Gamma}(c\phi_m^*, c),$$

for $m = 1, \ldots, M$, where $\boldsymbol{\mu}_Y, \boldsymbol{\Sigma}_Y$, a, σ_a, c and ϕ_m^*, $m = 1, \ldots, M$ are prespecified hyperparameters, and $\mathrm{Gamma}(g_1, g_2)$ denotes a gamma random variable with shape parameter g_1 and rate parameter g_2.

Joint Bayesian variable selection on $\boldsymbol{Z}$ in each of the submodels for $p(Y \mid R, \boldsymbol{Z}, X)$ and $p(R \mid \boldsymbol{Z}, X)$ is based on correlated latent covariate inclusion variables. In the joint likelihood given in equation (8.3), for each outcome $t = R$ or Y, let

$$\boldsymbol{\psi}_t = (\beta_{t,1}, \cdots, \beta_{t,p}, \gamma_{t,0}, \gamma_{t,1}, \cdots, \gamma_{t,p})$$

denote the regression coefficient vector, excluding the intercept. For each submodel, $t = Z$ or Y, variable selection is based on a spike-and-slab prior on $\boldsymbol{\psi}_t$, following Mitchell and Beauchamp (1988), George and McCulloch (1993) and Ishwaran et al. (2005). This procedure uses sparse posterior coefficient estimates to determine which variables to include in the submodel's linear component. For each outcome $t = Y$ and R, let $\boldsymbol{\lambda}_t = (\lambda_{t,1}, \ldots, \lambda_{t,2p+1})$ denote a vector of latent variable selection indicators corresponding to $(\boldsymbol{Z}, X, X\boldsymbol{Z})$ in the linear term. The jth variable in $(\boldsymbol{Z}, X, X\boldsymbol{Z})$ is included in the submodel for outcome t if $\lambda_{t,j} = 1$, and not included if $\lambda_{t,j} = 0$. The variable selection algorithm in each submodel is required to follow the *strong hierarchy interaction* constraint (Liu et al., 2015), which requires that, if the interaction term XZ_j is included, then the main effect term Z_j also must be included.

Because some covariates may be predictive of treatment effects on both R and Y, $\boldsymbol{\lambda}_R$ and $\boldsymbol{\lambda}_Y$ are assumed to follow a joint distribution, to borrow information between covariate effects on R and Y during the variable selection. To induce correlation, a bivariate Bernoulli distribution is assumed for $(\lambda_{R,j}, \lambda_{Y,j})$, for $j = 1, \ldots, 2p+1$. Denoting $p_{R,j} = \Pr(\lambda_{R,j} = 1)$ and $p_{Y,j} = \Pr(\lambda_{Y,j} = 1)$, which are the marginal probabilities that the jth variable is included in the submodel for $p(R \mid \mathbf{Z}, X)$ and $p(Y \mid R, \mathbf{Z}, X)$, respectively, the odds ratio for the jth pair of latent variables is

$$\rho_j = \frac{\Pr(\lambda_{Y,j} = 1, \lambda_{R,j} = 1)/\Pr(\lambda_{Y,j} = 0, \lambda_{R,j} = 1)}{\Pr(\lambda_{Y,j} = 1, \lambda_{R,j} = 0)/\Pr(\lambda_{Y,j} = 0, \lambda_{R,j} = 0)}.$$

Let $B(p_{R,j}, p_{Y,j}, \rho_j)$ denote the joint Bernoulli distribution of $(\lambda_{R,j}, \lambda_{Y,j})$. The spike-and-slab prior used for the joint variable selection is

$$\begin{aligned}
\psi_{R,j}|\lambda_{R,j} &\sim (1-\lambda_{R,j})N(0, v_{R,j}^2) + \lambda_{R,j}N(0, u_{R,j}^2 v_{R,j}^2), \quad j = 1, \ldots, 2p+1 \\
\psi_{Y,j}|\lambda_{Y,j} &\sim (1-\lambda_{Y,j})N(0, v_{Y,j}^2) + \lambda_{Y,j}N(0, u_{Y,j}^2 v_{Y,j}^2), \quad j = 1, \ldots, 2p+1,
\end{aligned}$$

where $u_{R,j}, v_{R,j}^2, u_{Y,j}, v_{Y,j}^2$, $j = 1, \ldots, 2p+1$ are hyperparameters. Large $u_{R,j}$ and small $v_{R,j}$ ensure that $\lambda_{R,j} = 1$ implies that a nonzero estimate of $\psi_{R,j}$ is included, while $\lambda_{R,j} = 0$ implies that the covariate corresponding to $\psi_{R,j}$ has a negligible effect on R. Similar choices are applied to obtain sparse vectors of coefficient estimates for Y. The latent indicator variables are assumed to follow the priors

$$(\lambda_{R,j}, \lambda_{Y,j}) \mid p_{R,j}, p_{Y,j}, \text{ and } \rho_j \sim B(p_{R,j}, p_{Y,j}, \rho_j), \quad j = 1, \ldots, p+1,$$

$$\lambda_{R,j} \sim \text{Bernoulli}(p_{R,j}) \text{ and } \lambda_{Y,j} \sim \text{Bernoulli}(p_{Y,j}), \quad j = p+2, \ldots, 2p+1.$$

The strong hierarchical property is ensured by the constraints

$$\begin{aligned}
p_{R,j} &= p_{R,j-p-1} p_{R,p+1} \min\{p_{R,j-p-1}, p_{R,p+1}\}, \quad j = p+2, \ldots, 2p+1 \\
p_{Y,j} &= p_{Y,j-p-1} p_{Y,p+1} \min\{p_{Y,j-p-1}, p_{Y,p+1}\}, \quad j = p+2, \ldots, 2p+1.
\end{aligned}$$

Priors for the remaining parameters are

$$\psi_{R,0} \sim N(u_0, \zeta_0^2), \ \alpha_Y \sim N(u_a, \zeta_a^2), \ \phi_m \sim \text{Gamma}(\tilde{c}\tilde{\phi}_m, \tilde{c}), \ m = 1, \ldots, M,$$

$$p_{R,j} \sim \text{Beta}(l_{R1,j}, l_{R2,j}), \ p_{Y,j} \sim \text{Beta}(l_{Y1,j}, l_{Y2,j}), \ \log \rho_j \sim N(r_{1j}, r_{2j}), \ j = 1, \ldots, p+1,$$

where

$$u_0, \zeta_0, u_a, \zeta_a, \ \tilde{c}, \tilde{\phi}_m, \ m = 1, \ldots, M, \ l_{Z1,j}, l_{Z2,j}, l_{Y1,j}, l_{Y2,j}, r_{1j}, \ r_{2j}, \ j = 1, \ldots, p+1,$$

are prespecified hyperparameters.

Joint variable selection performed at interim stages to obtain the subvectors $\mathbf{Z}_R$ and $\mathbf{Z}_Y$ may miss informative covariates early in the trial, due to an insufficient number of observed events for Y. As the trial progresses and survival data accumulate, the probabilities of identifying truly predictive covariates increase. This motivates repeatedly re-selecting $\mathbf{Z}_R$ and $\mathbf{Z}_Y$ as new data become available for each GS decision.

Adaptive Decisions

For each cohort $k = 1, \ldots, K$, let d_k denote the accumulated number of deaths ($Y_i = Y_i^o$) at the time when the kth adaptive enrichment is performed, $n_k = \sum_{j=1}^{k} c_j$ the total number of patients enrolled in the first k cohorts,

$$\mathscr{D}_k = \{(Y_i^o, \delta_i, R_i, X_i, \mathbf{Z}_i), i = 1, \ldots, n_k\}$$

the accumulated data, and $\mathbf{Z}_R^{(k)}$ and $\mathbf{Z}_Y^{(k)}$ the selected subvectors. The *personalized benefit index* (PBI) for a patient with biomarker vector $\mathbf{Z}$ is defined as

$$\begin{aligned} PBI(\mathbf{Z}_R^{(k)}, \mathbf{Z}_Y^{(k)} \mid \mathscr{D}_k) &= (1-\omega_k) \Pr\{\Delta_R(\mathbf{Z}_R^{(k)}, \boldsymbol{\theta}_R) > \varepsilon_1 \mid \mathscr{D}_k\} \\ &+ \omega_k \Pr\{\Delta_Y(\mathbf{Z}_Y^{(k)}, \boldsymbol{\theta}_Y) < \varepsilon_2 \mid \mathscr{D}_k\}, \end{aligned} \tag{8.4}$$

where $\omega_k = d_k/n_k$. The cutoffs ε_1 and ε_2 are design parameters that quantify minimal clinically significant improvements in response probability and the risk of death. Since larger $\Delta_R(\mathbf{Z}_R^{(k)}, \boldsymbol{\theta}_R)$ and smaller $\Delta_Y(\mathbf{Z}_Y^{(k)}, \boldsymbol{\theta}_Y)$ correspond to superiority of E over C, the statistic $PBI(\mathbf{Z} \mid \mathscr{D}_k)$ is defined as a weighted average of the posterior probabilities that a patient with covariates $\mathbf{Z}$ will benefit from E more than C in terms of a larger probability of response and a smaller risk of death. Early in the trial, when there are few observed deaths, ω_k will be smaller and $PBI(\mathbf{Z} \mid \mathscr{D}_k)$ will depend on R more than Y. As more events occur, the weight ω_k for the survival hazard ratio component in the PBI definition (8.4) will become larger and the weight $(1-\omega_k)$ for the response probability difference will become smaller, so $PBI(\mathbf{Z} \mid \mathscr{D}_k)$ will depend more on the survival time data.

The PBI is used as a criterion for doing adaptive enrichment by saying that a patient with biomarkers $\mathbf{Z}$ is eligible for enrollment into the $k+1^{st}$ cohort of the trial if E is sufficiently beneficial compared to C for those biomarkers. This is formalized by the inequality

$$PBI(\mathbf{Z}_R^{(k)}, \mathbf{Z}_Y^{(k)} \mid \mathscr{D}_k) > v \left(\frac{n_k}{N}\right)^g \tag{8.5}$$

for $k = 1, \ldots, K-1$, where $v > 0$ and $g > 0$ are prespecified parameters, calibrated by simulation to obtain a design with good properties. If the trial is not stopped early, when the final K^{th} cohort's outcomes have been evaluated at the end of follow-up, the final biomarker set $B(\mathbf{Z} \mid \mathscr{D}_K)$ is updated and used for the final tests. Bayesian posterior decision criteria are based on the treatment effect averaged over the enriched trial population. The kth interim decisions are based on $\mathscr{D}_k$, the accumulated data from the first k successive cohorts. Patients in each cohort are homogenous because they satisfy the same eligibility criteria, but patients may be heterogeneous between cohorts, since different cohorts may have different eligibility criteria due to the adaptive variable selection and refinement of the PBI during the trial.

The rules for GS decisions for superiority or futility based on treatment effects on Y follow the same logical structure as for a conventional GS test, with one important difference. Prior to each test, the set of E-sensitive patients first is determined and the most recently selected subvectors $\mathbf{Z}_R^{(k)}$ and $\mathbf{Z}_Y^{(k)}$ are used to define the test statistics, and also to determine the enrollment criteria for the next cohort if the trial is continued. Let

$$\mathbf{Z}_k^{sel} = \{\mathbf{Z} : PBI(\mathbf{Z}_R^{(k)}, \mathbf{Z}_Y^{(k)} \mid \mathscr{D}_k) > \nu(n_k/N)^g\}$$

denote the set of selected biomarkers satisfying the eligibility criteria used to define the E-sensitive patients who may be enrolled in the $k+1^{st}$ cohort. At this point in the trial, the covariate-averaged long-term outcome treatment effect is defined to be

$$T_{Y,k}(\boldsymbol{\theta}) = \int_{\mathbf{z}_Y^{(k)} \in \mathbf{Z}_k^{sel}} \Delta_Y(\mathbf{z}_Y^{(k)}, \boldsymbol{\theta}_Y)\hat{p}_k(\mathbf{z}_Y^{(k)})d\mathbf{z}_Y^{(k)},$$

where $\hat{p}_k(\mathbf{z}_Y^{(k)})$ denotes the empirical distribution of $\mathbf{Z}_Y^{(k)}$ on the set $\mathbf{Z}_k^{sel}$. Since these expectations are computed over the kth selected enrichment set, $\mathbf{Z}_k^{sel}$, which is the set of patients who are expected to benefit more from E than C in the kth cohort, $T_{Y,k}(\boldsymbol{\theta})$ is a subset-specific treatment effect, and thus it may be used as the basis for precision medicine. It follows from the definition of the hazard ration $\Delta_Y(\mathbf{z}_Y^{(k)}, \boldsymbol{\theta}_Y)$ that E is more effective than C for patients with $\mathbf{Z} \in \mathbf{Z}_k^{sel}$ if $T_{Y,k}(\boldsymbol{\theta})$ is sufficiently small.

To define GS test statistics, one must account for the fact that, due to adaptive enrichment, there are k heterogeneous cohorts at the kth analysis, and the empirical distribution $\hat{p}_k(\mathbf{Z}_Y^{(k)})$ changes with k as new data are obtained. Thus, denoting the number of events in the jth cohort by e_j, the test statistic at the kth analysis is defined as the event-weighted average of the treatment effects at stages $1, \cdots, k$,

$$\overline{T}_{Y,k}(\boldsymbol{\theta}) = \sum_{j=1}^{k} \left(\frac{e_j}{\sum_{l=1}^{k} e_l} \right) T_{Y,j}(\boldsymbol{\theta}).$$

Each $T_{Y,k}(\boldsymbol{\theta})$ is calculated using the data observed at the kth interim decision. Because Y is a time-to-event endpoint, $T_{Y,j}(\boldsymbol{\theta})$ must be updated at each later interim decision time. Let $b_1 < 1$ denote the hazard ratio under which E is considered to be superior to C in terms of Y, and let $b_2 > 1$ denote the hazard ratio under which E is considered inferior to C, with the values of (b_1, b_2) specified by the clinicians. Let (B_1, B_2) be probability cutoffs calibrated by preliminary simulations. For the interim analysis at each $k = 1, \ldots, K-1$, bearing in mind that superiority of E over C corresponds to smaller values of $\overline{T}_{Y,k}(\boldsymbol{\theta})$, the AED decision rules are as follows:

Superiority: Stop the trial for superiority of E over C in the E-sensitive patient subset $\mathbf{Z}_k^{sel}$ if

$$Pr\{\overline{T}_{Y,k}(\boldsymbol{\theta}) < b_1 | \mathscr{D}_k\} > B_1.$$

Futility: Stop the trial for futility of E over C in the E-sensitive patient subset $\mathbf{Z}_k^{sel}$ if

$$Pr\{\overline{T}_{Y,k}(\boldsymbol{\theta}) > b_2 | \mathscr{D}_k\} > B_2.$$

Final Decision: If the trial is not stopped early, at the final analysis, conclude that E is superior to C in the final E-sensitive patient subset $\mathbf{Z}_K^{sel}$ if

$$Pr\{\overline{T}_{Y,K}(\boldsymbol{\theta}) < b_1 | \mathscr{D}_K\} > B_1,$$

and otherwise conclude that E is not superior to C in $\mathbf{Z}_K^{sel}$.

An important practical issue during the process of constructing a design is deciding when to begin the adaptive enrichment. This depends on several factors, including the number of covariates, their information-to-noise ratio, the percentage of sensitive patients, the treatment difference between sensitive and insensitive patients, and the variances of the outcomes R and Y. In practice, logistical limitations usually will limit the number of interim decisions to 1, 2, or 3. Based on these considerations, as a rule of thumb, a reasonable time to initiate the adaptive enrichment is after 33% to 50% of the maximum number of patients have been enrolled.

If desired, at each interim analysis, an additional futility stopping rule may be included to account for the possibility that only a very small percentage of patients may benefit from E. For a prespecified lower threshold $0 < q < 1$ based on practical considerations, such a futility rule would stop the trial if the estimated proportion of E-sensitive patients in the trial is $< q$. The value $q = .10$ is used in the simulation study, and a value in the range $.01 - .10$ is reasonable for use in practice. At the end of the trial, identification of the final E-sensitive subset $\mathbf{Z}_K^{sel}$ based on the final PBI involves all covariates, because the Bayesian variable selection method based on the spike-and-slab prior may not drop covariates with little or no contribution for identifying $\mathbf{Z}_K^{sel}$. In practice, if desired, the E-sensitive subset identification rule may drop covariates having low posterior probability, say $< .10$, of being selected in the model of $p(Y,R \mid \mathbf{Z})$.

Park et al. (2022) reported a simulation study to evaluate OCs of the AED and compare it to several competing enrichment designs. A maximum trial sample size of 400 was assumed, with patients accrued according to a Poisson process with rate 100 per year, and randomized fairly between E and C. Up to two interim analyses were performed at 200 and 300 patients, with a final analysis one year after the last patient was enrolled. Ten covariates, $\mathbf{Z} = (Z_1,\ldots,Z_{10})$, were defined, each either with or without a treatment interaction effect, to define the E-sensitive subpopulation, with each covariate binary for simplicity. The simulation scenarios are given in Table 8.5. In scenario 1, no patients are E sensitive. In scenario 2, patients with $Z_1 = 1$ are E sensitive. In scenario 3, patients with $Z_1 = Z_2 = 1$ are E sensitive. In scenario 4, patients with $Z_1 = 1$ and $Z_2 = 0$ are E sensitive. In scenario 5, patients with $Z_1 = Z_2 = Z_3 = 1$ are E sensitive. In scenario 6, patients with $Z_1 = Z_2 = 1$ and $Z_3 = 0$ are E sensitive. In scenario 7, patients with $Z_1 = 1$ and $Z_2 = Z_3 = 0$ are E sensitive. For E-sensitive patients, several values of the hazard ratio $\Delta_Y < 1$ were evaluated, as well as several differences $\Delta_R > 0$ between the response rates of E and C. For non-E-sensitive patients, $\Delta_Y \geq 1$ and $\Delta_R \leq 0$. Values of R were simulated from a Bernoulli distribution with response probability given by

$$\pi(\mathbf{Z},X,\boldsymbol{\theta}_R) = \Phi\left\{\beta_{R,0} + \sum_{j=1}^{10} \beta_{R,j}Z_j + X\left(\gamma_{R,0} + \sum_{j=1}^{10} \gamma_{R,j}Z_j\right)\right\},$$

TABLE 8.5
Assumed covariates $\mathbf{Z}$ and true parameter values for simulation scenarios to evaluate the adaptive enrichment design and its competitors. $\mathbf{Z}$ determines the E-sensitive subgroup. π_E = response probability and $\tilde{\mu}_E$ = median survival time with E. $\Delta_Y = h_E(y|\mathbf{Z},\boldsymbol{\theta}_Y)/h_C(y|\mathbf{Z},\boldsymbol{\theta}_Y)$ and $\Delta_R = \pi(\mathbf{Z},E) - \pi(\mathbf{Z},C)$.

		E-sensitive pats			
Scenario	$\mathbf{Z}$ of E-sensitive pats	π_E	$\tilde{\mu}_E$	Δ_Y	Δ_R
1	No E-sensitive pats	0.50	0.817	1	0
2	$Z_1 = 1$	0.65	2.593	0.49	0.19
3	$Z_1 = Z_2 = 1$	0.65	2.768	0.60	0.23
4	$Z_1 = 1,\ Z_2 = 0$	0.65	2.342	0.59	0.19
5	$Z_1 = Z_2 = Z_3 = 1$	0.65	2.233	0.60	0.21
6	$Z_1 = Z_2 = 1,\ Z_3 = 0$	0.65	2.236	0.60	0.19
7	$Z_1 = 1,\ Z_2 = Z_3 = 0$	0.65	2.233	0.59	0.17

		Non-E-sensitive pats			
Scenario	$\mathbf{Z}$ of E-sensitive pats	π_E	$\tilde{\mu}_E$	Δ_Y	Δ_R
1	No E-sensitive pats	0.50	0.817	1	0
2	$Z_1 = 1$	0.46	0.585	1.220	−0.040
3	$Z_1 = Z_2 = 1$	0.40	0.586	1.300	−0.073
4	$Z_1 = 1,\ Z_2 = 0$	0.387	0.536	1.983	−0.073
5	$Z_1 = Z_2 = Z_3 = 1$	0.373	0.404	1.566	−0.141
6	$Z_1 = Z_2 = 1,\ Z_3 = 0$	0.300	0.225	2.247	−0.171
7	$Z_1 = 1,\ Z_2 = Z_3 = 0$	0.234	0.136	3.459	−0.236

and Y was simulated using the hazard function

$$h(y|R,X,\mathbf{Z},\boldsymbol{\theta}_Y) = h_0(y)\exp\left\{\sum_{j=1}^{10}\beta_{Y,j}Z_j + X\left(\gamma_{Y,0} + \sum_{j=1}^{10}\gamma_{Y,j}Z_j\right) + \alpha_Y R\right\},$$

where $h_0(y)$ follows a Weibull distribution with scale parameter 1 and shape parameter 0.6 to obtain a decreasing hazard. Simulation scenarios 2 – 7 considered three E-sensitive patient prevalences: 65%, 50% and 35%.

The designs were calibrated to have overall type I error probability of 0.05, with $b_1 = b_2 = 1$ for GS monitoring, using $\varepsilon_1 = 0$ and $\varepsilon_2 = 1$ to define the PBI, and the design parameters $v = 0.766$ and $g = 0.352$ for the eligibility criteria, based on preliminary simulations, and $q = 0.10$, so the trial would be stopped if less than 10% of patients were found to be E-sensitive.

The following four adaptive enrichment designs were included in the simulations as comparators to the AED.

GSED: The GS enrichment design of Magnusson and Turnbull (2013), which selects an E-sensitive subgroup at the first interim test based on one prespecified dichotomized biomarker.

InterAdapt: The GS design of Rosenblum et al. (2016), which allows interim early stopping based on tests for superiority or futility in the nominally "sensitive" subgroup, or for superiority in the entire population.

SSD: The adaptive enrichment design proposed by Simon and Simon (2013).

ACGSD: An all-comers GS design, included to focus on the contribution of the covariate-based adaptive enrichment method in the AED. The ACGSD is identical to the AED in all ways with the one difference that the ACGSD does not perform adaptive enrichment.

Because both GSED and InterAdapt require prespecified "sensitive" and "insensitive" subgroups based on one pre-chosen biomarker, in the simulations Z_1 was used to dichotomize the patient population into these two subgroups.The designs were compared using generalized power (GP), which is defined in this setting as the probability that the design correctly (1) identifies the sensitive subpopulation and (2) rejects $H_0 : \Delta_Y(\boldsymbol{Z}, \boldsymbol{\theta}) \geq 1$ when H_0 actually is not true with $\Delta_Y(\boldsymbol{Z}, \boldsymbol{\theta}) < 1$, i.e. when E is superior to C in the E-sensitive subgroup. As explained previously, GP is very different from conventional power, which ignores the adaptive E-sensitive subgroup identification process and is computed assuming that the E-sensitive subgroup is known *a priori*. The GP thus reflects the actual statistical decisions. Because the GP event is more demanding and thus smaller than the usual rejection event used to compute conventional power, it is more difficult to achieve a given large GP value, such as .80. In the extreme case where an E-sensitive subgroup that a design assumes is incorrect, the GP = 0. GP thus honestly quantifies how well a complex sequentially adaptive decision-making process performs to optimize a targeted therapy.

The simulation results are summarized in Table 8.6, where the proportion of sensitive patients for the first cohort is assumed to be 0.65, 0.50, or 0.35, while proportions for the second and third cohorts are statistics resulting from the adaptive enrichment, computed as means across the simulations. In scenario 1, E is ineffective for all patients, and all designs preserve the nominal type I error rate of 0.05, with InterAdapt having type I error rate of 0.06. In each of scenarios 2 – 7, E is only effective for a subgroup of patients. Each of these scenarios has three sub-cases, with 65%, 50%, or 35% sensitive patients in cohort 1. The tabled numerical percentages for cohorts 2 and 3 result from the adaptive enrichment decisions made by the AED, and consequently they are design operating characteristics and not assumed simulation study parameters.

Table 8.6 shows that, in most scenarios, the AED has much higher GP than all other designs. The many GP values of 0 for GSED and InterAdapt are due to the fact that both designs prespecify a nominally sensitive subgroup. If this assumed E-sensitive subgroup is incorrect, then the design's GP = 0, which is the case in each of scenarios 3 – 7. The all comers CGS design has GP = 0 in all scenarios because it does not pick a sensitive subgroup. The great advantage in GP and rejection probability for the AED stems from its adaptive enrichment of E-sensitive patients based on covariate signatures chosen using regression models for both short-term and long-term

TABLE 8.6

Generalized power of the AED and four comparator enrichment designs in a three-stage group sequential trial. Survival time follows a Weibull distribution with deceasing hazard. Generalized Power = Pr(Correctly identify the E sensitive subgroup and reject H_0 when $\Delta_Y < 1$).

	% Sensitive Patients in Cohort			Generalized Power				
Scenario	1st	2nd	3rd	AED	GSED	InterAdapt	SSD	ACGSD
1	0	0	0	NA	NA	NA	NA	NA
2	0.65	0.88	0.89	0.79	0.45	0.65	0.63	0.61
	0.50	0.81	0.81	0.73	0.41	0.65	0.36	0.46
	0.35	0.70	0.72	0.73	0.36	0.62	0.20	0.27
3	0.65	0.83	0.84	0.65	0	0	0.49	0.49
	0.50	0.77	0.77	0.63	0	0	0.28	0.43
	0.35	0.65	0.65	0.56	0	0	0.13	0.36
4	0.65	0.81	0.81	0.68	0	0	0.50	0.49
	0.50	0.66	0.68	0.66	0	0	0.32	0.39
	0.35	0.57	0.58	0.64	0	0	0.15	0.21
5	0.65	0.83	0.90	0.51	0	0	0.22	0.16
	0.50	0.80	0.81	0.50	0	0	0.18	0.15
	0.35	0.72	0.74	0.44	0	0	0.14	0.12
6	0.65	0.93	0.93	0.71	0	0	0.36	0.20
	0.50	0.88	0.89	0.63	0	0	0.22	0.06
	0.35	0.81	0.83	0.59	0	0	0.11	0.05
7	0.65	0.93	0.94	0.84	0	0	0.58	0.03
	0.50	0.90	0.91	0.81	0	0	0.32	0.01
	0.35	0.83	0.87	0.77	0	0	0.22	0.01

endpoints, and the fact that this is repeated throughout the GS process. In all scenarios, because the first cohort of the AED enrolls all comers, the percentage of sensitive patients enrolled in the first cohort is approximately its population prevalence. Since the AED enriches the identified sensitive patient subgroup in all subsequent cohorts, this results in an increasingly higher percentages of sensitive patients in cohorts 2 and 3. For example, in Scenario 2, the percentage of truly sensitive patient is the population value 65%, but thereafter, due to adaptive enrichment, the percentages of enrolled sensitive patients increase to 88% in cohort 2 and 89% in cohort 3. Because GSED takes a fixed enrichment approach, whether or not the prespecified biomarkers correctly identify sensitive patients plays a major role in detecting a treatment effect, and can easily lead to incorrect conclusions. If the subgroups are incorrectly specified by the GSED, the selected subgroup after the first interim test will include

TABLE 8.7
Ratio of median survival times of future patients for each design, using the all comers ACGSD median in the denominator, in simulation scenario 2 with 35% E-sensitive patients in the first cohort.

Design	Median Survival Time Ratio versus the All Comers Design
AED	2.04
GSED	1.22
InterAdapt	1.19
Simon-Simon	0.98

a substantial proportion of E-insensitive patients, which dilutes the estimated effect size and leads to a low probability of rejecting the null hypothesis in an assumed or identified sensitive subpopulation. For the SSD, the enrichment rule often is based on few events at early interim analyses, and it enriches 5% to 10% more than the percentage of sensitive patients in the first cohort. Compared to GSED and SSD, the AED is far more likely to conclude that the treatment is effective and stop the trial early for superiority. InterAdapt's power was calculated based on all patients in most cases, since on average InterAdapt terminates the trial earlier. The lower GP of both InterAdapt and ACGSD in Scenario 2 are due to the fact that they both mix substantial numbers of non-E-sensitive patients with E-sensitive patients. Compared to the AED, InterAdapt and ACGSD both are more likely to stop the trial early for futility.

A very important property of the AED is that it defines the Bayesian criterion $PBI(\mathbf{Z}_R^{(k)}, \mathbf{Z}_Y^{(k)} \mid \mathscr{D}_k)$ repeatedly for each $k = 1, \ldots, K$ as a mixture of posterior probabilities of covariate-specific treatment effects for the early outcome R and the late outcome Y. Using this criterion, the AED defines the final subgroup of E-sensitive patients using the rule $PBI(\mathbf{Z}_R^{(K)}, \mathbf{Z}_Y^{(K)} \mid \mathscr{D}_K) > \eta_K$ when the trial is completed.

Clinical benefit to future patients may be summarized by the median survival time (MST) for patients who are considered sensitive based on the rule specified by the design at the end of the trial. The MST is computed assuming that the patients will receive E if H_0 is rejected at the end of trial, or C if it H_0 is not rejected. For example, in simulation scenario 2 with 35% E-sensitive patients in the first cohort, for each design, the average of MST ratios for future patients, using the MST with the all comers ACGSD as the denominator, are given in Table 8.7. The table says that, in terms of future patient survival, the AED provides the greatest benefit among the four enrichment designs. For patients enrolled in an adaptive enrichment trial, the clinical benefit is the average effect from the mixture of treatment-sensitive patients and non-treatment-sensitive patients after the randomization. The MST of patients enrolled during an enrichment trial does not show a substantial gain in survival benefit from using any of the enrichment designs.

In summary, across a broad range of different scenarios, the AED has much higher GP compared to the enrichment designs of Magnusson and Turnbull (2013), Rosenblum et al. (2016), and Simon and Simon (2013), as well as the conventional all-comers GS design. This substantial improvement over existing adaptive enrichment designs may be attributed to the AED's repeated adaptive biomarker selection, and the effectiveness of its adaptive enrichment rule based on each patient's covariate-based personalized benefit index. By exploiting this structure, the AED greatly magnifies the signal in the patient covariate vector and increases the GP for detecting a treatment effect, if it exists, in an E-sensitive subgroup. The AED also may be considered more ethical than a GS design without enrichment, because the AED reduces the probability of enrolling non-E-sensitive patients who are unlikely to benefit from E, while providing longer expected survival time for future patients.

Computer Software A zip file of computer software *biom13421-sup-0001-SuppMat.zip* for implementing the adaptive enrichment design is available from Biometrics as Supporting Information for Park et al. (2022).

9

Bayesian Nonparametric Models

9.1 Regression Analysis in Medical Research

There is a saying that most of statistical data analysis is regression, or would like to be. In the preceding chapters, regression models for the distribution $p(Y \mid \mathbf{Z}, X)$ of an outcome Y as a function of covariates $\mathbf{Z}$ and a treatment or dose X were used as statistical tools to design experiments, choose or optimize treatments, explain relationships between observed variables, and provide a rational basis for making decisions and taking actions. The two main purposes of writing down a model for $p(Y \mid \mathbf{Z}, X)$ and fitting it to the available data are to explain the causal effects $\mathbf{Z} \rightarrow Y$ of covariates and $X \rightarrow Y$ of the treatments on the outcome variable, or to apply the fitted model so that future $(\mathbf{Z}^*, X^*)$ may be used to predict future potential outcome Y^*. These inferences may be used as the basis for making decisions, such as choosing X for a patient based on their $\mathbf{Z}$ vector, or designing a future experiment.

It is self-evident that, if data on Y alone are available, then whatever can be learned from estimating the distribution of Y *per se* is very limited. When data on both Y and $\mathbf{Z}$ are available, the underlying idea for regression analysis is that $\mathbf{Z}$ is observed first, then Y is observed. From a modeling viewpoint, if $\mathscr{Z}$ is the set of possible covariate vectors $\mathbf{Z}$, then $p_{\mathbf{Z}}$ is a set of probability distributions for Y that is indexed by $\mathbf{Z} \in \mathscr{Z}$, and in a medical setting treatments or doses X are included with $\mathbf{Z}$. Many of the examples given in Chapters 1 and 2, and all of the clinical trial designs that I have discussed, are based on regression models for one or more outcomes, a treatment or dose, and prognostic or predictive covariates or subgroups. Simpson's Paradox, which is a major issue in causal inference and science, may be framed generally as a set of known or unknown regression relationships. It arises in settings where covariates Z_1 and Z_2 are positively associated, Z_1 affects an outcome Y while Z_2 does not, but because Z_2 and Y are observed but Z_1 is not observed one may be misled by a fitted model for $p(Y \mid Z_2)$ to believe that Z_2 has a causal effect on Y. Before invention of the Salk vaccine for poliomyelitis, the observation that Y = incidence of polio cases increased when Z_2 = soda pop consumption increased led to the widespread belief that drinking soda pop caused polio. Then someone noticed that both Y and Z_2 went up with Z_1 = temperature, which makes sense from a mechanistic causal viewpoint since people drink more soda pop and the incidence of polio went up during the hotter summer months. Settings where the existence of a true cause for a disease was not known, or a treatment was not scientifically evaluated, have led to

DOI: 10.1201/9781003474258-9

some phenomenally bad medical practices. A patient having a heart attack may die because their chest pain from a blocked artery is misdiagnosed as acid indigestion. A patient with acid reflux may undergo emergency open heart surgery because their chest pain is misdiagnosed as a heart attack.

Abraham Maslow observed, "To a man with a hammer, everything looks like a nail". In the past, physicians routinely bled patients, sometimes fatally, in an attempt to treat a variety of different maladies. Smoking cigarettes was prescribed as a cure for asthma. People were given "shock therapy" or lobotomized to cure a wide variety of behaviors diagnosed as being caused by mental disorders. These activities have little or no basis in scientific thinking, and they illustrate the immense power of established conventions. It remains to be seen what will be said, in some future time, about infusing poisons into cancer patents in an attempt to cure them.

Returning to the practice of statistical science, the three main objects observed in medical statistics are patient covariates, $\mathbf{Z}$, treatments X, and clinical outcomes, Y. They characterize the information that the patient brings, what a physician may do to deal with a patient's disease, and what is observed and interpreted as a consequence of the covariates and the treatment. The idea underlying a regression model $p(Y \mid X, \mathbf{Z})$ is that it explains how $\mathbf{Z}$, which cannot be controlled, and X, which can be controlled, together cause Y. Without regression models, modern causal analysis would not exist, and natural phenomena would be explained by imagined whims of unseen gods, angels, or monsters. The discussion of causality in Chapter 2 centered around the trick of taking the treatment decisions out of the physicians' hands by randomizing patients between treatments $X = 0$ and $X = 1$ to remove any causal arrows $\mathbf{Z} \to X$, as well as any unobserved, unknown latent variables, $Z^{unknown} \to X$. Randomization ensures that one may compute an unbiased estimator of the between-treatment effect on Y from observed data, using the fitted regression model $p(Y \mid X, \mathbf{Z})$. Randomization is not done to benefit the patients being randomized, but rather to be able to make a fair treatment comparison. To achieve this futuristic statistical goal, randomization abrogates medical practice, relying on the ethical rationale that there is no reason to prefer one treatment over the other, but learning statistically whether or not one treatment is likely to be better than the other, and fairly estimating the magnitude of the difference in effects on Y, may benefit future patients.

Statisticians assume regression models reflexively to describe relationships that they wish to explore. How do toxicity, response, and time to disease worsening vary with dose and patient characteristics? If treatment effects may differ between subgroups, how should similar subgroups be combined to obtain more reliable subgroup-specific inferences? How can patient covariates be used to reduce bias if only observational, non-randomized data are available? Given a vector of candidate biomarkers, which entries in the vector actually show that a patient's biology may enable a targeted agent to kill cancer cells? To do medical statistics, the question is not whether regression models are useful. Rather, it is how to decide what regression models to use. Simple parametric models usually are assumed because they are computationally convenient and easy to understand. But the array of regression models that currently are available is immense. See, for example, Chipman et al. (2010b), De Iorio et al. (2009), Dey et al. (2000), Dobson and Barnett (2008), Harrell (2001), Ibrahim et al.

(2001), or Parmigiani (2002), among many others. In practice, aside from simulation studies where a data generating model $p^{true}(Y \mid X, \mathbf{Z})$ is assumed, it cannot be known whether a regression model is "correct" for a dataset in some mathematical sense. What really matters is that the assumed model should not be badly incorrect. A regression model should be chosen to provide a good fit to the data at hand, reveal and, ideally, explain relationships among the observed variables, and possibly make reliable predictions. This can be done in a practical way by using a robust regression model, constructed to provide a good fit to a wide array of possible datasets.

The wealth of available regression models notwithstanding, there often is a large gap between the large statistical toolkit that is available and what actually is used in medical research. Most regression analyses reported in the medical literature include no consideration of whether the assumed model actually fits the data being analyzed, and models with very restrictive assumptions often are used. If the published medical literature is to be believed, all that one needs to know about regression analysis is how to use canned statistical software to fit a logistic model for binary outcomes or a Cox model for survival time data, and how to construct tables with lots of p-values, bearing in mind that the cutoff .05 has magical properties. These conventional practices get at a central problem in medical research, and more generally all scientific research. If an assumed regression model is an oversimplification that ignores important properties of a dataset, then inferences based on the fitted model can be very misleading, which can lead to poor decisions that may cost patients their lives. There are a myriad of ways that this can happen. In Chapters 1 and 2, I described a wide variety of examples illustrating what can go wrong when specifying a statistical regression model. I wrote a book describing a wide variety of statistical practices commonly used in medical research that obviously are flawed, and for each flawed practice provided a reasonable alternative method (Thall, 2020).

A common example of oversimplification in the medical literature is the use of an estimated hazard ratio (HR) to characterize a between-treatment effect based on survival time data. As an estimand, a HR is very easy to understand, since numerical values that differ substantively from 1 in either direction correspond to a meaningful treatment difference. But for many datasets, a single HR does not exist. This is the case, for example, when two hazard functions that are being compared vary over time and have different shapes, so that the hazard ratio varies as a function of time. This often can be seen by looking closely at the Kaplan Meier plots of two survival distributions being compared. If the plots cross each other, for example, then a proportional hazards (PH) assumption clearly is wrong, and there is no single HR. The medical literature is full of papers in which a Cox PH regression model is used to analyze survival time data, without any consideration of whether the model's PH assumption is valid for the data being analyzed. If $h_X(t, \mathbf{Z})$ denotes the survival time hazard function for a patient with covariates $\mathbf{Z}$ given treatment X = 0 or 1, then the hazard ratio at time t is $HR(t, \mathbf{Z}) = h_1(t, \mathbf{Z})/h_0(t, \mathbf{Z})$. For a Cox model, $h_X(t(\mathbf{Z})$ $= \lambda(t)\exp(\alpha X + \boldsymbol{\beta}\mathbf{Z})$ where $\lambda(t)$ is an unknown baseline hazard function, but the treatment effect α and covariate effects $\boldsymbol{\beta}$ do not vary with t. This model has the convenient implication that $HR(t, \mathbf{Z}) = \exp(\alpha)$ for all t and $\mathbf{Z}$, so the convention of assuming a Cox model has led to the convention of estimating a single hazard ratio parameter, and using this estimate to compare treatments. This does not work, however, if $X = 0$ and $X = 1$ have different baseline hazards, $\lambda_1(t)$ and $\lambda_0(t)$, or if some of the covariate effects in $\boldsymbol{\beta}$ differ between treatments due to treatment-covariate interactions, e.g. if $\gamma_1 X Z_1$ should be included in the model's linear term. In these cases, the HR is more complicated than a single parameter, since it varies with t or Z_1, so taking about "the hazard ratio" makes no sense. For example, this may occur if a dataset is obtained from a heterogeneous sample of patients having different histologies and prognostic variables, so that the actual survival distribution is a mixture of distributions having different forms, and it may depend on treatments in ways that differ both qualitatively and quantitatively. For example, if the survival distribution for patients receiving salvage therapy for an advanced cancer has two modes because patients are likely to either die within three months (the salvage therapy did not work) or survive for at least a year (the salvage therapy worked), then talking about "median survival time" is very misleading. While this sort of observation may be made from simply inspecting a Kaplan-Meier curve, it is common practice to compute median survival time unthinkingly when reporting survival data. Similar problems arise in regression analysis with a real-valued variable that is not well approximated by a normal distribution, which may be seen if the histogram of the variable is not bell-shaped, is highly skewed, has multiple modes, or has spikes of positive probability at one or more numerical values.

9.2 Dirichlet Process Priors

Chapters 10, 11, and 12 will describe regression analyses of observational or randomized trial data from three different oncology settings, to evaluate either treatments or DTRs, each using a Bayesian nonparametric (BNP) regression model. To provide a conceptual basis for these applications, the rest of this chapter will provide a review of BNP models.

George Box stated, "All models are wrong, but some are useful". This famous quote may be extended by adding the observation that some models are more useful than others. BNP models are a very flexible family of statistical tools that can overcome the limitations of most conventional parametric statistical models. A major advantage of BNP models is that they can accurately approximate any distribution or function, a property known as *full support*. BNP models can be constructed to accommodate a very broad set of statistical problems, including density estimation, regression analysis (MacEachern and Müller, 1998; MacEachern, 1999), graphical modeling, machine learning, causal inference (Hill, 2011), classification, clustering, prediction, and spatiotemporal modeling (Reich and Fuentes, 2007). Fitted BNP models often identify unexpected structures in a dataset that are unlikely to be seen using conventional statistical models. Examples include clusters, complex treatment-covariate interactions, multiple modes, and unexpected patterns of change for a variable observed repeatedly over time. Reviews of BNP models are given by

De Iorio et al. (2009), Hjort et al. (2010), Gershman and Blei (2012), Müller and Mitra (2013), and Müller et al. (2016). A wide variety of applications are described by Müller and Rodriguez (2013), Mitra and Müller (2015), and Thall et al. (2017).

To explain BNP models, it is useful first to review conventional Bayesian models. Temporarily ignoring covariates, Classical Bayesian statistics starts by assuming a parametric probability model, $F_{\boldsymbol{\theta}}(\cdot)$, for an observable outcome Y characterized by a finite vector $\boldsymbol{\theta} = (\theta_1, \cdots, \theta_q)$ of parameters, including a prior distribution $p(\boldsymbol{\theta} \mid \tilde{\boldsymbol{\theta}})$, where $\tilde{\boldsymbol{\theta}}$ is a vector of fixed hyperparameters. A central idea underlying either a Bayesian or frequentist probability modeling is that the assumed functional form $F_{\boldsymbol{\theta}}$ and its q-dimensional parameter vector $\boldsymbol{\theta}$ do a good job of characterizing the unknown distribution of Y. A very common example is a normal model $Y_1, \cdots, Y_n \sim iid\ N(\mu, \sigma^2)$, which relies on the strong assumption that the symmetric bell-shaped form of the normal pdf does a good job of characterizing the unknown distribution F. Advantages of this model are that its parameters are easy to understand and easy to estimate. From a Bayesian viewpoint, assuming this particular form and putting priors on μ and σ^2 is a nicely structured but indirect way of putting a prior on F. While no model is correct, the practical issue is whether the model assumptions are a useful approximation that will lead to reliable statistical inferences using the fitted model. Since infinitely many functional forms are possible for F, a major disadvantage of assuming a normal distribution is that a given dataset fit to the model and subjected to goodness-of-fit assessments may show that a normal model fits the data poorly. Bimodal or skewed distributions, which can be seen by simply drawing a histogram, are common examples where a normal model obviously is wrong. For example, if a distribution has two modes, it may not make sense to talk about a single mean. If a model is wrong enough to be inadequate or inappropriate for making reliable inferences, then it should not be used. These issues apply to any parametric probability model, and they are central to statistical inference. Beyond simply assuming a convenient model, fitting it, and then assessing its fit, one may formulate a model so that it will be robust and thus likely to provide a good fit for a wide variety of possible datasets. The price of assuming a robust parametric model is that its higher parameter dimension and complexity may make it harder to fit, and sometimes harder interpret.

In a landmark paper in which he defined the first BNP model, Ferguson (1973) proposed a family of Bayesian probability models that are straightforward to interpret and very broadly applicable. Motivated by the desire to deal with nonparametric problems, Ferguson reasoned that the prior should have a large support in order to address the limitation that, conventionally, a probability distribution F, which may be infinite dimensional, is characterized by a finite dimensional vector $\boldsymbol{\theta}$ under an assumed parametric model. He avoided this limitation by defining a prior on F itself without assuming a parametric form $F(\cdot \mid \boldsymbol{\theta})$. To do this, he proposed the Dirichlet process (DP) prior.

In order explain a DP prior, it is useful to begin by reviewing the Dirichlet distribution, since it has very desirable properties that the DP exploits. Consider the experiment of drawing an object at random from a heterogeneous population that has been partitioned into k subpopulations having respective proportions $\boldsymbol{p} = (p_1, \cdots, p_k)$, with

$p_1 + \cdots + p_k = 1$. If this experiment is repeated n times, and R_j is the number of draws from the jth subpopulation, then $R_1 + \cdots + R_k = n$ and $\boldsymbol{R} = (R_1, \cdots, R_k)$ follows a multinomial distribution with parameters n and $\boldsymbol{p}$, denoted $\boldsymbol{R} \sim Mult(n, \boldsymbol{p})$. The counts have means $E(R_j) = np_j$, variances $var(R_j) = np_j(1 - p_j)$, and covariances $cov(R_j, R_\ell) = -np_j p_\ell$ for $1 \le j \ne \ell \le n$. A very tractable prior on the probability vector $\boldsymbol{p}$ is the *Dirichlet distribution*,

$$f(\boldsymbol{p} \mid \boldsymbol{\alpha}) = \frac{\Gamma(\alpha_+)}{\prod_{j=1}^k \Gamma(\alpha_j)} \prod_{j=1}^k p_j^{\alpha_j - 1} \tag{9.1}$$

where $\Gamma(\cdot)$ is the gamma function, $\boldsymbol{\alpha} = (\alpha_1, \cdots, \alpha_k)$, each hyperparameter $\alpha_j > 0$, and the sum is $\alpha_+ = \sum_{j=1}^k \alpha_j$. This is denoted by $\boldsymbol{p} \sim Dir(\boldsymbol{\alpha})$. The means are $E(p_j) = \alpha_j/\alpha_+$, the variances are $var(p_j) = \alpha_j(1 - \alpha_j)/(\alpha_+ + 1)$, and covariances are $\text{cov}(\alpha_j, \alpha_r) = -\alpha_j \alpha_r / \alpha_+$ for $1 \le j \ne r \le k$. One may think of α_j as the size and α_j/α_+ as the prevalence of the jth subpopulation. The posterior is $\boldsymbol{p} \mid \boldsymbol{R} \sim Dir(\boldsymbol{\alpha} + \boldsymbol{R})$, so the Dirichlet prior on $\boldsymbol{p}$ is conjugate for the multinomial distribution. An important special case is the beta distribution, when $k = 2$, denoted by $p \sim Beta(\alpha_1, \alpha_2)$, which has pdf

$$f(p) = \frac{\Gamma(\alpha_1 + \alpha_2)}{\Gamma(\alpha_1)\Gamma(\alpha_2)} p^{\alpha_1 - 1}(1 - p)^{\alpha_2 - 1}, \quad \text{for} \quad 0 \le p \le 1,$$

mean $E(p) = \mu = \alpha_1/(\alpha_1 + \alpha_2)$, and variance $var(p) = \mu(1 - \mu)/(\alpha_1 + \alpha_2 + 1)$. The beta is a conjugate prior for the $\text{Binomial}(p, n)$ distribution for R, with $[p \mid R, n] \sim Beta(\alpha_1 + R, \alpha_2 + n - R)$.

The Dirichlet distribution has the useful property that if some categories are combined, then the resulting distribution is Dirichlet with corresponding parameters summed. For example, if $k = 4$ and categories 3 and 4 are combined, then the multinomial vector becomes $(R_1, R_2, R_3 + R_4) \sim Mult(n, (p_1, p_2, p_3 + p_4))$, and starting with prior $(p_1, p_2, p_3, p_4) \sim Dir(\alpha_1, \alpha_2, \alpha_3, \alpha_4)$, it follows that

$$(p_1, p_2, p_3 + p_4) \sim Dir(\alpha_1, \alpha_2, \alpha_3 + \alpha_4).$$

This implies, in particular, that each entry of a Dirichlet is beta, formally if $\boldsymbol{p} \sim Dir(\boldsymbol{\alpha})$, then for any j the marginal is $p_j \sim Beta(\alpha_j, \alpha_+ - \alpha_j)$. The multinomial-Dirichlet model thus accommodates clustering categories, and it includes the binomial-beta model as a special case.

While most often the DP is assumed as a prior for the probability distribution of a sample of real-valued random variables, it is very general and can serve as the prior of a random variable taking on values in any measurable space. The random objects may be, for example, vectors, partitions, differential equations, matrices, types of T-cells, or hurricane wind fields. Ferguson defined a Dirichlet process in the context of the problem of estimating a probability density based on a sample $Y_1, \cdots, Y_n$ from an unknown distribution F. Rather than assuming a parametric model for F, he specified a Dirichlet process prior on F, treating it as an infinite dimensional parameter.

Let S be the space where a random variable Y takes on values, with $\mathscr{A}$ a sigma-field of measurable sets in S, and let F be a probability measure on $(S, \mathscr{A})$. The following structure for a DP prior on F provides a robust model for Bayesian analysis of a sample $Y_1,\cdots,Y_n \sim iid\ F$. In particular, assuming a DP prior avoids the conventional process of specifying a parametric model $F_{\boldsymbol{\theta}}(y)$ characterized by a finite dimensional $\boldsymbol{\theta}$ and specifying a prior $p(\boldsymbol{\theta} \mid \tilde{\boldsymbol{\theta}})$ with fixed hyperparameters $\tilde{\boldsymbol{\theta}}$. A DP prior is characterized by two hyper-parameters, a positive real-valued *total mass parameter* $M > 0$, and a *base probability measure* $F_0 = E(F)$ defined on $(S,\mathscr{A})$, so $F_0(S) = 1$. The definition of a DP prior on F indexed by sets in $\mathscr{A}$ is as follows.

DEFINITION. The probability measure F on $(S,\mathscr{A})$ is a *Dirichlet process with parameters M and F_0* if, for any measurable finite partition $\{A_1,\cdots,A_k\}$ of S,

$$F(A_1),\cdots,F(A_k) \mid M,F_0 \sim Dirichlet(MF_0(A_1),\cdots,MF_0(A_k)). \tag{9.2}$$

This is denoted $F \sim DP(M,F_0)$. In particular, definition (9.2) implies that, for any measurable set A,

$$F(A) \mid M,F_0 \sim Beta(MF_0(A),M[1-F_0(A)]),$$

so $F(A)$ has prior mean $F_0(A)$ and prior variance $F_0(A)[1-F_0(A)]/(M+1)$. This shows that a larger prior total mass M corresponds to greater prior precision.

A very useful property of the DP is that it is conjugate for any distribution F. Denote the degenerate probability distribution with mass 1 on the point Y by δ_Y. Exploiting the properties of the Dirichlet distribution, given a sample $Y_1,\cdots,Y_n \sim iid\ F$, if $F \sim DP(M,F_0)$ then the posterior also is a Dirichlet process,

$$\begin{aligned} F \mid Y_1,\cdots,Y_n,M,F_0 &\sim DP\left(M+n,\ \frac{M}{M+n}F_0 + \frac{1}{M+n}\sum_{i=1}^{n}\delta_{Y_i}\right) \\ &= DP\left(M+n,\ \frac{M}{M+n}F_0 + \frac{n}{M+n}\hat{F}_n\right), \end{aligned} \tag{9.3}$$

where $\hat{F}_n = \frac{1}{n}\sum_{i=1}^{n}\delta_{Y_i}$ denotes the empirical distribution function. This says that, given the sample, *a posteriori*, the prior total mass parameter M is increased by adding the sample size, and the DP posterior has a base measure that is a weighted average of the prior probability F_0 and the empirical probability distribution $\hat{F}_n$, with the shrinkage toward F_0 decreasing with larger sample size n.

Sethuraman (1994) provided an equivalent, constructive definition of a DP prior, called the *stick-breaking* representation, that is very useful in practice. Let $\theta_1,\theta_2,\cdots$ be a sequence of *iid* elements of S sampled from the base probability measure F_0, and let $\upsilon_1,\upsilon_2,\cdots$ be a sequence of random probabilities that are $iid{\sim}Beta(1,M)$. Denote

$$F = \sum_{h=1}^{\infty} w_h\delta_{\theta_h}, \tag{9.4}$$

where the weights are $w_1 = \upsilon_1$ and $w_h = \upsilon_h\prod_{\ell<h}(1-\upsilon_\ell)$ for each $h \geq 2$. Sethuraman (1994) proved that, using this construction, $F \sim DP(M,F_0)$. The steps of the construction are as follows. First, sample $\theta_1 \sim F_0$, sample $\upsilon_1{\sim}Beta(1,M)$, and give

weight $w_1 = \upsilon_1$ to δ_{θ_1}. This may be thought of as breaking a piece of length w_1 from a probability stick of length 1, which gives the proportion $1 - w_1$ of the stick remaining. Next, sample $\theta_2 \sim F_0$, sample $\upsilon_2 \sim Beta(1,M)$, and break off a second piece of length $w_2 = \upsilon_2(1 - \upsilon_1)$ from the remaining stick, which leaves the proportion $1 - w_1 - w_2$ of the stick and gives weight w_2 to δ_{θ_2}. Continue this process for probabilities $w_3, w_4, \cdots$, successively breaking off a piece of the remaining stick at each step, so that by construction $\sum_{h=1}^{\infty} w_h = 1$. This construction shows how, given M and F_0, the weights and point mass locations used to define F may be generated in sequence, which facilitates computation. It also shows that the prior has a positive probability of multiple values of Y in the support of F being equal to each other. One may think of each set of identical Y values in a sample as a cluster. When carrying out this construction in practice, the sum over h does not actually go to ∞, but is stopped at some convenient finite number, depending on the dataset being analyzed.

A finite approximation to a $DP(M, F_0)$ distribution may be obtained by fixing an integer H, generating a finite set of H weights from a uniform Dirichlet distribution,

$$(w_1, \cdots, w_H) \sim Dirichlet\left(\frac{M}{H}, \cdots, \frac{M}{H}\right)$$

and specifying any H fixed locations $\theta_1, \cdots, \theta_H \in S$. If G_H is the distribution with probability mass w_h at θ_h for each $h = 1, \cdots, H$, denoting the empirical distribution $\hat{G}_H = \frac{1}{H}\sum_{h=1}^{H} \delta_{\theta_h}$, it follows that, for any finite partition $\{A_1, \cdots, A_J\}$ of S,

$$G_H(A_1), \cdots, G_H(A_J) \sim Dirichlet\big(M\hat{G}_H(A_1), \cdots, M\hat{H}_H(A_J)\big).$$

But by definition, this implies that $G_H \sim DP(M, \hat{G}_H)$. Elaborating this construction by generating the locations as random quantities $\theta_1, \cdots, \theta_H \sim$ *iid* F_0, since the empirical distribution $\hat{G}_H \to F_0$ uniformly as $H \to \infty$, this is a heuristic proof that G_H is an approximation of a $DP(M, F_0)$ distribution.

The desirable properties of the DP notwithstanding, it has the limitation that any $F \sim DP$ has discrete support, which is shown by the representation in equation (9.4). Since this may be inappropriate for applications with continuous variables $Y \sim F$, a useful practical extension is to elaborate the DP by defining a *DP mixture (DPM) distribution*, given by

$$F(y) = \int_{\boldsymbol{\theta}} \phi(y \mid \boldsymbol{\theta}) dG(\boldsymbol{\theta}), \tag{9.5}$$

where the kernel function $\phi(y \mid \boldsymbol{\theta})$ is a continuous probability density indexed by the parameter vector $\boldsymbol{\theta}$ and the mixing distribution G follows a $DP(M, F_0)$ prior. The DPM is a continuous prior on the probability measure F. The DPM may be represented equivalently as a hierarchical model, in which the indices $\boldsymbol{\theta}_1, \boldsymbol{\theta}_2, \cdots$ of the kernel function are endowed with the hyperprior G,

$$\begin{aligned} Y_i \mid \theta_i &\sim \ indep\ \phi(y_i \mid \theta_i), \quad \text{for } i = 1, 2, \cdots \\ \theta_i &\sim \ iid\ G \\ G &\sim \ DP(M, F_0). \end{aligned} \tag{9.6}$$

This hierarchical representation gives G as the level 1 prior of the θ_i's and the $DP(M, F_0)$ as the level 2 prior of G. Marginalizing with respect to the $\boldsymbol{\theta}_i$'s gives

$$Y_1, Y_2, \cdots \mid \phi, G \sim iid \int_{\boldsymbol{\theta}} \phi(\cdot \mid \boldsymbol{\theta}) dG(\boldsymbol{\theta}) = F(\cdot),$$

which is equivalent to expression (9.5). The DPM may be denoted compactly as $F \sim DPM(M, F_0, \phi)$.

A very popular choice for ϕ is a normal pdf, with additional priors assumed on its mean and variance. Denote a sequence of normal means by $\{\mu_h\}$ and variance $\sigma^2 > 0$. With a slight abuse of notation, denote the $N(\mu, \sigma^2)$ pdf by $\phi(y \mid \mu, \sigma^2)$. A normal DPM model is completed by assuming the hyperpriors $\mu \sim N(\tilde{\mu}, \tilde{\sigma}^2)$ and $\sigma^2 \sim IG(\tilde{a}, \tilde{b})$. This gives a DPM as a weighted average of normal distributions,

$$F(y) = \int \phi(y \mid \mu, \sigma^2) dG(\theta) = \sum_{h=1}^{\infty} w_h \phi(y \mid \mu_h, \sigma^2), \tag{9.7}$$

where $G = \sum_{h=1}^{\infty} w_h \delta_{\mu_h} \sim DP(M, G_0)$. Expression (9.7) shows that G puts probability mass w_h on the mixture component $\phi(\cdot \mid \mu_h, \sigma^2)$. Since F is a weighted average of normal distributions, it has smooth support. The DPM with a normal kernel is extremely useful as a practical tool for density estimation. This model has the desirable property that any smooth pdf can be approximated to any specified degree of error by a mixture or normals. DPM models were introduced by Ferguson (1983), and have been applied widely. See, for example, Escobar and West (1995), Müller and Rodriguez (2013), and Müller et al. (2016).

The following example illustrates how a DPM with a normal kernel may be used to do density estimation. Xu et al. (2017) used a slight extension of a DPM to estimate a probability distribution in the context of constructing a design for a randomized trial to compare a gel sealant, Progel, to standard of care for resolving intra-operative air leaks (IALs) after pulmonary resection. IALs are a severe problem in patients who have undergone a pleurectomy decortication, an extensive surgical procedure in which the lining surrounding one lung is removed and then tumors growing inside the chest cavity are removed. A prolonged IAL may lead to multiple adverse events, including longer chest tube drainage, greater postoperative pain, and increased risk of infection, leading to longer hospitalization. Standard of care is to suture visible leaks and use staples as reinforcement, but this gives unpredictable results. The outcome variable used as a basis for the design was Y = time to resolve an IAL. To accommodate the possibility that no air leak develops, $Y = 0$, Xu et al. (2017) assumed a slightly elaborated DPM distribution in each treatment arm, taking the form

$$F(y) = \nu \delta_0(y) + (1 - \nu) \sum_{h=1}^{\infty} w_h \phi(y, \mu_h, \sigma^2), \tag{9.8}$$

where $\nu = \Pr(Y = 0)$. Equation (9.8) generalizes (9.7) by including a point mass $Y = 0$. The left side of Figure 9.1 gives a histogram of Y obtained from historical data, showing a long tail, possibly multiple modes, and a spike at $Y = 0$. The right side of the figure shows the DPM normal mixture model-based density estimate of Y

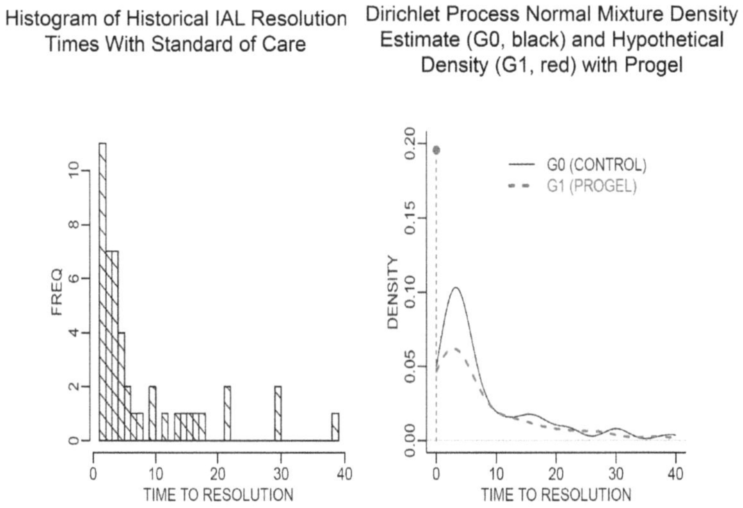

FIGURE 9.1
Left: Histogram of the time to resolve an intraoperative air leak, including the value 0 if no air leak developed, for standard of care. Right: Dirichlet Process Mixture Model fit to the historical data, with an additional point mass at 0 if no air leak develops, and a hypothetical distribution corresponding to using Progel.

(in black), and also includes a hypothetical distribution (in red) corresponding to the use of Progel rather than sutures and staples. This shows that the slightly elaborated mixture of normal pdfs (9.8) gives quite a good smooth estimate of an extremely irregular distribution, with shrinkage toward the prior to prevent overfitting the data. This is typical of DPM-based density estimates.

Returning to a general DP, to derive the predictive distribution of a future value Y_{n+1} given a sample $Y_1, \cdots, Y_n$ from F following a $DP(M, F_0)$ prior, note that Y_{n+1} is sampled from the posterior

$$F \mid Y_1, \cdots, Y_n \sim DP\left(M+n, \frac{M}{M+n}F_0 + \frac{n}{M+n}\hat{F}_n\right).$$

This implies that, for any measurable set A, the predictive distribution of a future observation Y_{n+1} is given by

$$\begin{aligned}
\Pr(Y_{n+1} \in A \mid Y_1, \cdots, Y_n) &= E\{F(A) \mid Y_1, \cdots, Y_n\} \\
&= \frac{M}{M+n}F_0(A) + \frac{1}{M+n}\sum_{i=1}^{n} \delta_{Y_i}(A),
\end{aligned}$$

which says that the predictive distribution is the weighted average

$$Y_{n+1} \mid Y_1, \cdots, Y_n \sim \frac{M}{M+n}F_0 + \frac{1}{M+n}\sum_{i=1}^{n} \delta_{Y_i}. \tag{9.9}$$

Thinking of a sample as it is accumulated, for $n = 1,2,,\cdots$, the sequence of predictive distributions is called a *Polya urn scheme* (Blackwell and MacQueen, 1973), which can be explained using the following metaphor about successively placing colored balls into an urn based on the sampled values of Y. Assume that $Y \sim F$ with prior $F \sim DP(M, F_0)$, and also that each element in the space S of observed Y values has its own unique color. At the start, the urn is empty. One begins by drawing the first sample value $Y_1 \sim F$, paints a ball the color of Y_1, and places the colored ball into the urn. The posterior of F is now the mixture $DP(M+1, \frac{M}{M+1}F_0 + \frac{1}{M+1}\delta_{Y_1})$ of the base measure F_0 and a point mass at Y_1. At step 2, one draws a second observation Y_2 from the current posterior, $Y_2 \sim F \mid Y_1$. A second ball is painted the color of Y_2 and placed into the urn. According to the DP posterior, either $Y_2 \neq Y_1$, and thus Y_2 will be a different color than Y_1, which occurs with probability $M/(M+1)$, or $Y_2 = Y_1$ and therefore Y_2 will be the same color as Y_1, which occurs with probability $1/(M+1)$. The urn now contains two balls that are either different colors with probability $M/(M+1)$ or the same color with probability $1/(M+1)$. Continuing this process, at step $n+1$, after $Y_1, \cdots, Y_n$ have been observed and a total of n colored balls have been placed into the urn, the $n+1^{st}$ ball is either a new color with probability $M/(M+n)$ or is one of the first n colors with probability $n/(M+n)$. An important property of the predictive distribution (9.9) and the urn scheme is that, since the $n+1^{st}$ draw has a point mass of size $1/(M+n)$ at each previous sample value $Y_1, \cdots, Y_n$, there is a positive probability that the urn will have multiple balls of the same color. This is the case regardless of the base measure F_0 that was assumed for the DP prior.

One way to think about the urn scheme for a DP is to consider each set of balls having the same color to be a cluster, with the clusters corresponding to a partition of the integers $\{1, \cdots, n\}$. If the clusters of balls having the same color are indexed by $j = 1,2,\cdots,k_n$ for $H \leq n$, then all Y_i's in the jth cluster are the same value. where k_n is the number of induced clusters from a sample of size n. Recall that the ith observation Y_i may be either a new value different from $\{Y_1, \cdots, Y_{i-1}\}$ and thus increase the number of clusters by 1, or the same as one of the previous values. Consequently, since $k_1 = 1$, for each $i \geq 2$, the distribution of the number of clusters is

$$\begin{aligned} \Pr(k_i = k_{i-1}) &= \frac{i-1}{M+i-1} \\ \Pr(k_i = k_{i-1}+1) &= \frac{M}{M+i-1}. \end{aligned}$$

Denoting the digamma function by $\psi(z) = \Gamma'(z)/\Gamma(z)$, since $\Gamma(n+1) = n\Gamma(n)$, this recursion gives

$$E(k_n) = \sum_{i=1}^{n} \frac{M}{M+i-1} = M\{\psi(M+n) - \psi(M)\} \approxeq M\log\left(1+\frac{n}{M}\right).$$

Similarly, the variance is

$$var(k_n \mid n) = M\{\psi(M+n) - \psi(M)\} + M^2\{\psi'(M+n) - \psi'(M)\} \approxeq M\log\left(1+\frac{n}{M}\right)$$

for $n > M$. This shows that the number of clusters increases at a rate approximately proportional to the logarithm of the sample size.

Using a more culinary metaphor, the probability distribution of the partition induced by the urn scheme, equivalently induced by sampling values from a distribution with a $DP(M,F_0)$ prior, may be called a *Chinese restaurant process*. The Chinese restaurant process requires the assumptions that there is a Chinese restaurant with an infinite number of tables, each table can accommodate an unlimited number of customers, customers enter the restaurant one at a time, and they are not only willing but more likely to sit with a large group of complete strangers. So it is an infinitely large restaurant, which suggests that the food must be really good. Under these assumptions, here is the Chinese restaurant process. The first customer enters the restaurant and sits at any table. Then a second customer enters the restaurant and randomly either sits at the first customer's table with probability $1/(M+1)$, or sits alone at a new table with probability $M/(M+1)$. If the first two customers end up sitting at the same table, then the induced partition has one set, $\{1,2\}$, but if they sit at different tables then the induced partition is $\{1 \mid 2\}$. If the first two customers have sat at different tables, then the third customer sits with one of them with probabilities $1/(M+2)$ each, creating a two-set partition that is either $\{1,\ 3 \mid 2\}$ or $\{1 \mid 2,\ 3\}$, or sits at a different, third table, inducing the three-set partition $\{1 \mid 2 \mid 3\}$. Continuing in this manner, the $n+1^{st}$ customer sits at a new table with probability $M/(M+n)$ and at a table with, say, R_j customers with probability $R_j/(M+n)$. This is known as the *rich get richer* property of the Chinese restaurant process, since it is more likely that a new customer will sit at a table with more customers, that is, a table with larger R_j. The underlying assumptions may be made somewhat more realistic by assuming that there are a finite number of tables, possibly by following the finite approximation of a $DP(M,F_0)$ given earlier, and only considering fixed values of n.

The slow rate of growth for k_n, proportional to $\log(n)$, is a consequence of the *rich-gets-richer* property of the sampling scheme, in which the larger the number of customers sitting at a table, the more likely it is that a new customer will sit at that table. In the Polya urn scheme metaphor, the larger the number of balls of a given color, the more likely it is that a new ball will be that color. In terms of sampling sequentially from a distribution F, this says that, given $Y_1,\cdots,Y_n$, not only does Y_{n+1} have a positive probability of equaling one of the previous values, but it is more likely to equal previous values that have appeared repeatedly in the sample, and the probability of equaling a particular Y_j is proportional to the number of times that the value of Y_j has been repeated in the current sample.

Teh et al. (2004) provided a hierarchical DP mixture model called the *Chinese restaurant franchise*, which consists of J Chinese restaurants, each a subpopulation, with n_j customers $Y_{j,1},\cdots,Y_{j,n_j}$, entering restaurant $j = 1,\cdots,J$, according to latent factor variables $\phi_{j,1},\cdots,\phi_{j,n_j} \sim$ *iid* G_j which themselves have DP hyperpriors, $G_1,\cdots,G_J \mid M,F_0 \sim$ *iid* $DP(M,F_0)$. Thus, the distribution $F(\cdot \mid \phi_{j,i})$ of $Y_{j,1}$ is indexed by the latent factor $\phi_{j,i}$, which itself is indexed by both the customer index i and the restaurant index j. This hierarchical DP structure may be written as follows:

$$\begin{aligned} Y_{j,i} \mid \phi_{j,i} &\sim F(\cdot \mid \phi_{j,i}) \\ \phi_{j,i} \mid G_j &\sim G_j \\ G_1, \cdots, G_J &\sim iid\ DP(M_0, G_0) \\ G_0 \mid M, F_0 &\sim DP(M, F_0) \end{aligned} \tag{9.10}$$

The hierarchical DP mixture model has a restaurant-specific likelihood for the observed Y_{ij}'s, restaurant-specific DP priors for the latent $\phi_{j,i}$'s, a DP prior for the DP distributions of the latent factors, and a DP hyperprior for the base measure G_0 of the DP priors on the latent factors. The distribution G_j corresponds to the jth restaurant, which is the metaphor for a subgroup. Given a likelihood for the observed Y_{ij}'s, the two basic ideas of the Chinese restaurant franchise are to introduce latent subgroup (table) variables for each restaurant (subpopulation), and assume DP priors for everything, including the latent variables and all distributions.

The discreteness of G_0 allows the DP mixture components to be shared between restaurants in the franchise. In the jth restaurant, the way that the n_j customers sit corresponds to a partition of $\{\phi_{j,1}, \cdots, \phi_{j,n_j}\}$, so the latent variables determine clusters (tables). The Chinese restaurant franchise is an example of a class of hierarchical Bayesian models in which latent random subgroup membership variables are included to determine clusters, with an additional prior or hyperprior on the latent variables. Such latent variable structures are extremely useful in practice. Latent variables are components of the models, described in earlier chapters, for the Subtite design of Chapple and Thall (2018), the precision phase I-II design of Lee et al. (2021), the basket trial design of Psioda et al. (2019), the precision randomized phase II screen-and-select design of Lee, Thall and Msaouel (2022), and the BAGS design of Lin et al. (2021a). To compute posteriors, a very useful aspect of MCMC algorithms is that they treat latent variables like additional parameters, with a key issue being how and when the latent variables and parameters are sampled in the posterior computation algorithm.

9.3 Dependent Dirichlet Processes

MacEachern (1999) proposed a very flexible class of Bayesian regression models by extending the stick-breaking representation of a DP for a random probability distribution. Given a set $\mathscr{Z}$ of covariates $\mathbf{Z}$, he defined a *dependent Dirichlet process* (DDP) to be a family $\{G_{\mathbf{Z}} : \mathbf{Z} \in \mathscr{Z}\}$ of random probability measures for $Y \mid \mathbf{Z}$ where each $G_{\mathbf{Z}}$ follows a DP. This gives the prior distribution of $Y \mid \mathbf{Z}$ as a DP indexed by $\mathbf{Z}$, formally

$$Y \mid \mathbf{Z} \overset{indep}{\sim} G_{\mathbf{Z}}.$$

A DDP is obtained by extending the stick-breaking construction of a DP so that the weight and location parameters of $G_{\mathbf{Z}}$ are indexed by $\mathbf{Z}$. For each $\mathbf{Z} \in \mathscr{Z}$, the weights are

$$w_h(\mathbf{Z}) = \upsilon_h(\mathbf{Z}) \prod_{r<h} (1 - \upsilon_r(\mathbf{Z})) \quad \text{with} \quad \upsilon_h(\mathbf{Z}) \overset{iid}{\sim} beta(1, q_{\mathbf{Z}}), \tag{9.11}$$

and the atoms in the space of outcomes of the DP base probability measure $H_{\mathbf{Z}}$ are independent stochastic processes $\eta_1(\mathbf{Z}), \eta_2(\mathbf{Z}), \cdots$. The weights $\{w_h(\mathbf{Z})\}$ and atoms $\{\eta_h(\mathbf{Z})\}$ are assumed to be independent. The stick-breaking representation gives the prior as a mixture

$$G_{\mathbf{Z}} = \sum_{h=1}^{\infty} w_h(\mathbf{Z}) \delta_{\eta_h(\mathbf{Z})}.$$

A DDP characterizes regression of Y on $\mathbf{Z}$ through the parameter $q_{\mathbf{Z}}$ in the beta distribution used to generate the weights via the stick breaking algorithm, and by the distribution of the locations of the prior point masses $\eta_h(\mathbf{Z})$ of Y. Due to its immense usefulness, the family of DDPs has grown substantially over the years. Computational methods for DDPs are given by MacEachern and Müller (1998). Barrientos et al. (2012) provide an explanation of full support properties, and Quintana et al. (2022) provide a comprehensive review.

In most applications of DDPs, regression of Y on $\mathbf{Z}$ is characterized using only one of the two functions, either $q_{\mathbf{Z}}$ in the beta distributions or the atom locations $\eta_h(\mathbf{Z})$, with the other function specified as a sequence of random variables that do not vary with $\mathbf{Z}$. The most commonly used DDP is the *single weights* version. This is constructed using a weight distribution that does not depend on $\mathbf{Z}$, with $\{w_h\}$ obtained using the usual beta in the stick-breaking construction, and the $\eta_h(\mathbf{Z})$'s independent stochastic processes. The mixture representation of a DDP is

$$G_{\mathbf{Z}} = \sum_{h=1}^{\infty} w_h \delta_{\eta_h(\mathbf{Z})}. \tag{9.12}$$

The *ANOVA DDP*, proposed by DeIorio et al. (2004) is a version of a single weights DDP in which the entries of $\mathbf{Z}_i$ represent factors in an analysis of variance model for a designed experiment, including main effects, interactions, and identifiability constraints. In this case, the set $\mathscr{Z}$ of covariate vectors is discrete and represents elements of a design matrix, with linear terms $\eta_h(\mathbf{Z}) = \boldsymbol{\beta}'_h \boldsymbol{d}_{\mathbf{Z}}$ for the atoms in (9.12), where $\{\boldsymbol{\beta}_h, h = 1, 2, \cdots\}$ is a sequence of iid random vectors with distribution G_0 and $\boldsymbol{d}_{\mathbf{Z}}$ is a design vector defined in terms of factorial covariates in $\mathbf{Z}$. Each $G_{\mathbf{Z}}$ has a DP prior with atoms at the $\eta_h(\mathbf{Z})$'s. To extend this to an analysis of covariance model that also includes continuous covariates, an ANOVA DDP may be defined as a DPM with normal kernels,

$$Y_i \mid G_{\mathbf{Z}_i} \sim indep \int \phi(\cdot \mid \boldsymbol{\beta}' \boldsymbol{d}_{\mathbf{Z}}, \sigma^2) dG(\boldsymbol{\beta}).$$

For example, one may assume the prior $\sigma^2 \sim IG(a,b)$. This DPM may be summarized in the hierarchical form

$$\begin{aligned} Y_i \mid \boldsymbol{\beta}_i, \sigma^2 &\sim indep\, N(\boldsymbol{\beta}_i' \boldsymbol{d}_{\mathbf{Z}_i}, \sigma^2) \\ \boldsymbol{\beta}_1, \boldsymbol{\beta}_2, \ldots &\sim iid\ G \\ G &\sim DP(M, G_0). \end{aligned}$$

A *single atoms* DDP is obtained if the atoms in a DDP are assumed to be the same sequence $\eta_1, \eta_2, \cdots$ for all $\mathbf{Z}$, while the weights $w_h(\mathbf{Z})$ vary with $\mathbf{Z}$ according to the covariate dependent stick-breaking construction given in (9.11). Thus, for a single atoms DDP, regression of Y on $\mathbf{Z}$ is characterized by the beta parameters $q_{\mathbf{Z}}$ in the distribution $\upsilon_h(\mathbf{Z}) \sim iid\ beta(1, q_{\mathbf{Z}})$ that determines $w_h(\mathbf{Z}) = \upsilon_h(\mathbf{Z}) \prod_{r<h}(1 - \upsilon_r(\mathbf{Z}))$ in the stick-breaking process. This says that the prior on the partition of [0, 1] induced by the weights varies with $\mathbf{Z}$, which may be useful when the focus is on the partition and it is desired to reflect heterogeneity by allowing the partition to vary with the characteristics $\mathbf{Z}$ of each observation. There are numerous other formulations of DDP models, tailored to address a wide variety of regression problems. A comprehensive review is provided by Quintana et al. (2022).

Another useful proposal made by MacEachern (1999) is to assume a Gaussian process (GP) as a prior on the set of $\eta_h(\mathbf{Z})$'s in a DDP. First, to define a GP, for simplicity temporarily suppress the subscript h, and let $\mathbf{Z}_1, \cdots, \mathbf{Z}_n$ be a sample of n covariate vectors in the p-dimensional reals, $\mathscr{R}^p$. A GP prior for a random function f on $\mathscr{R}^p$ is characterized by a real-valued mean function $\mu(\mathbf{Z})$ on $\mathscr{R}^p$ and a covariance function $C(\mathbf{Z}_j, \mathbf{Z}_r)$ on $\mathscr{R}^p \times \mathscr{R}^p$. The definition of a GP is as follows.

DEFINITION A random function f is a *Gaussian process with mean function* μ *and covariance matrix* C if, for any sample $\mathbf{Z}_1, \cdots, \mathbf{Z}_n$ of p-vectors, $(f(\mathbf{Z}_1), \cdots, f(\mathbf{Z}_n))$ follows a multivariate normal distribution with mean vector $(\mu(\mathbf{Z}_1), \cdots, \mu(\mathbf{Z}_n))$ and positive definite covariance matrix C having (j,r) entry defined by the kernel function κ as $C_{j,r} = \kappa(\mathbf{Z}_j, \mathbf{Z}_r)$.

Thus, κ determines C in a GP. This prior may be denoted by either $f \sim GP(\mu, C)$ or $f \sim GP(\mu, \kappa)$. In practice, a useful version of this model is obtained by defining the means of a GP prior as parametric linear combinations $\mu(\mathbf{Z}) = \boldsymbol{\beta}'\mathbf{Z}$, while also specifying a prior on $\boldsymbol{\beta}$, and defining the covariance kernel as the squared exponential form

$$\kappa(\mathbf{Z}_j, \mathbf{Z}_r) = \exp\left(-\frac{||\mathbf{Z}_j - \mathbf{Z}_r||^2}{2\sigma^2}\right)$$

for any $1 \le j, r \le n$. A general application of a GP is to specify a traditional nonparametric regression model, $Y_i = f(\mathbf{Z}_i) + \varepsilon_i$, with $E(\varepsilon_i) = 0$ and $\text{var}(\varepsilon_i) = \sigma_n^2$, for $i = 1, \cdots, n$, while assuming that the mean function f follows a Gaussian process prior, $f \sim GP(0, \kappa(\cdot,\cdot))$. This implies that $Y_i \mid f_i \sim N(f(\mathbf{Z}_i), \sigma_n^2)$, so unconditionally the covariance matrix of the observed outcome vector $\boldsymbol{Y}_n = (Y_1, \cdots, Y_n)$ is $cov(\boldsymbol{Y}_n) = C + \sigma_n^2 I_n$, where I_n is the $n \times n$ identity matrix. This provides a robust model for a very broad class of random functions.

To use the GP as a prior for a DDP, a single weights DDP mixture model is assumed using a normal distribution as the kernel function. This model often is called a *DPP-GP*. The usual stick-breaking weight representation with all $\upsilon_h \sim beta(1, M_h)$ and $w_h = \upsilon_h \prod_{r<h}(1 - \upsilon_r)$ gives the distribution of Y as the normal mixture

$$F(y \mid \mathbf{Z}) = \sum_{h=1}^{\infty} w_h \phi(y \mid \theta_h(\mathbf{Z}), \sigma^2), \tag{9.13}$$

with the mean of the normal summand indexed by h given by the covariate function $\theta_h(\mathbf{Z})$. It is assumed that the normal means $\theta_1(\mathbf{Z}), \theta_2(\mathbf{Z}), \cdots$ follows GP priors, denoted $\theta_h(\mathbf{Z}) \sim GP(\mu_h, C)$. Thus, each normal mean in the sum (9.13) has its own GP prior. For each h, the GP mean may be specified by assuming that it equals the linear combination $\theta_h(\mathbf{Z}) = \boldsymbol{\beta}_h' \mathbf{Z}$. The covariance matrix may be specified, for example, by

$$C(\mathbf{Z}_i, \mathbf{Z}_j) = \exp\left\{ -\sum_{m=1}^{q} (Z_{i,m} - Z_{j,m})^2 \right\} + I(i = j)J^2$$

for $i, j = 1, \cdots, n$, where q is the dimension of $\mathbf{Z}$. To stabilize numerical computations, jitter J^2 is added to the diagonal terms of the covariance matrix, usually a small value such as $J = .10$. The model is completed by assuming the hyperpriors $\boldsymbol{\beta}_h \sim N(\tilde{\beta}, \tilde{\Sigma})$ and $M_h \sim Gamma(\lambda_1, \lambda_2)$ for all h, and $\sigma^2 \sim IG(\lambda_3, \lambda_4)$. The DDP-GP may be written in the compact form

$$\begin{aligned} Y \mid \mathbf{Z} &\sim F(\cdot \mid \mathbf{Z}) \\ F &\sim DDP - GP\{\{\mu_h\}, C, \{\boldsymbol{\beta}_h\}, \sigma^2\} \end{aligned}$$

Because different parameter vectors $\boldsymbol{\beta}_1, \boldsymbol{\beta}_2, \cdots$ are specified for the means of the normal components of the mixture (9.13), the DDP-GP allows a wide variety of possible forms for the regression of Y on $\mathbf{Z}$, including multiple modes and nonlinear interactions that cannot be revealed by more conventional regression models.

The family of DDP-GPs is quite general, since the mean functions $\mu_h(\mathbf{Z})$ can take any form, although in most applications a linear DDP is used. This provides a very flexible basis for discovering a wide variety of regression structures in a dataset, including multiple modes for the outcome variable distribution, clusters, and interactions among the elements of $\mathbf{Z}$ that are not of the simple conventional form $\beta_{r,k} Z_r Z_k$. These are patterns that a conventional regression model typically fails to identify. Two applications the DDP-GP will be described below. The first is an analysis of a leukemia chemotherapy dataset, in Chapter 10, including a simulation study that shows using G-estimation under a DDP-GP may provide excellent bias correction for observational data. The second application is an analysis of a stem cell transplantation dataset, in Chapter 11, where the fitted DDP-GP identified a third-order interaction between each patient's age, disease status at transplant, and the optimal pharmacokinetically guided dose of a key agent in their preparation regimen.

Computer Software There are many computer programs and software packages for implementing Bayesian nonparametric methods, including DPpackage for fitting many DP models, BNPdensity for density estimation, and more generally the OpenBUGS, JAGS, and STAN packages. See, for example, Karabatsos (2016).

10

Evaluating Multistage Treatment Regimes for Leukemia

10.1 Chemotherapy for Acute Leukemia

Leukemia is a family of rare cancers of the blood and lymphatic system, with a global rate of about 32 cases per 100,000 people annually, and most cases occurring in older adults. AML presents with a variety of symptoms, including fatigue, shortness of breath, bruising, bleeding in the gums, infections, bone or joint pain, swollen lymph nodes, and weight loss. AML can be diagnosed by taking a sample of cells from the bone marrow, which typically has $\geq 20\%$ blast cells for a patient with AML, compared to about 5% in a healthy person. AML subtypes with different prognoses may be defined by chromosomal or genetic abnormalities, including rearrangements or duplications of genetic material in chromosomes 5, 7, 8, 11, 15, or 16 or particular genes, including the BCR-ABL fusion gene. For example, AML patients with an inversion of chromosome 16 or translocations between chromosomes 8 and 21, or between 15 and 17, have a comparatively favorable prognosis. A closely related disease, myelodysplastic syndrome (MDS), is characterized by malformed, dysfunctional blood cells that also have chromosomal abnormalities. In practice, AML and MDS often are treated the same way, with some combination of chemotherapy, stem cell transplantation, or a targeted agent.

Chemotherapy (chemo) for acute AML/MDS begins with remission induction, in which a combination of chemicals is infused onto the patient's circulating blood intravenously. The goal is to achieve a complete remission (CR), defined as the patient having less than 5% blast cells, a platelet count greater than $10^5/mm^3$, and a WBC count greater than $10^3/mm^3$, based on a bone marrow biopsy. These are considered normal functional levels for these three types of cells. Achieving a CR is necessary for long-term survival in AML/MDS, but it is far from being a cure since most AML patients who achieve a CR subsequently suffer a disease recurrence. In some cases, a patient who achieves a CR is given additional treatment, known as consolidation therapy, to reduce the risk of recurrence. If the induction chemo does not achieve a CR, so that the patient's disease is considered *resistant*, or if a CR is achieved but the patient later suffers a relapse, then a *salvage* chemo generally is given in a second attempt to achieve a CR. For a patient who achieves CR but later relapses and is treated with a salvage therapy, the duration of the initial CR is a very strong prognostic variable for their subsequent survival time following salvage therapy.

DOI: 10.1201/9781003474258-10

The most effective chemo for AML/MDS is cytosine arabinoside (ara-C, cytarabine), given by continuous intravenous infusion. ara-C is used as a component of induction, consolidation, and salvage therapies. High-dose ara-C (HDAC) acts by penetrating leukemic cells and inhibiting DNA synthesis and repair. However, HDAC also is associated with high rates of severe toxicity that may require extensive supportive care. Generally, ara-C is given in combination with an anthracycline, such as daunorubicin or idarubicin, possibly a targeted therapy such as an FLT3 or IDH inhibitor, and other cytotoxic agents. The aim is to enhance effectiveness by targeting different aspects of AML biology, possibly with synergistic interactions. Adverse events associated with HDAC include myelosuppression of bone marrow function leading to a decrease in WBCs, RBCs, and platelets, and mucositis, which is inflammation and ulceration of the mucous membranes lining the mouth and gastrointestinal tract that causes pain, diarrhea, nausea, and vomiting.

10.2 A Randomized Trial of Frontline Chemotherapies

All-trans retinoic acid (ATRA) is a nutrient manufactured from vitamin A that helps cells to grow and develop. ATRA was found to be effective as a treatment for acute promyelocytic leukemia (APL), which is a subtype of AML involving abnormalities of chromosomes 15 and 17 that make up 10% to 15% of cases. ATRA can induce CR's and improve overall survival in APL patients. Granulocyte colony-stimulating factor (GCSF) is a natural growth factor that stimulates the bone marrow to produce more WBCs. Motivated by the desirable effects of ATRA seen in APL and of GCSF in all patients, Estey et al. (1999) designed a clinical trial to compare the effects of four remission induction chemotherapy (chemo) combinations on clinical outcomes in newly diagnosed AML/MDS patients with poor prognosis. The trial used a 2×2 factorial design that studied the effects of adding ATRA, GCSF, or both to the baseline chemo combination of fludarabine + ara-C + idarubicin (FAI), which at the time of the trial was a standard chemo combination used as an induction regimen in AML. Patients were randomized fairly among four treatment arms, FAI, FAI+ATRA (FAI+A), FAI+GCSF (FAI+G) , and FAI+ATRA+ GCSF (FAI+A+G). The trial was designed to compare the probability of success, defined as the patient being alive and in CR six months from the start of treatment, between the baseline FAI group and each of the other treatment groups, FAI+A, FAI+G, and FAI+A+ F. Each comparison was done using a two-sided test with type I error rate .05 and power .80. The method of Pocock and Simon (1975) was used to balance the randomization on patient baseline covariates, including age, hemoglobin, WBC count, platelet count, performance status, and type of cytogenetic abnormality. While this design is straightforward, at the time of the trial it was quite unusual since, conventionally, randomization seldom was used to evaluate any new agent or combination for treating AML. Rather, the most common approach was to conduct a single-arm phase II trial of a new induction regimen (Gehan and Freireich, 1974).

Estey et al. (1999) used conventional statistical methods to report the results of this iconoclastic clinical trial. Various between-treatment comparisons were done using parameter estimates and p-values to quantify the strength of evidence. Methods included a Cox model for event time regressions, a logistic model for regression of the probability of CR, and logrank tests. The empirical CR rates in the four arms were 21/53 (40%) for FAI, 29/53 (55%) for FAI+G, 28/55 (51%) for FAI + A, and 32/54 (59%) for FAI+G+A. These values suggest an early benefit if ATRA, G-CSF, or both are added to FAI. To compare survival distributions, Figure 10.1 gives the Kaplan-Meier plots for the four arms, showing the same general pattern of improved survival with any of the three combination arms compared to FAI. The four Kaplan-Meier plots indicate that, ignoring covariate effects, the treatment effects on survival were qualitatively similar to their effects on the CR rates. That is, adding one or both of the ATRA or G-CSF was beneficial.

Log rank tests comparing survival with FAI to survival with each of the other three treatment arms gave p-value .15 for FAI vs FAI + G, .023 for FAI vs FAI + A, and .055 for FAI versus FAI + A +G. As an information-based replacement for a p-value, Greenland et al. (2016) defined $S = -\log_2(\text{p-value})$, rounded to the nearest integer, called the *S value*. An S value quantifies the surprise that one would feel if they saw all heads in S flips of a coin that they believed to be fair. Thus,

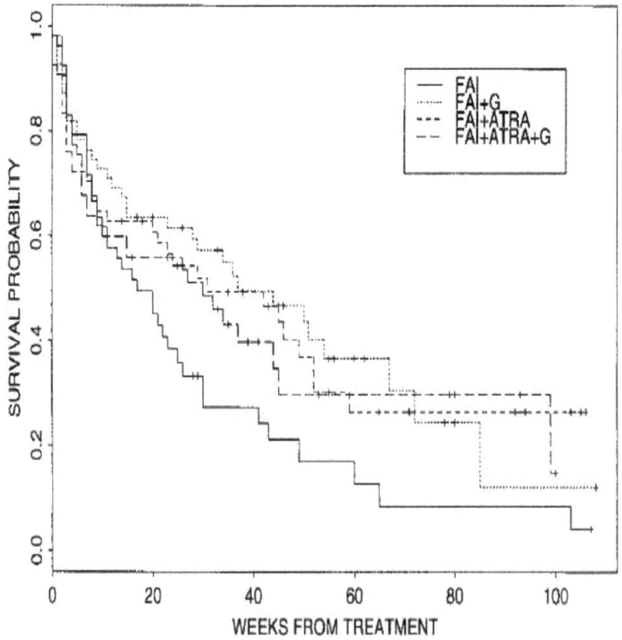

FIGURE 10.1
Kaplan-Meier plots of survival time distributions for the four treatment arms in the leukemia trial.

an S value may be interpreted as the information, in terms of bits of information, provided by the p-value against the null hypothesis. The p-values given above for the between-treatment logrank tests give respective values S = 3, 5, and 4, which together give at most moderately surprising evidence that any of the three combinations is different from FAI in terms of survival. In light of the four empirical CR rates and Kaplan-Meier plots, the trial's data suggest that adding ATRA, G-CSF, or both to FAI improves survival, but by a small amount.

To evaluate the effects of the treatments and patient baseline prognostic variables on the outcomes, Estey et al. (1999) reported two fitted regression models, a logistic model for the probability of CR, and a Cox model for OS. Both fitted regression models showed, as expected, that each of a set of well-established prognostic variables for AML/MDS had a strong effect on each of these clinical outcomes. They concluded that G-CSF had a "significantly" beneficial effect on the probability of CR, but that neither G-CSF nor ATRA had any "significant" effects on either OS time or event-free survival time among patients who achieved CR. However, the most significant aspect of these analyses is that they ignored a much deeper issue.

10.3 Accounting for Multistage Therapy

A problematic limitation of the trial design and statistical analyses reported by Estey et al. (1999) is that they only accounted for frontline treatments. This makes evaluation of outcomes beyond disease progression, such as survival time, problematic because therapy of leukemia is a multistage process. Randomization among the four frontline treatments did not account for later choices of salvage therapies made subjectively by the attending physicians, at the times when either a patient's disease was declared resistant to frontline therapy or when progressive disease was seen following CR. Consequently, for any outcomes following salvage therapy, effects of frontline therapy were confounded with salvage therapy effects. Like most clinical trial designs in oncology, this trial's design was myopic in that it failed to recognize this fact. The relevant objects to evaluate were not frontline therapies alone, but rather were multistage dynamic treatment regimes (DTRs). The remainder of this chapter will present the structure of the DTRs that actually were used in the leukemia trial and describe two re-analyses of the data, one frequentist and one Bayesian, that account for the DTRs. Both analyses assume regression models for seven different transition times between disease states, use these to define overall survival time, and apply G-estimation to evaluate the effects of DTRs and prognostic covariates when estimating mean overall survival time. The main difference is between the assumed frequentist and Bayesian regression models.

To identify the DTRs used to treat patients in the trial, it is useful to begin by noting that the multistage structure of AML therapy accommodates the fact that CR may or may not be achieved by the frontline therapy. Rather, three possible outcomes for frontline therapy may be defined. The first possible outcome is that a CR is achieved. The second outcome is that a patient's disease is *resistant* to the frontline treatment, in that a CR could not be achieved. In this case, the attending physician routinely gives the patient a *salvage therapy*, in a second attempt to achieve CR. The third possible outcome is that the patient dies during induction therapy, due to the combined effects of the patent's disease, prognostic covariates, and adverse effects of the chemo combinations, which are very aggressive and often cause the severe toxicities described above. For example, death during induction is more likely for patients who are older or who have comorbidities (Walter et al., 2011; Malfuson et al., 2008). If a patient achieves CR but later relapses, then a salvage therapy may be given in an attempt to achieve a second CR. Thus, a patient's chemotherapy for AML/MDS may include three treatments, consisting of a frontline therapy A and, if necessary, a salvage therapy B_1 to treat disease that is resistant to induction, or a salvage therapy B_2 to treat disease recurrence following a CR. The nontrivial likelihood of this last event is what motivated the generalized phase I-II design described in Chapter 3.

These considerations show that a major problem when analyzing the survival data from this trial is that, while patients were randomized among the four frontline treatments A using a 2×2 factorial design, all salvage treatments, B_1 or B_2, were chosen subjectively by the attending physicians. This introduced selection bias for estimating the rates of any events that occurred after the initial outcomes CR or resistance from induction chemo. In nearly all oncology clinical trials studying frontline treatments, patients are not randomized among salvage treatments given subsequently. Moreover, in reported data analyses from oncology trials, whether they are single arm or randomized, a correction seldom is made for selection bias due to the subjective choice of salvage or consolidation treatments. Ignoring these effects renders many reported statistical analyses of overall survival time based on data from many oncology trials either misleading or useless.

To address this problem when analyzing the AML/MDS trial data, Wahed and Thall (2013) first identified each patient's possible outcomes, defined the transition times between possible disease states, and constructed a likelihood function that provided a basis for defining DTRs. They treated the possible combinations of frontline and salvage therapies as DTRs, characterized by the triple (A, B_1, B_2). They defined two indicators. For stage 1, $\zeta_1 = 0$ if death occurred during induction, $\zeta_1 = 1$ if disease was declared resistant to frontline therapy, and $\zeta_1 = 2$ if CR was achieved. For patients who achieved CR, they defined the second indicator $\zeta_2 = 0$ if the patient died in CR and $\zeta_2 = 1$ if progressive disease occurred prior to death. Using these treatment identifiers and indicators, the possible treatment sequences and patient outcomes in the leukemia trial are illustrated in Figure 10.2, which shows that, depending on their sequence of outcomes, each patient could receive either B_1 or B_2, but not both.

In addition to the indicators ζ_1 and ζ_2, Figure 10.2 identifies a total of seven different transition times between disease states. For the three possible induction chemo outcomes, the respective transition times are denoted by T^D for death, T^R for resistant disease, and T^C for CR. These are semi-competing risks, since at most one of the three events can occur, and death censors both resistant disease and CR, but

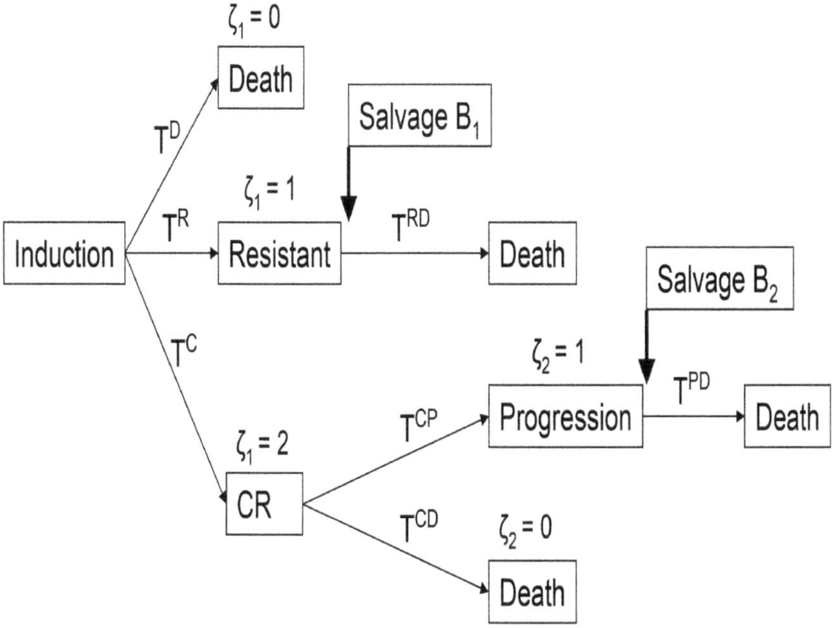

FIGURE 10.2
Schematic of the possible treatment and outcome paths for each patient in the AML/MDS chemotherapy trial. For patients who died during induction chemo, T^D is the transition time from induction to death. For patients who achieved CR, T^{CP} is the transition time from CR to disease progression, and so on. A dynamic treatment regime is the triple (A, B_1, B_2).

not conversely. Thinking of T^D, T^R, and T_C as potential outcomes,

$$\begin{aligned} \zeta_1 = 0 \quad &\text{if} \quad T^D < \min\{T^C, T^R\} \\ \zeta_1 = 1 \quad &\text{if} \quad T^R < \min\{T^D, T^C\} \\ \zeta_1 = 2 \quad &\text{if} \quad T^C < \min\{T^D, T^R\}. \end{aligned}$$

The figure also shows that CR can be followed by either death (ζ_2=0) or disease progression (ζ_2=1), where ζ_2 is defined similarly to ζ_1, in terms of the transition times. While it may not appear to make sense that a patient may achieve CR and then die without disease recurrence, death in CR may occur due to the combined effects of poor prognosis and toxicities from the frontline chemo, which may include organ failure. Figure 10.2 shows that, depending on a patient's induction chemo outcome (CR, disease resistant to induction, died during induction), they could receive either salvage B_1 for resistant disease or salvage B_2 for disease progression after *CR*. For convenience, the seven possible transition times are indexed by $j \in \{D, R, C, RD, CD, CP, PD\}$, bearing in mind that, indexing the patient's state at the start of frontline chemo by 0, $j = C$ refers to the transition $0 \rightarrow C$, $j = CP$ refers to the transition $C \to P$, and so on. All seven transition times are potential outcomes since, for each patient, each time may or may not be observed depending on how their DTR plays out.

Since each salvage therapy B_1 or B_2 in a given DTR (A, B_1, B_2) was chosen adaptively using the patient's baseline covariates and current history, this was a form of personalized therapy. In general, both B_1 and B_2 include adaptive decision rules, but these depended on how the attending physicians chose salvage treatments, and these rules were not available in the dataset. This illustrates the general fact that a DTR in a medical therapy includes personalized treatment decisions. Since the overall posterior mean survival time for each DTR is computed as an average over means for each $\mathbf{Z}$ in the generalized estimation scheme, described below, this process includes estimation of mean survival, or prediction of survival time, for each possible (DTR, $\mathbf{Z}$) combination. Thus, the analysis is based on choosing a personalized therapy using $\mathbf{Z}$ in the context of a DTR, rather than choosing a single frontline treatment. A key practical difference is between individual physicians choosing B_1 or B_2 subjectively, and the statistical approach of using a fitted regression model to determine an optimal DTR among the 16 possibilities. In the latter case, a physician would choose A, and if necessary B_1 or B_2, for a patient with a given $\mathbf{Z}$ by following the optimal DTR determined by the data analysis.

By accounting for all possible paths of treatments and outcomes, Figure 10.2 shows that each patient's survival time could take one of four possible forms, each expressed as a sum of transition times. Overall survival time is given by

$$\begin{aligned} T &= T^D\, I[\zeta_1 = 0] \\ &+ (T^R + T^{RD})\, I[\zeta_1 = 1] \\ &+ (T^C + T^{CD})\, I[\zeta_1 = 2,\ \zeta_2 = 0] \\ &+ (T^C + T^{CP} + T^{PD})\, I[\zeta_1 = 2,\ \zeta_2 = 1]. \end{aligned} \tag{10.1}$$

This shows that the seven transition times are potential outcomes, with several semi-competing risks. If $T = T_D$, then none of the other six transition times could be observed, if $T = T^R + T^{RD}$ then none of the other five transition times could be observed, and so on. To interpret the DTRs illustrated in Figure 10.2, it is useful to draw a DAG that shows possible causal effects. This is given in Figure 10.3. While the DAG is rather complex, it still simplifies the actual causal structure in that it does not distinguish between the two different types of salvage therapies B_1 and B_2. Figure 10.3 shows that, while randomization among the four frontline chemo combinations removed the causal arrow from [Prognostic Covariates] to [Frontline Chemo], other covariate effects on later outcomes and choices of salvage therapies remained. Additionally, there may have been direct effects of the frontline chemo on any later event time, including T^{CP} = time from CR to progressive disease as well as the time of death. Since the relevant therapeutic object for evaluating effects on overall survival (OS) time is not only the frontline treatment A, but rather is the DTR (A, B_1, B_2), there was selection bias due to the subjective choice of salvage treatments. Consequently, the initial randomization among induction regimens was not enough to provide data that would give fair between-DTR comparisons of effects

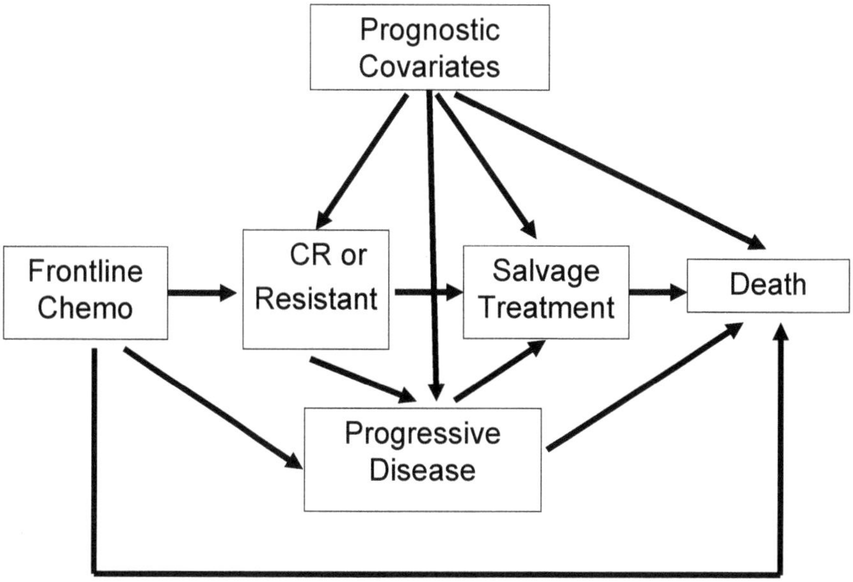

FIGURE 10.3
Causal diagram of relationships between baseline prognostic covariates, frontline treatments, salvage treatments, interim outcomes, and survival time in the dynamic treatment regime for AML/MDS.

on OS, so some form of bias correction is needed to compare mean survival times associated with the DTRs.

Salvage treatments that were chosen by the attending physicians took many forms in the AML/MDS dataset. This is commonplace in therapy of most cancers, since there is a constant flow of new agents to be tested, and most often they first are evaluated in patients who have advanced disease, have relapsed, or were treated previously and had resistant disease. Consequently, in order to make progress analytically, each salvage therapy was categorized using a binary indicator of whether it contained HDAC, motivated by the fact that HDAC is the most effective agent against AML/MDS. Thus, B_1 and B_2 each was classified as containing HDAC or not. Since there were four induction chemos, this produced a total of 16 possible DTRs. In the regression analysis described below, Y = log overall survival time is a function of covariates $\mathbf{Z}$, and the DTR (A, B_1, B_2). This notation represents the possible treatments appearing in the branching structure of Figure 10.2, and it replaces the notation used in Chapter 2 that represents regimes more simply as a sequence of treatments given in consecutive stages.

Many methods have been proposed for evaluating DTRs arising from observational data or longitudinal studies. Seminal papers by Robins (1986a) and Robins et al. (2000) present G-estimation of structural nested models. Additional references include IPTW estimation of marginal structural models (Robins et al., 2008), augmented IPTW (AIPTW) (Zhao et al., 2015), and G-estimation (Murphy, 2003). A review is given by Moodie et al. (2007). A wide variety of methods have been developed to evaluate and compare DTRs from clinical trials or observational data (Murphy, 2005a; Murphy and Bingham, 2009; Murphy, 2005b; Thall et al., 2000, 2002). A comprehensive review is given by Tsiatis et al. (2021).

In the frequentist analysis of the DTR reported by Wahed and Thall (2013), the distribution of the $\log(T^j)$ for $j \in \{D,R,C,RD,CD,CP,PD\}$ in Figure 10.2 was assumed to follow a parametric accelerated failure time model, of the form

$$Y_i^j = \log(T_i^j) = \mathbf{Z}_i'\boldsymbol{\beta}^j + \sigma^j\, \varepsilon_i \tag{10.2}$$

for patient $i = 1,\cdots,n$, where $\sigma^j > 0$, the $Y_i^{j'}$ values of previous transition times were included as covariates along with $\mathbf{Z}_i$, and the ε_i's are iid. To define a likelihood function, for each j, let T^o denote the time from the start of induction chemo to the last follow-up, $U^j = \min\{T^j, T^o\}$ and $\delta_j = 1$ if $U^j = T^j$, and $\delta_j = 0$ if $U^j < T^j$. Denoting the pdfs by f^j and survival functions by $\bar{F}^j$, and suppressing notation for $\mathbf{Z}_i$ and $\boldsymbol{\beta}^j$, this gives the likelihood for the stage 1 transition times of a patient treated with induction chemo A as

$$\mathscr{L}_1 = \prod_{j=C,R,D} \left\{ f^j(T^0 \mid A)^{\delta^j} \bar{F}^j(T^0)^{1-\delta^j} \right\}^{I[\zeta_1 = j]} \tag{10.3}$$

For patients with resistant disease in stage 1, $j = RD$, indicated by $\zeta_1 = 1$, define $U^{RD} = \min\{T^{RD}, T^0 - T^R\}$ and $\delta^{RD} = I[U^{RD} = T^{RD}]$. The conditional likelihood of U^{RD} given T^R, the frontline chemo A, and salvage chemo B_1 for resistant disease is

$$\mathscr{L}_{2,R} = \{f^{RD|R}(U^{RD} \mid T^R, A, B_1)\}^{\delta^{RD}} \{\bar{F}^{RD|R}(U^{RD} \mid T^R, A, B_1)\}^{1-\delta^{RD}}. \tag{10.4}$$

Defining the likelihood contributions of T^{CP}, T^{PD}, and T^{CD} similarly to account for the possible events after CR, a tedious but straightforward probability computation shows that the likelihood contribution of patients for whom induction chemo achieves a CR is

$$\begin{aligned}
\mathscr{L}_{2,C} &= \left\{ f^{CD|C}(U^{CD} \mid T^C, A) \right\}^{\delta^{CD} I[\zeta_2 = 0]} \left\{ f^{CP|C}(U^{CP} \mid T^C, A) \right\}^{\delta^{CP} I[\zeta_2 = 1]} \\
&\times \left\{ \bar{F}^{CD|C}(U^{CD} \mid T^C, A) \bar{F}^{CP|C}(U^{CP} \mid T^C, A) \right\}^{1-\delta^{CD}-\delta^{CP}} \\
&\times \left\{ f^{PD|CP}(T^{PD,0} \mid T^C, T^{CP}, A, B_2) \right\}^{\delta^{PD}} \\
&\times \left\{ \bar{F}^{PD|CP}(T^{PD,0} \mid T^C, T^{CP}, A, B_2) \right\}^{(1-\delta^{PD}) I[\zeta_2 = 1]}.
\end{aligned} \tag{10.5}$$

Combining terms, since the indicator ζ_1 keeps track of the possible stage 1 outcomes and hence which branch of Figure 10.2 is followed by the patient, the overall likelihood is given by the equation

$$\mathscr{L} = \mathscr{L}_1 \times \{\mathscr{L}_{2,R}\}^{I[\zeta_1=1]} \times \{\mathscr{L}_{2,C}\}^{I[\zeta_1=2]},$$

bearing in mind that $\mathscr{L}$ is a function of the seven transition times, the baseline covariates, and the 16 DTRs.

To analyze the leukemia data, Wahed and Thall (2013) evaluated Weibull, exponential, log-logistic, and log-normal distributions for each Y^j in (10.2), and chose a distribution by minimizing the Bayes information criterion. To construct a G formula for estimation, first consider the induction stage, and denote the mean log transition times of the three possible observed outcomes following induction chemo A by

$$\theta^j(A,\mathbf{Z}) = E(Y^j \mid A,\mathbf{Z}) \quad \text{for} \quad j = D,R,C.$$

Denote

$$\theta^{RD}(A,B_1,\mathbf{Z},\mathbf{Z}^{(R)}) = E(Y^{RD} \mid A,B_1,\mathbf{Z},\mathbf{Z}^{(R)})$$

for the mean log time from resistance to death after induction with A and salvage with B_1, where $\mathbf{Z}^{(R)}$ includes Y^R, and possibly updated elements of $\mathbf{Z}$. Denote

$$\theta^{CD}(A,\mathbf{Z},\mathbf{Z}^{(C)}) = E(Y^{CD} \mid A,\mathbf{Z},\mathbf{Z}^{(C)})$$

for the mean log time to death while in CR induced with A, where $\mathbf{Z}^{(C)}$ includes Y^C. Denote

$$\theta^{CP}(A,\mathbf{Z},\mathbf{Z}^{(C)}) = E(Y^{CP} \mid A,\mathbf{Z},\mathbf{Z}^{(C)})$$

for the mean log time to progression while in CR induced with A. Denote

$$\theta^{PD}(A,B_2,\mathbf{Z},\mathbf{Z}^{(C)},\mathbf{Z}^{(P)}) = E(Y^{CP} \mid A,B_2,\mathbf{Z},\mathbf{Z}^{(C)},\mathbf{Z}^{(P)})$$

for the mean log time from progression to death after induction with A and salvage with B_2 following progression after CR, where $\mathbf{Z}^{(P)}$ includes Y^P,

Denote the mean of the potential log survival time $\log(T)$ for a future patient treated with DTR (A,B_1,B_2) by $\mu(A,B_1,B_2)$, and let m denote the covariate distribution. Wahed and Thall (2013) applied the G formula of Robins (1986b) to derive the following G estimation equation for the DTRs,

$$\mu(A,B_1,B_2) = \int_{\mathbf{Z}} \Bigg\{ \Pr(\zeta_1 = 0 \mid A,\mathbf{Z})\theta^D(A,\mathbf{Z})$$

$$+ \Pr(\zeta_1 = 1 \mid A,\mathbf{Z}) \Bigg\{ \theta^R(A,\mathbf{Z}) + \int_{\mathbf{Z}^{(R)}} \theta^{RD}(A,B_1,\mathbf{Z},\mathbf{Z}^{(R)})dm(\mathbf{Z}^{(R)}) \Bigg\}$$

$$+ \Pr(\zeta_1 = 2 \mid A,\mathbf{Z}) \Bigg(\theta^C(A,\mathbf{Z}) + \int_{\mathbf{Z}^{(C)}} \Bigg[\Pr(\zeta_2 = 0 \mid \zeta_1 = 2,A,\mathbf{Z},\mathbf{Z}^{(C)}) \Bigg)$$

$$\times\ \theta^{CD}(A,\mathbf{Z},\mathbf{Z}^{(C)}) + \Pr(\zeta_2 = 1 \mid \zeta_1 = 2, A, \mathbf{Z}, \mathbf{Z}^{(C)}) \Big\{ \theta^{CP}(A,\mathbf{Z},\mathbf{Z}^{(C)})$$

$$+ \int_{\mathbf{Z}^{(P)}} \theta^{PD}(A, B_2, \mathbf{Z}, \mathbf{Z}^{(C)}, \mathbf{Z}^{(P)}) dm(\mathbf{Z}^{(P)}) \Big\} \Big] dm(\mathbf{Z}^{(C)}) \Big) \Big\} dm(\mathbf{Z}). \quad (10.6)$$

This equation may be derived by taking the expected values of the terms on the right hand side of equation 10.1 that expresses survival time in terms of the seven transition times and the indicators ζ_1 and ζ_2, and then averaging over the respective covariate distributions.

For each (A, B_1, B_2), once the component regression models for the transition times in equation (10.6) have been fit to the data, estimates of the seven conditional means θ^C, θ^R, θ^D, θ^{RD}, θ^{CD}, θ^{CP}, and θ^{PD} can be plugged into the right hand side of (10.6) to compute an approximately unbiased estimator of $\mu(A, B_1, B_2)$. Wahed and Thall (2013) used this to compute both likelihood-based G estimators and IPTW estimators. They used age and type of cytogentic abnormality as prognostic covariates in the model for each transition time, and also in logistic regression models for propensity scores, specifically, the probabilities of receiving each type of salvage treatment, B_s = HDAC (s=1) or not (s = 2), used in the IPTW estimation.

For IPTW estimation, let T_i^o denote the observed time from start of induction chemo to death or censoring at the last follow-up of the ith patient. Denote δ_i =1 if T_i^o is the time of death and 0 otherwise. Let $I_i(A)$, $I_i(B_1)$, and $I_i(B_2)$ be the respective indicators of whether each per-stage treatment was given. Let $\hat{K}$ be a consistent estimator of the censoring time survival distribution, typically a Kaplan-Meier estimator. Thus, $\hat{K}(T_i)$ consistently estimates $\Pr(\delta_i = 1)$, the probability that T_i is the time of death. The IPTW estimator is

$$\hat{\mu}^{IPTW}(A, B_1, B_2) = \frac{\sum_{i=1}^n w_i(A, B_1, B_2) T_i^o}{\sum_{i=1}^n w_i(A, B_1, B_2)} \quad (10.7)$$

where the ith weight is

$$w_i(A, B_1, B_2) = \frac{I_i(A)\delta_i}{\hat{K}(T_i^o)} \Bigg[I(\zeta_{1,i} = 0) + \frac{I(\zeta_{1,i} = 1) I_i(B_1)}{\hat{P}(B_1 \mid \zeta_{1,i} = 1, A, \mathbf{Z}_i, \mathbf{Z}_i^{(R)})}$$

$$+ I\{\boldsymbol{\zeta}_{1,i} = (2,0)\} + \frac{I\{\boldsymbol{\zeta}_i = (2,1)\} I_i(B_2)}{\hat{P}\{B_2 \mid \boldsymbol{\zeta}_i = (2,1), A, \mathbf{Z}_i, \mathbf{Z}_i^{(C)}, \mathbf{Z}_i^{(P)}\}} \Bigg].$$

Their results showed that, across the 16 DTRS, the likelihood-based estimators were much larger than the corresponding IPTW estimators, all estimators had wide 95% confidence intervals and, for both methods, the four DTRs with frontline chemo A = FAI + ATRA had slightly higher estimated mean survival times compared to the other 12 DTRs. For patients with disease resistant to FAI + ATRA induction, it was irrelevant whether their salvage therapy contained HDAC, but for patients who achieved CR with FAI + ATRA salvage with HDAC was superior. Some transition time samples included comparatively large values, which Wahed and Thall (2013) considered outliers that might be excluded when computing the estimators.

This point will be important in the Bayesian DDP-GP-based estimation described in the next section, since the DDP-GP mixture model naturally accommodates relatively large sample values.

10.4 Bayesian Regression Analysis of the Leukemia Data

Xu et al. (2016a) provided a third analysis of the leukemia data, mimicking the approach of Wahed and Thall (2013) by focusing on the effects of the 16 DTRs, defining overall survival time as the sum given in equation (10.1) and estimating each DTR's mean survival time using the G equation (10.6). The main difference was that Xu et al. (2016a) did Bayesian estimation, assuming a DDP-GP regression model for each log transition time conditional on the patient's DTR, baseline covariates **Z**, and all preceding transition times. The motivation for this approach was the well-established fact that DDP regression models are robust and have full support, and provide much better fits to complicated data structures than can be obtained using conventional parametric regression models.

For the DDP-GP of each transition time distribution, a DDP mixture model with a Gaussian kernel was assumed. Let $\mathbf{Z}_i^j$ denote the covariate vector of the ith patient augmented to include all Y^r for previous transition times in the DTR. Denote $\phi(\cdot;\mu,\sigma)$ as a normal pdf with mean μ and variance σ^2 for a smoothing kernel. For each transition time T^j, define the stick breaking prior for the weights by $\upsilon_1^j, \upsilon_2^j, \sim iid$ beta$(1,\alpha^j)$, $w_1^j = \upsilon_1^j$ and $w_h^j = \upsilon_h^j \prod_{\ell<h}(1-\upsilon_\ell^j)$ for each $h \geq 2$ for all h and j. The distribution of each Y^j is the DDP mixture

$$F^j(y \mid \mathbf{Z}^j) = \sum_{h=0}^{\infty} w_h^j \, \phi(y;\theta_h^j(\mathbf{Z}^j),\sigma_h^j).$$

An important practical aspect of this form for distribution of the jth transition time is that it naturally accommodates relatively large values, which might be considered outliers in a more conventional analysis, as done by Wahed and Thall (2013).

For the Gaussian process prior, temporarily suppressing the indices h and j, given any n covariate vectors $\mathbf{Z}_1,\cdots,\mathbf{Z}_n$, the means $(\theta(\mathbf{Z}_1),\cdots,\theta(\mathbf{Z}_n))$ follow an n-variate normal distribution with mean vector $(\mu(\mathbf{Z}_1),\cdots,\mu(\mathbf{Z}_n))$ and $n \times n$ covariance matrix having (i,j) entry $C(\mathbf{Z}_j,\mathbf{Z}_j)$, denoted $\theta \sim GP(\mu,C)$. Reintroducing the indices j for transition time and h for the DDP summand, this may be written

$$\theta_h^j(\mathbf{Z}^j) \sim GP(\mu_h^j,C^j), \quad \text{for } j \in \{C,R,D,CP,CD,RD\}, \ h=1,2,\cdots.$$

For the GP mean function, they assumed the linear form

$$\mu_h^j(\mathbf{Z}^j,\boldsymbol{\beta}_h^j) = \mathbf{Z}^j\boldsymbol{\beta}_h^j.$$

For the covariance matrix, they assumed

$$C^j(\mathbf{Z}_i^j,\mathbf{Z}_r^j) = \exp\left\{ -\sum_{m=1}^{M^j} (W_{im}^j - W_{rm}^j)^2 \right\} + J^2 I[i=r]$$

for $i,r = 1,\cdots,n$, where M^j is the number of covariates in each $\mathbf{Z}_i^j$ at the jth transition and J is the variance on the diagonal reflecting the amount of jitter. For the hyperparameter priors, for each transition time T^j, they assumed

$$\boldsymbol{\beta}_h^j \sim \mathrm{N}(\boldsymbol{\beta}_0^j, \Sigma_0^j), \ (\sigma^j)^2 \sim iid\ \mathrm{IG}(\lambda_1,\lambda_2) \ \text{ and } \ \alpha^j \sim iid\ \mathrm{Gamma}(\lambda_3,\lambda_4).$$

This DDP-GP model may be summarized as

$$\begin{aligned} Y^j \mid \mathbf{Z}^j, F^j &\sim F^j(\cdot \mid \mathbf{Z}^j) \\ F^j &\sim \mathrm{DDP-GP}\{\{\mu_h^j\}, C^j, \alpha^j, \{\boldsymbol{\beta}_h^j\}, \sigma^j\}. \end{aligned}$$

For each transition time T^j, to determine the fixed hyperparameters $\boldsymbol{\beta}_0^j, \Sigma_0^j$, and $\lambda_1,\lambda_2,\lambda_3,\lambda_4$, Xu et al. (2016a) took an empirical Bayes approach, by using preliminary fits of the lognormal distributions $\log(T^j) \sim N(\mathbf{Z}^j\boldsymbol{\beta}_0^j, \sigma_0^2)$, assuming a diagonal matrix for Σ_0^j. The empirical estimates of σ^j were used to obtain approximate values of λ_1 and λ_2, and since nothing was known about the α_j's they assumed $\lambda_3 = \lambda_4 = 1$. This approach is practical because the parameters $\boldsymbol{\beta}_1^j, \boldsymbol{\beta}_2^j, \cdots$ determine the prior means $\mu_1^j, \mu_2^j, \cdots$ for the GP prior, which characterize the regression of T^j on $\mathbf{Z}^j$ and the per-stage treatment selections. The imputed treatment effects rely on the predictive distribution under this regression model. Similarly to empirical Bayes priors used in hierarchical models, this empirical Bayes prior avoids excessive prior shrinkage, which might result from an arbitrary choice of $\boldsymbol{\beta}_0^j$ and Σ_0^j, that might smooth away the treatment effects, which are the primary focus of the analysis. In this regard, the total mass parameter α^j for the stick-breaking construction of the DDP determines the number of clusters in the underlying Polya urn, but because most clusters are small the posterior predictive values used for the inferences are insensitive to the prior on α^j. This type of empirical Bayes approach for determining hyperparameters is used quite commonly when a full prior elicitation is impossible or impractical.

Figure 10.4 gives notched box-and-whisker plots summarizing the estimated posterior predicitve distributions of OS time for each DTR in the leukemia dataset, computed under the DDP-GP. Estimates of predicted OS were smallest for the four DTRs with FAI as induction, regardless of salvage treatment, with smaller 90% credible intervals for these inferior regimes. FAI plus ATRA as induction gave the largest predicted survival, except for the regimes (FAI+GCSF, HDAC, other) and (FAI+GCSF, other, other), and the best regime was (FAI+ATRA, other, other). The (FAI + ATRA, $Z^{2,1}$, other) regime was superior to (FAI + ATRA, $Z^{2,1}$, HDAC) regardless of $Z^{2,1}$. These results imply that in general, (1) FAI + ATRA was the best induction therapy, (2) if the patient's disease was resistant to FAI + ATRA, then it did not matter whether salvage therapy contained HDAC or not, and (3) if patients experienced progression after achieving CR with FAI + ATRA, then salvage therapy with non HDAC was

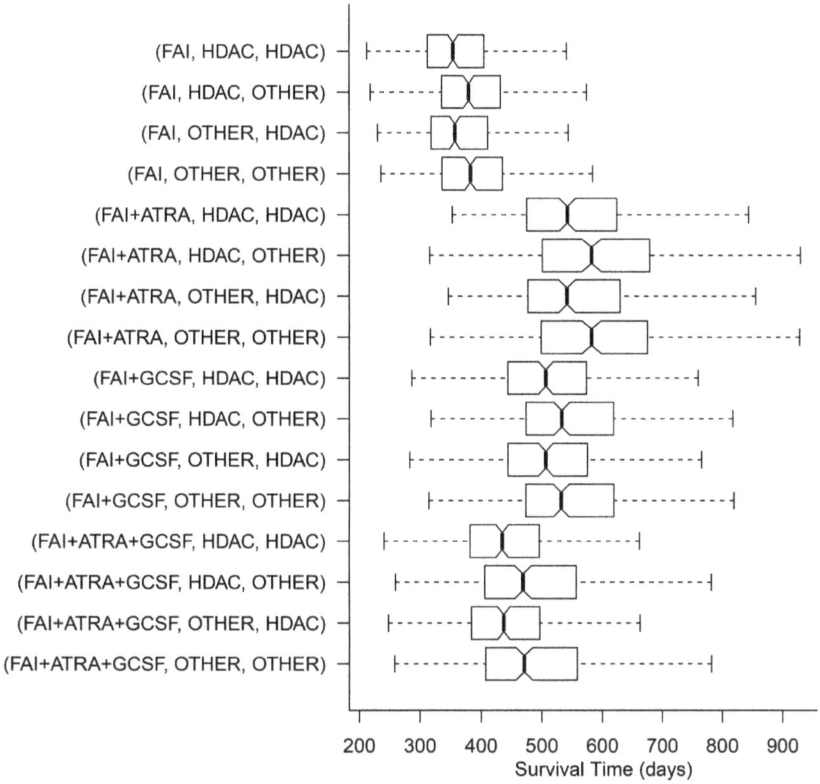

FIGURE 10.4
Box and whisker plots of estimated marginal posterior distributions of overall survival time under the Bayesian nonparametric DDP-GP, for each of the 16 possible DTRs in the leukemia trial.

superior. However, all of these comparative effects were small. The fact that adding ATRA to FAI as induction improved predicted survival suggests that adding ATRA to salvage B_1 or B_2 also might be beneficial.

To estimate overall mean survival time as a function of DTR, Xu et al. (2016a) also used IPTW as a comparator. The IPTW and DDP-GP G-estimation based estimators are summarized in Table 10.1. The most prominent overall results are that the two methods give nearly the same rankings of the 16 DTRs, but they give very different numerical estimates, with the DDP-GP likelihood-based G estimation values much larger than the corresponding IPTW estimators. This is not surprising, given the properties of the two procedures. The IPTW estimator uses covariates to estimate the regime propensity scores for B_1 and B_2, while the DDP-GP estimator computes posterior mean survival time based on a highly structured likelihood that accounts for patient baseline covariates, all transition times and per-stage treatments, with the

TABLE 10.1
IPTW and DDP-GP G-estimation-based estimators of mean survival time for each of the 16 possible dynamic treatment regimes in the AML/MDS trial. Each Bayesian estimator is a posterior mean, with a 90% posterior credible interval (CrI). For the frontline chemotherapies, 'A' denotes ATRA and 'G' denotes G-CSF added to FAI.

Regime	IPTW	DDP-GP	90% CrI
(FAI, HDAC, HDAC)	191.7	390.4	(286.5, 546.6)
(FAI, HDAC, Other)	198.2	416.3	(295.8, 581.7)
(FAI, Other, HDAC)	216.6	394.2	(287.2, 538.6)
(FAI, Other, Other)	222.4	420.2	(296.5, 579.1)
(FAI+A, HDAC, HDAC)	527.4	572.9	(416.6, 829.1)
(FAI+A, HDAC, Other)	458.8	617.2	(434.4, 905.8)
(FAI+A, Other, HDAC)	532.3	573.5	(413.6, 830.4)
(FAI+A, Other, Other)	464.4	617.7	(434.5, 900.3)
(FAI+G, HDAC, HDAC)	326.2	542.1	(393.5, 725.2)
(FAI+G, HDAC, Other)	281.8	578.2	(419.7, 781.0)
(FAI+G, Other, HDAC)	327.7	542.5	(392.8, 726.1)
(FAI+G, Other, Other)	283.4	578.7	(421.5, 781.3)
(FAI+A+G, HDAC, HDAC)	337.4	458.3	(327.9, 651.2)
(FAI+A+G, HDAC, Other)	285.6	502.5	(360.3, 727.4)
(FAI+A+G, Other, HDAC)	362.6	459.4	(328.1, 651.6)
(FAI+A+G, Other, Other)	309.6	503.6	(358.8, 726.9)

final estimate obtained by marginalizing over the empirical covariate distribution, following the G-computation formula. Additionally, the IPTW estimate is based on the overall sample, whereas the likelihood-based DDP-GP method models each transition time's conditional distribution separately, which reduces the effective sample size for each fitted transition time model and thus increases overall variability.

To validate the DDP-GP model-based approach and compare it to conventional frequentist bias correction methods, Xu et al. (2016a) performed a simulation study in which treatments were chosen based on patient covariates, using multistage adaptive rules that mimicked those used by physicians in the leukemia trial. In this way, selection bias was intentionally built into the simulated regimes, to reflect how physicians choose salvage therapies. Details of the simulation study design are given by Xu et al. (2016a). As comparators, they included IPTW and "doubly robust" augmented IPTW (AIPTW). To compare two treatments, $X = 0$ or 1, denoting the estimated propensity score $\hat{\pi}_i = \hat{P}(X_i = 1 \mid \mathbf{Z}_i)$, the AIPTW estimator is

$$\widehat{ATE}^{AIPTW} = \frac{1}{n}\sum_{i=1}^{n}\left\{\left[\frac{I(X_i=1)Y_i}{\hat{\pi}_i} - \frac{I(X_i=0)Y_i}{1-\hat{\pi}_i}\right]\right.$$

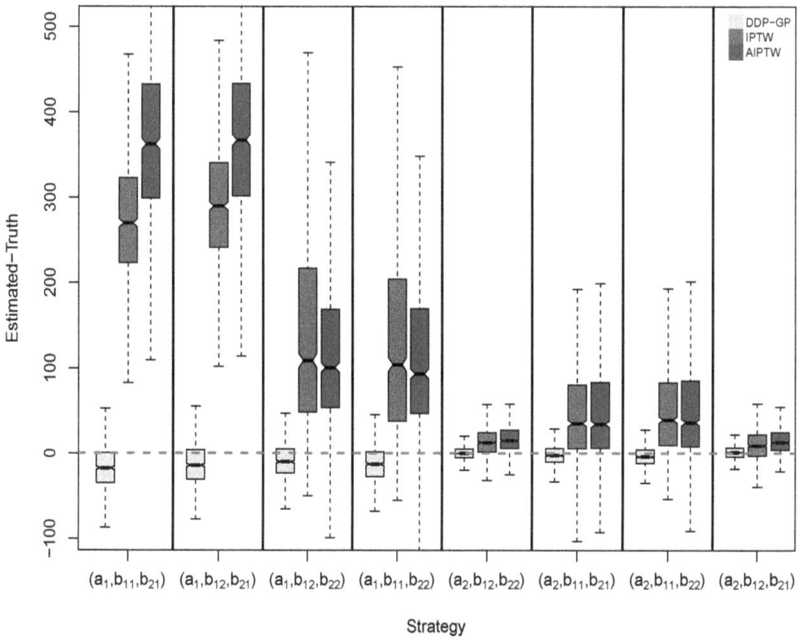

FIGURE 10.5
Box and whisker plots of estimated mean overall survival times for eight adaptive strategies, with treatments chosen based on patient covariates, based on simulated data. Bayesian estimates were obtained as posterior means from a fitted dependent Dirichlet process with a Gaussian process prior, applying a G computation formula (DDP-GP, yellow boxes). Inverse probability of treatment weighted (IPTW) estimates are given by green boxes), and augmented IPTW (AIPTW) estimates are given by blue boxes

$$-\left[\frac{I(X_i=1)-\hat{\pi}_i}{\hat{\pi}_i(1-\hat{\pi}_i)}\right]\left[(1-\hat{\pi}_i)\hat{E}(Y_i \mid X_i=1,\mathbf{Z}_i)\right.$$
$$\left.\left.+\hat{\pi}_i\hat{E}(Y_i \mid X_i=0,\mathbf{Z}_i)\right]\right\}.$$

Figure 10.5 gives estimates obtained from a fitted DDP-GP model based model G estimation using posterior means, IPTW, and AIPTW. The vertical axis gives the difference between each estimated mean survival time and the true mean in each simulation, so values closer to 0 are more desirable. Each notched box-and-whisker plot gives the interquartile range (IQR) from the 25th percentile to 75th percentile, with a mid-line at the median. The top whisker goes to Q3+1.5 IQR, and the bottom whisker goes to Q1 - 1.5 IQR. The IPTW estimates are represented by green boxes, and the A-IPTW estimates by blue boxes. The plots show that, for eight different simulated multistage adaptive strategies, the DDP-GP model-based G estimation method (yellow boxes) gives far more accurate and far more reliable estimates.

A practical advantage of assuming a DDP-GP, or any DDP regression model, is that it does away with the often tedious process of fitting a set of different candidate parametric models for each transition time distribution and choosing the that gives the best fit, as done by Wahed and Thall (2013). Moreover, in settings where a distribution has multiple modes, a conventional accelerated failure time or other parametric model cannot accommodate this, while a DDP with a Gaussian smoothing kernel incorporates such structures very naturally. Given these advantages, the rather involved process of specifying and calibrating prior hyperparameters certainly appears to be worthwhile.

The results in Figure 10.5 are quite striking, since IPTW and AIPTW are very commonly used bias correction methods. These simulation results suggest that it is quite advantageous to use a Bayesian nonparametric DDP regression model, when implementing likelihood-based G estimation to correct for selection bias. This may be attributed to the facts that the DDP-GP model has the desirable properties of shrinkage, robustness, and full support. Since G estimation relies on the assumption that the regression model is correct, and the DDP-GP is as close to having a "correct" model that one may hope for in practice, this seems like an ideal marriage of causal estimation and Bayesian nonparametric regression modeling. From a practical viewpoint, these simulation results strongly suggest that this combination of causal and Bayesian methodologies is very effective for estimating mean survival time based on data obtained from complex multistage treatment settings.

11

Personalizing Dose in Stem Cell Transplantation

11.1 Allogeneic Stem Cell Transplantation

This chapter describes a Bayesian nonparametric regression analysis that led to a greatly improved method for determining an optimal personalized dose of an agent that is part of a preparative regimen for allogeneic stem cell transplantation (allosct). Given the complexity of this treatment modality, it is useful to review some medical background before describing the statistical model and method.

Allosct is a very aggressive treatment for severe hematologic diseases, primarily acute leukemia, myelodysplastic syndrome, or lymphoma. To prepare for this treatment, the patient first undergoes human leukocyte antigen (HLA) typing, and a suitable donor, being an HLA-matched sibling or matched unrelated donor, or related haploidentical or volunteer cord blood donor is identified, to provide stem cells for the allosct. Since blood stem cells are made in the bone marrow as part of the hematopoietic process, the bone marrow also is the origin of the patient's cancer cells. Consequently, once a donor has been identified, the first therapeutic step is to ablate the patient's bone marrow with high-dose chemotherapy, or possibly chemoradiation therapy, in an attempt to greatly reduce or eradicate the source of the cancer cells. The chemotherapy combination used for this purpose is called the pretransplant *preparative* or *conditioning regimen*. The next step is to infuse the donor stem cells into the recipient's circulating blood. The goal is that the infused cells will repopulate the patient's bone marrow, re-establish hematopoiesis to produce functional blood cells, and reconstitute the immune system. This is known as *engraftment*, most commonly defined as three consecutive days of a peripheral white blood cell count $> 500 \times 10^6/L$, with platelet engraftment defined as independence from the need for platelet transfusion for at least seven days and a platelet count $> 20 \times 10^9/L$. Allosct carries several severe risks, including graft failure, infection due to the fact that the patient does not have a functioning immune system between bone marrow ablation and engraftment, disease recurrence, and graft-versus-host disease (GVHD), where the engrafted cells attack the patient's normal cells. Consequently, preventing GVHD is critical to the success of allosct, and GVHD prophylaxis always is given as a key component of treatment. Because graft failure has a nearly 100% mortality rate without an effective salvage treatment, if it occurs then a second allosct is necessary

DOI: 10.1201/9781003474258-11

to save the patient's life. Disease recurrence requires some form of salvage therapy, which may be a second allosct, chemotherapy, a targeted therapy, or immunotherapy.

The preparative regimen's drugs, and their doses and schedules, are critically important and have received a great deal of attention over the years, as methods for allosct have been developed and refined (Luger et al., 2012; Gyurkocza and Sandmaier, 2014). A potential complication of the preparative regimen is that it may cause hepatic veno-occlusive disease (VOD), characterized by blockage of the small blood vessels leading into the liver and inside the liver. This was found to be a very important dose-limiting toxicity of preparative regimens including oral busulfan (Bu). Because a general problem with oral administration of any drug is that it first must pass through the gastrointestinal tract, absorption of a given dose of oral Bu and the amount of the drug delivered to the bone marrow are highly variable. A dose of oral Bu may not be sufficiently immunosuppressive to achieve engraftment reliably, because hepatic first-pass effects produce a reduced concentration of the active drug in the bone marrow. On the other hand, a higher dose of oral Bu given to increase the probability of achieving engraftment also is more likely to produce hepatic VOD and other severe adverse events. These problems motivated the development of intravenous Bu (IV-Bu), which greatly reduces the risk of VOD and improves patient survival. This is, in principle, not a new or surprising finding. When precision, accuracy, and highly reproducible bioavailability are important for achieving a good treatment outcome, intravenous administration should always be favored (Benet and Sheiner, 1985).

Andersson et al. (2000) showed that, compared to oral Bu, administering Bu intravenously reduces the variability in busulfan systemic exposure dose, characterized by the area under the fitted plasma concentration versus time curve (AUC) obtained from a fitted pharmacokinetic (PK) model for data from an individual patient, by a multiplicative factor of 10 to 20. To determine an individual patient's optimal administered IV-Bu dose prior to ablation, the patient often is given a low "test" dose for one or two infusions, and a PK analysis is done of the Bu in their circulating blood over time. Their computed AUC then is used to characterize how quickly they metabolize Bu, and this is used to calculate their optimal therapeutic dose, based on a database comparison. Details of PK-guided individualized dosing are provided by Andersson et al. (2017). Using this approach, Andersson et al. (2000) identified an optimal IV-Bu dosing interval of approximately $950 - 1520$ μMol-min per administration, for a total AUC dose of $15{,}200 - 24{,}400$ μMol-min for a 16-dose regimen. The units μMol-min quantify the integrated area under the PK concentration versus time curve, used to quantify the Busulfan exposure in the patient over time. For a given dose, if a patient who metabolizes slowly the AUC will be relatively large, whereas a fast metabolizer produces a smaller AUC. Thus, a fast metabolizer should be given a comparatively higher dose in order to achieve a given target AUC.

In a retrospective analysis of observational data from seven medical centers performing allosct, Kashyap et al. (2002) compared the probabilities of VOD and 100-day mortality (D100) for oral and IV busulfan. Their data are summarized in Table 11.1. Assuming independent beta(.5, .5) priors for the four probabilities $\pi_Y(X)$, where Y = VOD or D100 and X = IV-Bu or fixed-dose Bu, given the observed count

TABLE 11.1
Comparisons of the rates reported by Kashyap et al. (2002) of veno-occlusive disease (VOD) and 100-day mortality (D100) between allosct patients treated with X = oral versus intravenous (IV) busulfan. For each combination of treatment and outcome Y = VOD or D100, the probability $\pi_Y(X)$ is assumed to follow a beta(.5, .5) prior, $\hat{\pi}$ denotes the posterior mean, and CrI denotes a posterior credible interval.

	Oral Busulfan		IV Busulfan	
Outcome	# cases / n	$\hat{\pi}$ (95% CrI)	# cases / n	$\hat{\pi}$ (95% CrI)
VOD	10 / 30	.34 (.19 - .51)	5 / 61	.08 (.03 - .17)
100-day Mortality	6 / 30	.21 (.09 - .37)	2 / 61	.04 (.01 - .10)

data in Table 11.1,

$$\Pr\{\pi_{VOD}(Oral) > \pi_{VOD}(IV) \mid data\} > .99$$

and

$$\Pr\{\pi_{D100}(Oral) > \pi_{D00}(IV) \mid data\} > .99. \tag{11.1}$$

That is, based on the observational data analyzed by Kashyap et al. (2002), oral busulfan almost certainly had higher rates of both VOD and 100-day mortality than IV-Bu. While the patients were not randomized between oral Bu and IV-Bu, some sensitivity analyses are informative. In the 61 IV-Bu patients, 48% (29/61) were heavily pretreated, defined at ≥ 3 prior chemotherapy regimens, prior radiation, or prior HSCT, compared to 33% (10/30) of the oral-Bu patients. There were no severe imbalances in any other key prognostic variables, which included an indicator of whether or not the patient had active disease at transplant. The fact that the heavy pretreatment rate was much higher in the IV-Bu group makes its much lower VOD and D100 rates shown in equation (11.1) and Table 11.1 quite striking. Although these statistical comparisons of VOD and D100 rates based on non-randomized, observational data would have benefited from IPTW or G-estimation to correct for bias, the following sensitivity analyses are useful. Given the observed 33% VOD rate for the 30 oral Bu patients, in order to see the same VOD rate in the 61 IV-Bu patients would have required 20 VOD cases. Since only 5 IV-Bu patients with VOD were seen, imbalances in covariates other than being heavily pretreated would need to explain why, in the observed data, 15 IV-Bu patients among the hypothetical 20 did not experience VOD. Similarly, since only 2 IV-Bu patients suffered D100, to see a 20% D100 rate in the 61 IV-Bu patients would have required 12 deaths, so similar covariate imbalances would need to explain why, in the observed data, 10 IV-Bu patients among the hypothetical 12 did not die within 100 days.

Andersson et al. (2002) refined the practice of using each patient's PK data to individualize their IV-Bu dose based on analysis of survival time as a function of AUC by analyzing a small sample of data from 36 CML patient who underwent allosct. To estimate their AUC, each patient was given the dose of Bu at 1.0 mg/kg IV over two hours for two infusions. The patient's estimated AUC then was used to individualize their dose, with a PK-guided dose adjustment performed at the next infusion, and that dose was used for the remaining scheduled infusions of the IV-Bu regimen. An optimal AUC interval was identified by fitting a Cox proportional hazards model for overall survival time and constructing a plot of the death rate as a function of AUC, estimated using smoothing splines, with a 95% confidence band for the estimated death rate obtained by bootstrapping. The plot of the estimated hazard of death as a function of AUC, given in Figure 3 of Andersson et al. (2002), was a U-shaped. When this estimated curve was derived, it provided the first graphical validation of the well-established medical fact that a systemic exposure that is too low leads to a high rate of relapse, while a systemic exposure that is too high leads to life threatening toxicities including, but not limited to, VOD. Essentially, the smoothed hazard function was the first statistical analysis to quantify what "too low" and "too high" meant for the AUC of IV-Bu. It provided a qualitative validation of the belief that an intermediate targeted AUC is best, and it triggered a global change in therapeutic practice, including many later refinements of the optimal AUC interval. This hazard function estimate implied that there was an optimal interval for the AUC of IV-Bu that minimizes the hazard of death. The optimal AUC interval to maximize survival was found by an exhaustive search among intervals that included the value AUC = 1232 μMol-min that minimized the estimated hazard function, and the optimal interval was found to be approximately 950 $-$ 1520 μMol-min per administration in a 16-dose course, for a total systemic exposure of 15,200 $-$ 24,320 μMol-min.

In a separate study, Bredeson et al. (2013) showed that IV administration of busulfan improved patient survival substantially. In a multi-center retrospective study of survival and toxicity in 674 patients ranging from infants to 30-year-old adults using IV-Bu in the preparative regimen for allosct, Bartelink et al. (2016) evaluated different methods for estimating AUC. They reported a daily AUC in the interval 78 $-$ 101 mg $\times$ h/L, corresponding to 22,000 $-$ 24,800 μMol-min for a four-day treatment course. that maximized event-free survival, with an event defined as graft failure, relapse, or death.

To validate the PK-based method for individualizing the dose of IV-Bu, Andersson et al. (2017) conducted a randomized trial of allosct for AML/MDS patients that compared PK-guided individualized IV-Bu dosing targeting an average daily AUC of 6000 μM-min $\pm$ 10%, that is, the targeted daily interval 5400 - 6600, versus a fixed IV-Bu dose of 130 mg/m^2, which corresponds to a average daily AUC of about 5000 μM-min, with an estimated range of 3000 to 8000. The hypothesis motivating the trial was that PK-guided IV-Bu dosing would provide more precise, better standardized Bu systemic exposure, and thus yield better leukemia control and reduce transplant-related mortality. A total of 218 patients were randomized between fludarabine (Flu) with PK-guided IV-Bu (n=111), versus Flu with a fixed-dose of IV-Bu (n=107). A Bayesian piecewise exponential time regression model was fit to the trial's data to assess the relationship between survival time and patient covariates and treatment arm, assuming N(0,100) priors on all treatment and covariate parameters.

Covariates included age, and indicators of whether the patient had MDS versus AML, poor cytogenetic risk, disease in CR at transplant, and whether the patient was FLT-3+, which is a receptor for a tyrosine kinase expressed by immature hematopoietic cells. The fitted survival time regression model, summarized in Table 3 of Andersson et al. (2017), showed that, for the coefficient β of the indicator that IV-Bu dose was PK-guided, the posterior probability of a beneficial effect, defined as a lower risk of death, was $\Pr(\beta < 0 \mid data) = .922$, which corresponds to odds of 12 to 1 in favor of PK-guided over fixed IB-Bu dosing for survival time.

Andersson et al. (2017) also reported an additional analysis of the data from this trial, focusing on the subset 34 patients who received fixed-dose IV-Bu of 130 mg/m^2 with observational PK analyses showing that they had a systemic exposure, represented by AUC levels ranging between 5400 and 6600 μM-min, which was the PK guided targeted interval. Survival and PFS of these patents then were compared to a matched set of patients in the PK-guided IV-Bu dose group. These comparisons were motivated by the idea that, if patients had approximately the same systemic exposure then, after matching on prognostic covariates, from a causal viewpoint there should be little difference between the groups in terms of either OS or PFS. The comparisons were done as a 2-to-1 matched-pair analysis that obtained 68 PK-guided IV-Bu patients as matches, using propensity scores computed from the prognostic variables diagnosis (MDS versus other), age, and whether or not the patient had active disease at transplant (CR = no active disease versus NCR = active disease). Logrank tests comparing the matched groups had p-values .84 for OS and .61 for PFS, which have corresponding S-values of 0 and 1. These S-values say that, after PK guided and fixed-dose IV-Bu patients were matched for both delivered dose and covariates, the data provide no surprise whatsoever if one believes the null hypotheses that survival time and PFS time both are distributed identically between the two groups. Considered together, these comparisons imply that differences between the survival time and PFS time distributions in the entire sample were due to the causal effects of PK guided versus fixed dosing, and that PK-guided dosing is greatly superior.

11.2 A DDP with a Refined Gaussian Process Prior

The following regression analysis, reported by Xu et al. (2019), addressed the question of whether PK-guided IV-Bu could be personalized further by using well established patient prognostic covariates to refine the optimal targeted AUC interval. Xu et al. (2019) performed a Bayesian statistical regression analysis of an observational dataset of 151 allosct patients for whom PK analyses of systemic busulfan had been performed, whether the patient received a fixed or PK guided IV-Bu dose. They focused on survival time as a function of AUC, age, and whether the patient was in CR or had active disease (NCR) at the time of allosct. Age was included as a covariate because older age is positively associated with severity of comorbidities in allosct patients, and patients in CR at the time of allosct have much better expected survival

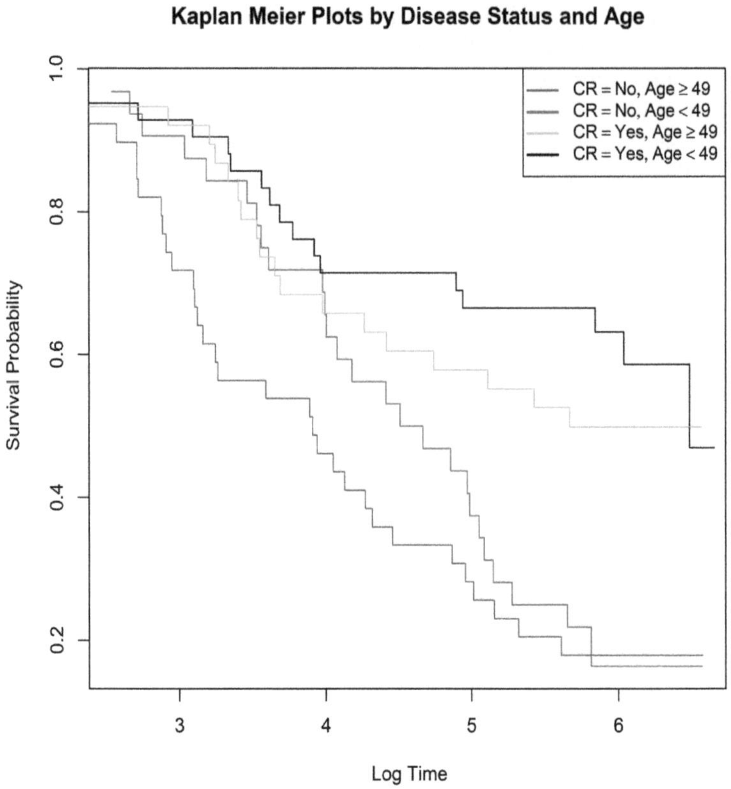

FIGURE 11.1
Kaplan Meier estimates of survival time distributions for four combinations of patient disease status (CR = Yes if disease in complete remission, No if not) and age on date of transplant)

than patients not in CR. Figure 11.1 illustrates the estimated survival distributions for four subgroups defined by CR status and age.

The strong dependence of survival time on age and CR status as prognostic variables in this dataset, which reflects what has been seen historically in allosct, is shown by Figure 11.1. This gives Kaplan-Meier estimates of the survival time distributions for the four subgroups of patients determined by CR = Yes or No and whether the patient's age was < 49 or ≥ 49 years. The goal of the statistical analyses reported by Xu et al. (2019) was to develop a method for computing an optimal targeted AUC interval tailored to each patient's Age and CR status. They did this by formulating and fitting a DDP-GP survival time regression model and using the fitted model to predict survival as a function of Age, CR, and AUC.

Denote Y = log time to death, V = censoring on the log(time) domain, T = $\min\{Y,V\}$, and $\delta = I(Y \leq V)$, so the observed data for patient $i = 1,\cdots,151$ are the outcome variables $(T_i,\ \delta_i)$, and the three covariates $\mathbf{Z}_i$ =(Age$_i$,CR$_i$,AUC$_i$). Xu et al. (2019) assumed a DDP-GP mixture model for the survival distribution $F(\cdot \mid \mathbf{Z})$, using a Gaussian kernel to obtain a continuous random distribution. Suppressing i, the DDP mixture model can be summarized as the usual weighted average of normals,

$$F(y \mid \mathbf{Z}) = \sum_{h=0}^{\infty} w_h\ \phi(y;\ \theta_h(\mathbf{Z}), \sigma^2) \tag{11.2}$$

where the weights are obtained from the stick-breaking prior $\{\upsilon_h\} \sim iid$ beta$(1,\alpha)$, with $w_1 = \upsilon_1$ and $w_h = \upsilon_h \prod_{\ell<h}(1-\upsilon_\ell)$ for each $h \geq 2$. A GP prior is assumed for each normal mean, $\theta_h(\mathbf{Z}) \sim\ GP(\mu_h, V)$, for $h = 1, 2, \cdots$, with the covariance function assumed by Xu et al. (2016a) replaced by the refined form

$$Cov(\mathbf{Z}_i, \mathbf{Z}_\ell) = \sigma_0^2\ \exp\left\{-\sum_{j=1}^{p} \frac{(Z_{ij}-Z_{\ell j})^2}{\lambda_j^2}\right\} + \delta_{i\ell} J^2, \tag{11.3}$$

where $p = 3$ in the allosct application. The covariance function (11.3) includes $p+1$ additional scale parameters, $\lambda_1, \cdots, \lambda_p$ for the p covariates, and also an overall multiplicative scale parameter σ_0^2. Including the covariate-specific λ_j's overcomes the limitation that assuming a single λ gives the same weight to all covariates when quantifying dependence between patients. The overall scale parameter σ_0^2 accounts for variability not accommodated by the variance σ^2 of the normal component distributions in the DDP mixture in the covariance function. These parametric extensions of the GP covariance matrix provide a more robust model.

Denoting the model parameter vector by $\boldsymbol{\theta}$ and the data by $\mathscr{D}_n = \{T_i, \delta_i, \mathbf{Z}_i\}_{i=1}^n$, the likelihood function is

$$\mathscr{L}(\boldsymbol{\theta}, \mathscr{D}_n) = \prod_{i=1}^{n} f(y_i \mid \mathbf{Z}_i, \boldsymbol{\theta})^{\delta_i} \{1 - F(y_i \mid \mathbf{Z}_i, \boldsymbol{\theta})\}^{1-\delta_i},$$

where $f(y \mid \mathbf{Z}, \boldsymbol{\theta})$ denotes the pdf and $F(y \mid \mathbf{Z}, \boldsymbol{\theta})$ denotes the cdf of Y for a patient with covariates $\mathbf{Z}$. Given the DDP-GP prior, the model is completed by assuming priors on the regression parameters,

$$\boldsymbol{\beta}_h \sim \text{iid}\ N(\boldsymbol{\beta}_0, \Sigma_0), \quad 1/\sigma^2 \sim \text{Gamma}(a_1, b_1), \quad \alpha \sim \text{Gamma}(a_2, b_2),$$

and the covariate scale parameters

$$\sigma_0 \sim N(0, \tau_\sigma^2), \quad \lambda_j \sim \text{iid}\ N(0, \tau^2), \quad j = 1, \cdots, p.$$

Collecting terms, the hyperparameter vector is $\boldsymbol{\theta}^*$ = $(\boldsymbol{\beta}_0, \Sigma_0, a_1, b_1, a_2, b_2, , \tau_\sigma^2, \tau^2)$. As done by Xu et al. (2016a), an empirical Bayes approach was used to obtain the regression hyperparameter vector $\boldsymbol{\beta}_0$ by fitting a preliminary normal distribution $log(Y) \sim N(\mathbf{Z}\tilde{\boldsymbol{\beta}}, \tilde{\sigma}^2)$ to the data and using the fitted normal model estimate $\hat{\tilde{\boldsymbol{\beta}}}$ = $\boldsymbol{\beta}_0$, while assuming that Σ_0 is a diagonal matrix with all diagonal values 10. The hyperparameters (a_1, b_1) were tuned to obtain a prior mean of $\tilde{\sigma}^2$ equal to the empirical estimate. Setting $a_2 = b_2 = 1$ ensured a vague prior on the beta stick-breaking parameter α, and assuming $\tau = \tau_\sigma$ = 10 ensured that the ranges of the λ_r's and σ_0 were sufficiently large to account for variability in the data.

11.3 Optimizing Personalized Dose Intervals

The objective of the data analysis was to estimate optimal AUC intervals, personalized using age and CR status, so that the best targeted IV-Bu dose could be given to each future patient. The optimal intervals were obtained as posterior predictive quantities from the fitted DDP-GP survival time regression model. For this computation, the DDP model given by (11.2) first was marginalized with respect to $F(y \mid \mathbf{Z})$ and expressed in its equivalent hierarchical form by defining iid latent indicators $\{\zeta_1, \cdots, \zeta_n\}$ having a distribution determined by the weights $\{w_h\}$. This was done by assuming that

$$\begin{aligned} Y_i \mid \zeta_i = h, \mathbf{Z}_i &\sim indep\, N(\theta_h(\mathbf{Z}_i), \sigma^2) \\ \Pr(\zeta_i = h) &= w_h, \end{aligned} \quad (11.4)$$

for patients $i = 1, \cdots, n$ and DDP mixture indices $h = 1, 2, \cdots$. Let $\rho_n = \{S_1, \ldots, S_H\}$ denote the partition of the n patients determined by H clusters induced by the ζ_i's in the underlying Polya urn scheme representation. The fitted BNP model was used to derive the posterior predictive distribution $p(Y_{n+1} \mid \mathbf{Z}_{n+1}, \mathscr{D}_n)$ of the log survival time Y_{n+1} of a future $n+1^{st}$ patient with covariates $\mathbf{Z}_{n+1}$. To implement the DDP-GP model, the computations were performed using the R package `DDPGPSurv`, which can be downloaded from CRAN.

The cluster representation may be exploited to compute the posterior predictive distribution of a future, $n+1^{st}$ patient's outcome, obtained by averaging over the partition ρ_n and the posterior of $\boldsymbol{\theta}$,

$$p(Y_{n+1} \mid \mathbf{Z}_{n+1}, \mathscr{D}_n) = \sum_{\rho_n} p(\rho_n \mid \mathscr{D}_n) \int p(\boldsymbol{\theta} \mid \mathscr{D}_n)$$

$$\left\{ \sum_{h=1}^{H+1} p(Y_{n+1} \mid n+1 \in S_h, \theta_h(\mathbf{Z}_{n+1}), \boldsymbol{\theta})\, p(n+1 \in S_h \mid \mathbf{Z}_{n+1}, \mathscr{D}_n, \rho_n, \boldsymbol{\theta}) \right\} d\boldsymbol{\theta}. \quad (11.5)$$

The term indexed by $H+1$ in (11.5) accounts for the possibility that the $n+1^{st}$ patient may form their own singleton cluster. The averages with respect to the posterior distribution $p(\rho_n \mid \mathscr{D}_n)$ of the partition, and the predicted distribution of Y_{n+1} averaged over the $H+1$ partition sets, were computed by averaging over the MCMC posterior sample of $\boldsymbol{\theta}$ values.

The posterior predictive distribution of Y_{n+1} given by (11.5) may be used to compute the optimal AUC, given the patient's Age and CR status, to maximize the patient's expected log survival time, as

$$\widehat{\text{AUC}}_{n+1} = \text{argmax}_{\text{AUC}} E(Y_{n+1} \mid \mathbf{Z}_{n+1}, \mathscr{D}_n). \quad (11.6)$$

Since laboratory-based PK evaluation methods for determining a median daily Bu systemic exposure may have about a 3% to 6% error, the above optimal AUC +/- 10% was considered a reasonable interval to target, given as

$$\left[0.9 \times \widehat{\text{AUC}}_{n+1}(\mathbf{Z}_{n+1}), \quad 1.1 \times \widehat{\text{AUC}}_{n+1}(\mathbf{Z}_{n+1}) \right].$$

11.4 Simulation Study

To evaluate the DDP-GP model's performance for estimating the survival distribution and optimal AUC intervals, Xu et al. (2019) conducted a simulation study assuming a data structure that mimicked the allost dataset, with $p = 3$ covariates sampled from the allosct dataset, and a sample of $n = 200$ patients either with survival times uncensored or with 25% censoring. They assumed that $[Y_i \mid \mathbf{Z}_i] \sim N(\mu(\mathbf{Z}_i), \sigma_0^2)$ where, suppressing the patient index i for brevity, the mean took the highly nonlinear form

$$\mu(\mathbf{Z}) = 4 - .1Z_1 + .7Z_2 + .3Z_3 - .07Z_2^2 + .1Z_1Z_2 + .2Z_2Z_3 - .18Z_1Z_2Z_3.$$

Indexing the simulated trials by $b = 1, \cdots, B = 100$, the bth computed posterior predictive probability that a patient with covariates $\mathbf{Z}$ survives beyond time t is

$$\hat{S}_b(t \mid \mathbf{Z}) = \Pr(Y_{n+1} > t \mid \mathbf{Z}, \mathscr{D}_n),$$

and an overall predicted survival probability estimate was computed by averaging over the empirical covariate distribution, as

$$\hat{S}(t) = \frac{1}{B}\sum_{b=1}^{B}\frac{1}{n}\sum_{i=1}^{n}\hat{S}_b(t \mid \mathbf{Z}_i).$$

Six models were considered as comparators. Two were accelerated failure time (AFT) models of the form, again suppressing the index i,

$$Y = \beta_0 + \beta_1 Z_1 + \beta_2 Z_2 + \beta_3 Z_3 + \beta_4 Z_2^2 + \beta_5 Z_1 Z_2 + \beta_6 Z_2 Z_3 + \beta_7 Z_1 Z_3 + \sigma\varepsilon, \quad (11.7)$$

where a normal distribution for ε gave lognormal Y or an extreme value distribution for ε gave Weibull Y. They also evaluated four semiparametric models for the baseline hazard function, including a Polya Trees (PT) prior (Hanson and Johnson, 2002), and a transformed Bernstein polynomials (TBP) prior (Zhou and Hanson, 2018), assuming an AFT regression model as the frailty model with the same form as (11.7) in both the PT and TBP methods. The simulations showed that the DDP-GP model-based estimate and the RF and BART estimates each reliably recovered the shape of the true survival function, while the log-normal, Weibull, TBP, and PT methods each produced a substantially biased estimate.

Xu et al. (2019) performed an additional simulation study to compare the DDP-GP with the more elaborate covariance matrix (11.3) for the GP prior to what is obtained assuming the simpler form with all $\lambda_j = \lambda$ and $\sigma_0^2 = 1$. The goal was to compare the two DDP-GP models in terms of their ability to estimate optimal personalized AUC ranges reliably. The simulation study, provided in a supplement to the paper, was based on data for 200 patients, either with or without 25% censoring, generated assuming a hypothetical version of the allosct data, with Y distributed as a highly nonlinear function of $p = 8$ covariates. Figure 11.2 shows that the DDP-GP model assuming the refined form for $Cov(\mathbf{Z}_i, \mathbf{Z}_\ell)$ given by (11.3) did a substantially better job of estimating the true mean survival time curve μ_Y(log AUC) for each age, compared to the DDP-GP model assuming the simpler form of V for the GP prior.

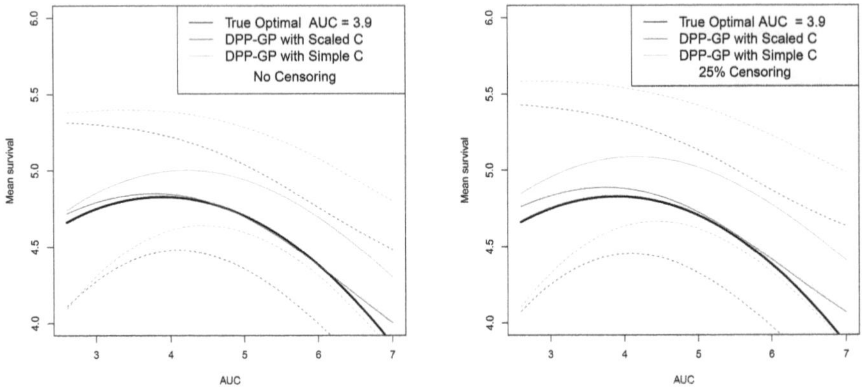

FIGURE 11.2
Comparison of DDP-GP estimates of log mean survival as a function of AUC on the log scale, based on simulated samples of n = 200 patients with no censoring (left) or 25% censoring (right). Estimates are based on the GP prior with either simple covariance matrix V (green line), or a refined V with additional covariate-specific scale parameters (red line). Pointwise 95% credible interval bands as given as dotted lines, and the true curve is given as a solid black line.

11.5 Analyses of the Allosct Data

Analyses of the allosct data reported by Xu et al. (2019) identified, for each combination of CR status and Age, an optimal AUC interval that yields higher expected survival times than any AUC that is either below or above the interval. Figure 11.3 gives predicted posterior mean log(survival time) as a function of log(AUC) for each of the eight combinations of CR status = Yes or No and age = 30, 40, 50, or 60 years. The figure shows that the predicted optimal AUC intervals differ substantially between the (Age, CR status) combinations. This has extremely important implications for a physician choosing an individual allosct patient's personalized targeted AUC based on their Age and CR status. For example, the optimal AUC interval for a 50-year-old patient not in CR is $4.7 \pm 0.47 = [4.23, 5.17]$, while the optimal interval for a 40-year-old patient in CR is $5.8 \pm 0.58 = [5.22, 6.38]$ Because these intervals do not overlap, these two patients should have very different targeted AUC values to maximize their expected survival times.

Figure 11.4 gives a striking illustration of the optimal targeted AUC intervals as functions of Age and CR status, based on predictions obtained from the fitted DDP-GP regression model. The figure shows that there was a nonlinear three-way interaction between the predicted optimal AUC, Age and CR status. There was a strong negative association between optimal AUC and Age, and the rates of decrease

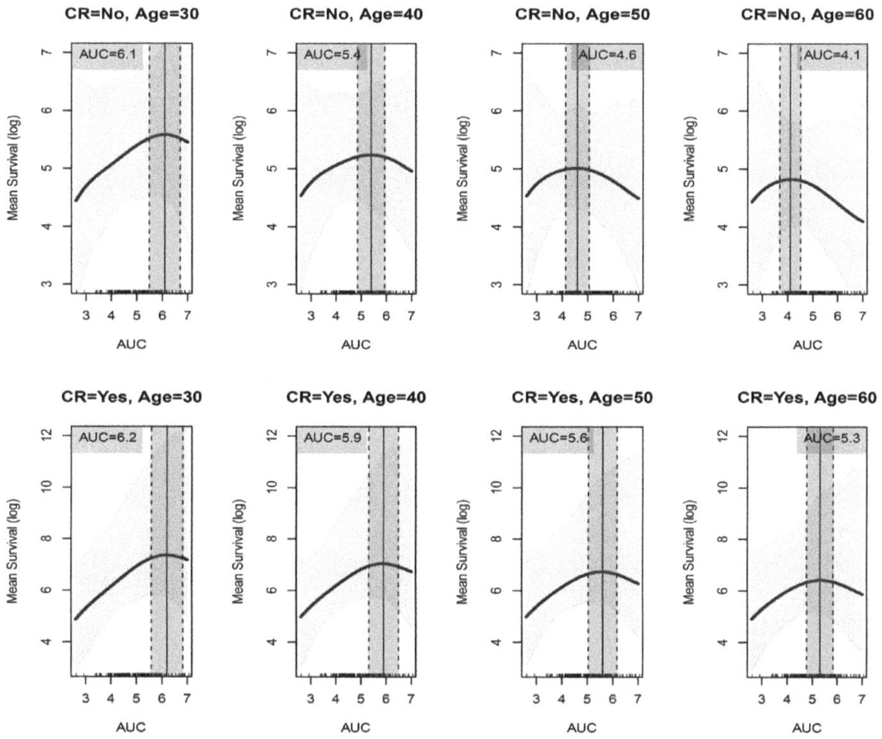

FIGURE 11.3
Estimated log mean survival time for a patient either in complete remission (CR) or with active disease (No CR) at transplant, Age = 30, 40, 50, or 60 years, as a function of AUC, with the optimal targeted AUC interval for each (CR status, Age) combination.

for the optimal AUC with older Age were different for patients in CR and not in CR at transplant, with a nonlinear rate in the no CR subgroup. While CR status had little or no effect on the estimated optimal AUC interval for younger patients with Age ≤ 28, the optimal AUC intervals decreased with Age at different rates for the CR = Yes versus CR = No subgroups, with the intervals becoming completely disjoint for patients older than 55 years. For Age ≤ 28, the curves for CR and No CR in Figure 11.4 agree with the conclusion of Bartelink et al. (2016) for pediatric and adolescent patients, while the curves for Age > 28, show that CR status mattered a great deal for older patients. These results provide a concrete empirical basis for planning a targeted AUC for an AML/MDS patient undergoing allosct based on their Age and CR status, that is, for doing precision medicine.

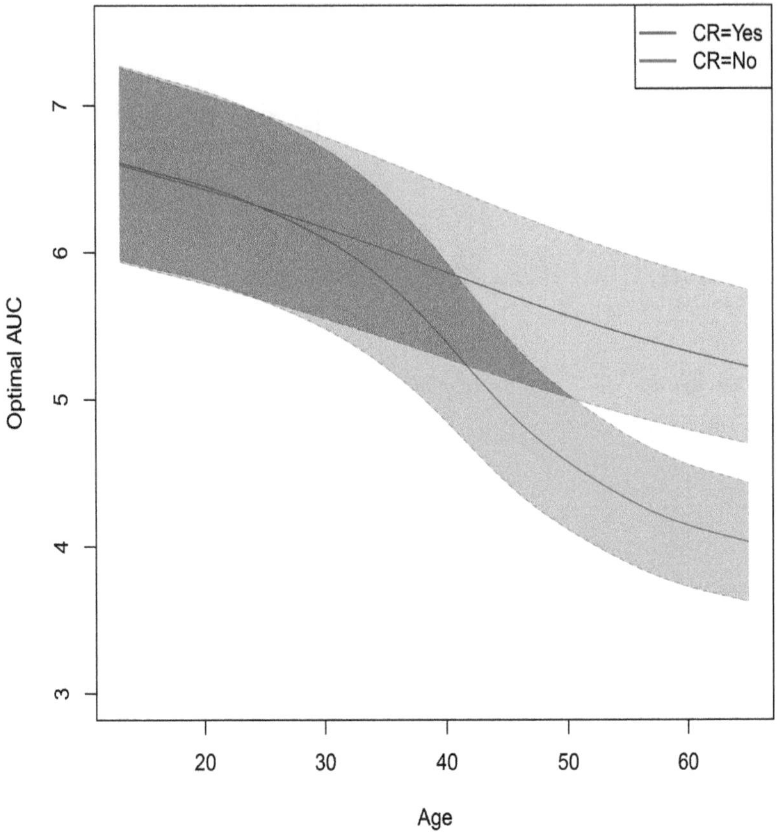

FIGURE 11.4
Estimated pharmacokinetically guided optimal targeted AUC interval to maximize expected survival time for patients receiving intravenous busulfan as part of a conditioning regimen for an allogeneic stem cell transplant, as a function of Age and disease status at transplant, CR (complete remission) or not .

Computer Software The data analysed by Xu et al. (2019) and computer programs used for the analyses can be obtained from
https://rss.onlinelibrary.wiley.com/hub/journal/14679876/series-c-datasets

12

Choosing a Breast Cancer Treatment

12.1 A Limitation of the Medical Literature

Papers reporting clinical trial results in medical journals focus primarily on statistical inferences about effects of treatments on clinical outcomes. In oncology, outcomes often include an early anti-disease effect represented by a binary or ordinal "response" variable observed relatively soon after initiation of therapy, a list of treatment-related adverse events, PFS time, survival time, and immunological or other biological variables that often are measured repeatedly over time. Conventional practice dictates that one element of $\boldsymbol{Y}$ must be designated as the "primary" outcome, with the other entries considered "secondary" or "exploratory" outcomes. Each patient's data may be summarized generally as an outcome vector, $\boldsymbol{Y}$, a vector $\boldsymbol{Z}$ of prognostic covariates, and an element from a set $\{x_1, \cdots, x_K\}$ of treatments studied in the trial, possibly including doses or different therapeutic modalities. In most reported data analyses, the outcome variables are considered one at a time, with each variable Y_j in the vector $\boldsymbol{Y}$ analyzed by computing summary statistics, possibly with graphics such as Kaplan-Meier plots of an event time distributiona or a histogram of Y_j by treatment arm, and a fitted regression model $f(Y_j \mid \boldsymbol{Z}, X, \boldsymbol{\theta})$, where $\boldsymbol{\theta}$ denotes the model's parameter vector including treatment and covariate effects.

Most commonly, Cox proportional hazards models are fit for a time-to-event time outcomes and logistic models are fit for binary outcomes. There seldom is any consideration of how well, or whether, each assumed model fits the trial's data, or whether the sample size is large enough for the statistical methods underlying distribution theory, which may be asymptotic, to be valid, or reliable. Inferences about $\boldsymbol{\theta}$ are reported, with strength of evidence quantified most often using p-values. These statistical analyses often are accompanied by medical or biological interpretations including hypothetical mechanistic causal explanations for estimated relationships. Scientific flaws with these conventions include using assumed regression models without any goodness-of-fit analyses to determine whether they fit the trial's data, using p-values to quantify the strength of evidence while artificially dichotomizing a trial's results as being "significant" or not and, not infrequently, analyzing treatment effects in subgroups selected by data dredging in search of p-values that are $<$ .05. These foolish, sometimes fatal conventions are discussed extensively by Thall (2020).

DOI: 10.1201/9781003474258-12

In this chapter, I will describe a re-analysis of data from a published clinical trial. This was motivated, in part, by the desire to overcome the limitations imposed by the convention of identifying one primary outcome and reporting a trial's results on that basis. This may be very misleading when two or more outcomes are important. The convention of identifying one primary outcome makes a trial appear to be simple and focused, and it facilitates power and sample size computations, which appear to provide a halo of scientific respectability. This practice can become extremely problematic when there are risk-benefit trade-offs between desirable and undesirable outcomes that are ignored. This was the case in a randomized two-arm trial of targeted treatments for advanced breast cancer reported by (Dickler et al., 2016), described below. In this trial one treatment combination gave longer expected PFS time, but it also had a much higher rate of severe toxicities. This problem arises quite frequently in oncology, where adverse treatment effects are ubiquitous. It is a truism in cancer therapy that more advanced disease requires more aggressive therapy, which is likely to increase the rates of both desirable anti-disease effects and undesirable regimen-related adverse effects. Because designating one Y_j as primary ignores these fundamental facts, this common practice frequently is misleading, and it may serve patients poorly.

Aside from failing to address risk-benefit trade-offs, another important limitation of most medical papers is due to the fact that statistical inference and medical decision-making are not the same thing. The conventional practice is simply to report statistical analyses of a clinical trial's data. This of course can be very useful, but it falls short of addressing the real reason that clinical trials are conducted in the first place, which is to improve medical practice. An implicit assumption in published medical papers reporting a clinical trial's results is that, assuming the trial was designed sensibly and conducted properly, the data analyses were done reasonably well, and the results are described clearly, decision-making by practicing physicians based on the trial results should be straightforward. In many cases, this simply is not true. A typical medical paper reporting a trial's results does not provide a structure that practicing physicians can use to incorporate a trial's results when making their day-to-day decisions for individual patients seen in the clinic. While fitted statistical regression models and plots of a trial's results provide a scientific basis for making inferences about model parameters $\boldsymbol{\theta}$, they do not include a formal structure for the next step, which is using such inferences to make personalized treatment decisions based on each new patient's characteristics, $\mathbf{Z}$. Interpreting published statistical inferences about $\boldsymbol{\theta}$ is the penultimate step in a process that, ideally, should end with practicing physicians using the published results to make informed treatment decisions for their patients.

Choosing an individual patient's treatment based on their covariates and published clinical trial results often is not straightforward. Making treatment decisions is especially difficult if there are two important but qualitatively different outcomes, such as a desirable anti-disease effect and an undesirable adverse effect, each of which may vary with each patient's covariates. If separate consideration of two such outcomes produces conflicting treatment choices, as in the trial reported by Dickler et al. (2016), then inevitably this leads to some sort of consideration of the risk-benefit trade-off between the outcomes. Almost invariably, this is done informally since, unavoidably, any formal statistical structure for evaluating treatment and covariates effects on two or more outcomes, such as PFS time and a list of toxicities, is complex. The issue is how making treatment decisions while accounting for trade-offs may be done in a way that is both scientifically rigorous and practical. The key to a statistical analysis that achieves these goals is explicit specification of a risk-benefit trade-off. The methodology for doing this described in this chapter requires close collaboration between physicians and statisticians.

12.2 Evaluating Targeted Agents for Breast Cancer

Dickler et al. (2016) reported results of a randomized phase III study to compare letrozole plus bevacizumab (*L*+*B*) to letrozole plus placebo (*L*) as first-line therapy for patients with hormone receptor-positive advanced breast cancer. From a therapeutic perspective, the trial's results were less than ideal. This section describes how a medical risk-benefit decision analysis may be done using a Bayesian nonparametric regression model and a utility function that quantifies the trade-off between PFS time and severe toxicity.

Letrozole is a nonsteroidal aromatase inhibitor that works by decreasing the amount of estrogen produced by the body. This slows the growth of breast cancer cells because they require estrogen to reproduce. Letrozole is one type of estrogen therapy (ET). Because the vascular endothelial growth factor (VEGF) receptor for angiogenesis is a potential mechanism of a cancer's resistance to an ET, a therapeutic strategy is to augment an ET with an anti-disease agent that attacks the VEGF signaling pathway. In the study, this was done by combining letrozole with bevacizumab, which is a humanized monoclonal antibody to VEGF. The desired anti-disease effect of bevacizumab is that it reduces blood supply to the tumor. Previous trials had investigated effects of combining an ET with bevacizumab, but the comparative effects of *L*+*B* and *L* on PFS in advanced breast cancer were not known.

Patients were randomized between *L*+*B* and *L* within four strata defined by two binary covariates recorded at enrollment. One was an indicator of whether the patient had measurable disease, and the other was an indicator of whether the length of their previous disease-free interval prior to entering the trial was $\leq$ 24 months. The primary efficacy endpoint was PFS time, defined as the time from study entry to first disease progression or death, subject to independent administrative right censoring, with progression defined using RECIST v1.0 criteria, A total of 21 qualitatively different types of toxicity were monitored, including the grade of each toxicity occurrence in each patient, using NCI Common Toxicity Criteria version 3, with possible grades being none, mild, moderate, severe, life-threatening, and fatal, coded as 0, 1, 2, 3, 4, 5.

In a conventional frequentist data analysis, Dickler et al. (2016) reported that *L*+*B* had significantly longer PFS time compared to *L*, based on the one-sided p-value = 0.016, and that this effect did not vary substantially with age or any other covariates. They also reported that 46.8% of patients treated with $L+B$ experienced at least one severe (grade ≥ 3) toxicity, compared to only 14.2% with L. Thus, adding bevacizumab to letrozole tripled the severe toxicity rate. A closer look at the PFS data shows that, despite the nominally "significant" p-value of the statistical test comparing PFS between $L+B$ and L, the average PFS times of the two treatment arms actually were not very different. The estimated median PFS was 20.2 months (95% CI 17.0 to 24.1) with $L+B$ compared to 15.6 months (95% CI 12.9 to 19.7) with L. An even smaller difference was seen for estimated median survival times, which were 47.2 months (95% CI 39.0 to 56.8) with $L+B$, compared to 43.9 months (95% CI 37.6 to 49.6) with L. So the estimated median survival times differed by 3.3 months and the confidence intervals had a large degree of overlap. This is an example of the well known fact that whether or not the p-value of a statistical test is $< .05$ often has little to do with whether the difference in the outcome distributions between two treatments is meaningful for a patient. For this trial, as often is the case, the median survival time estimates and accompanying confidence intervals are far more informative than a p-value.

From a patient's perspective, the small estimated increases of 4.6 months in median PFS time and 3.3 months in median survival time with $L+B$ are accompanied by great uncertainty as to which treatment may actually have longer PFS. This is because the pairs of confidence intervals for each outcome have a high degree of overlap. In terms of adverse effects, a very worrying result was that $L+B$ had a much higher estimated severe toxicity rate than L. Consequently, if one wishes to consider both PFS time and the risk of severe toxicity, it is very unclear how a physician and patient might decide which treatment is preferable.

Lee et al. (2022) addressed this problem by accounting explicitly for subjective trade-offs between PFS time and the total amount of toxicity experienced by a patient. They constructed a Bayesian statistical framework for making personalized treatment decisions based on data having the form obtained from the breast cancer trial. To construct this framework, they first computed a summary statistic from the 21 toxicity grades called the scaled *total toxicity burden* (TTB). For a vector $\boldsymbol{W}$ of K toxicities, each having J possible grades, the scaled TTB was defined as

$$Q = Q(\boldsymbol{W}) = \frac{\sum_{k=1}^{K} W_k}{K \times (J-1)}.$$

This takes on values between 0 (no toxicity of any kind) and 1 (all toxicities at the highest grade). Because the definition of Q relies entirely on the grade of each type of toxicity, it assigns the same weight, $1/\{K \times (J-1)\}$, to any observed grade W_k of any type of toxicity. For example, since $J = 6$ and $K = 21$ in the breast cancer dataset, the contributions to TTB of grade 3 toxicities of two qualitatively different types both were $3/(21\times 5) = 3/105$. This is simpler than the more general definition of TTB given by Bekele and Thall (2004), which requires qualitatively different types of toxicity to be assigned possibly different subjective "severity scores" for each grade. This definition is much more demanding, because it requires these subjective scores to be elicited from physicians familiar with the disease, toxicities, and

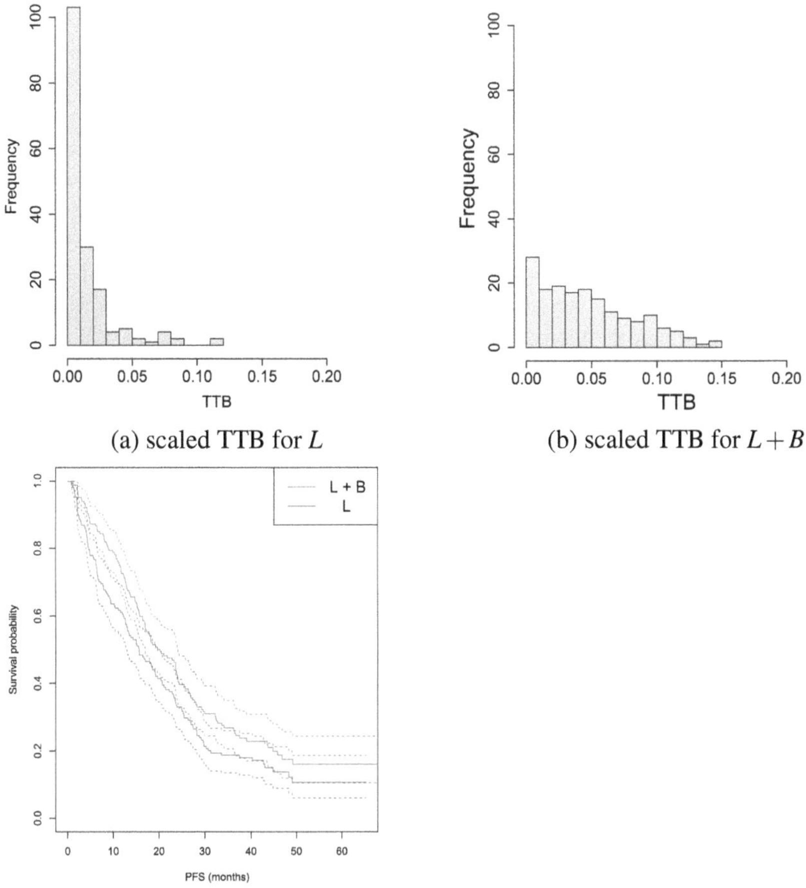

(a) scaled TTB for L

(b) scaled TTB for $L+B$

(c) Kaplan-Meier estimates of $S(t)$

FIGURE 12.1
Histograms of scaled total toxicity burden for the breast cancer data are given in panels (a) for L = letrozole and (b) for L+B = letrozole plus bevacizumab. Panel (c) gives Kaplan-Meier estimates and 95% confidence intervals of the PFS distributions for L (red) and $L+B$ (blue).

treatment. In this sense, TTB as defined by Bekele and Thall (2004) is similar to a utility function. The scaled TTB used here has the advantage that no subjective scores must be elicited, since grades are used to quantify severity regardless of type of toxicity, which makes it much more practical. Figure 12.1 illustrates the observed TTB distributions for the two treatment arms, and gives Kaplan-Meier estimates of the survival functions of PFS for each treatment, where L+B is represented by blue and L by red.

Considered together, the results reported by Dickler et al. (2016) and the summary TTB distributions, given here in Figure 12.1, are problematic. Since L+B has slightly longer PFS but L has much lower TTB, it is unclear how to decide which treatment is better, either overall or for individual patients. This example may be considered a prototype for the general problem with the way that conventional methods are used for reporting PFS or overall survival time distributions and toxicity data from a randomized comparative trial. The first major issue is that a nominally "significant" difference between key parameters based on a computed p-value being $<$ 0.05 or not for a two-sample test comparing PFS was accompanied by estimated parameters that do not differ by a clinically meaningful amount. Because the overlap between the two CIs for median PFS time based on the breast cancer trial data was substantial, the evidence for superiority of L+B over L in terms of PFS was weak. This is an example of the well-known problem that a nominally "significant" p-value often is inadequate or misleading for the purpose of clinical decision-making, since the sizes of estimated effects often are at odds with whether a computed p-value is below the arbitrary cut-off .05. The common practice of using .05 as a p-value cut-off to make a dichotomous decision reduces a great deal of information to an oversimplified "significant" versus "not significant" conclusion. Since this corresponds to four bits of information against the null hypothesis that the median PFS times are equal, it hardly is convincing evidence that L+B is superior.

Since the L+B arm had a slightly larger average PFS and a much larger average TTB than L, this is a "win-lose" scenario that leads, inevitably, to consideration of a risk-benefit trade-off between these two outcomes. This type of setting is very common in oncology, since most anti-cancer treatments have adverse effect. This problem also arises when evaluating treatments for many other diseases. In nearly all similar settings, quite often some mention is made of a "tradeoff" or "risk benefit" when reporting trial results, but beyond paying lip service to this issue no actual decision analysis is provided.

Another problem is the absence of estimated median PFS and mean TTB values that account for important prognostic variable effects. This does not refer to treatment-subgroup interactions, but rather to estimated posterior distributions of treatment effects under a regression model that accounts for how PFS and TTB each varies with additive prognostic covariate effects and treatment. Additive prognostic covariate effects can lead to substantive changes in one's conclusions about comparative treatment effects, as $\mathbf{Z}$ varies from patient to patient.

If either PFS or TTB is considered alone, then selecting an optimal treatment is straightforward because smaller TTB and longer PFS each is more desirable. This is of no practical use, however, because L+B is preferable in terms of Y, but L is preferable in terms of Q. For a practicing physician who desires to choose between treatments for an individual patient, these two outcomes must be considered together, and the patient's prognostic covariates also may play a key role. How to make this decision for an individual breast cancer patient depends on the fact that there is a risk-benefit trade-off between the desire to limit toxicity and the desire to improve PFS. Moreover, different patients may not have the same risk-benefit trade-off. Consequently, to address this problem formally, a key question that must be resolved at the start is precisely what one means by "risk-benefit trade-off".

The Bayesian statistical analysis given by Lee et al. (2022) extends a usual regression-based data analysis by using a precision utility function to connect the fitted regression model with subsequent medical decision-making by a practicing physician. This requires construction of a family of precision utility functions, $U(\boldsymbol{Y},\boldsymbol{Z})$, that assign numerical desirability scores to all $\boldsymbol{Y} = (PFS,TTB)$ pairs for a patient with prognostic variables $\boldsymbol{Z}$. To represent their subjective trade-offs between PFS and TTB, a physician and patient may choose a particular utility function from a family of functions. Because $U(\boldsymbol{Y},\boldsymbol{Z})$ varies with $\boldsymbol{Z}$, two patients with different prognostic covariates may have different treatment preferences for the same outcome pair $\boldsymbol{Y}$.

Decision analysis based on utility functions and individual preferences has been studied and applied in many areas, including business (Pennings and Smidts (2003); Loewenstein et al. (1989)), engineering (Chen et al. (1998); Bagočius et al. (2014)), and operations research (Walsh et al. (2004); Roy et al. (2017)). Utility-based decision procedures are almost never included in papers reporting statistical data analyses from medical studies, however. The analysis of the breast cancer dataset that will be given below illustrates how a utility function and a Bayesian statistical regression model can be used together to choose between two treatments for a patient with given prognostic variables. This was done using Bayesian nonparametric (BNP) multivariate regression model, $f(\boldsymbol{Y} \mid X,\boldsymbol{Z},\boldsymbol{\theta})$, that accounts for association among between PFS and TTB, and how each outcome depends on the patient's treatment, $X = L+B$ or L, and baseline covariates, $\boldsymbol{Z}$. The BNP model, which is a linear DDP, includes a vector of continuous unobserved latent variables that are used to induce a joint distribution for the observed ordinal toxicity outcomes. This regression model is extremely robust, and inferences based on it are very reliable.

12.3 Bayesian Regression Analysis of the Breast Cancer Data

The following analyses rely on two key components, a flexible Bayesian nonparametric regression model $f(\boldsymbol{Y} \mid X,\boldsymbol{Z})$ and a utility function $U(\boldsymbol{Y},\boldsymbol{Z})$. Specifying and fitting the DDP may be considered a modern version of a conventional regression analysis. It enjoys the important advantage that, rather than assuming a Cox model, or possibly an accelerated failure time model, as is done most often, the linear DDP provides an extremely flexible basis for fitting the data and making inferences. An important aspect of the DDP model is that the outcome $\boldsymbol{Y}$ is a high dimensional vector including PFS time and the grades of 21 different toxicities. While constructing a multivariate regression model in this case using conventional methods may be quite difficult, it can be done quite naturally in the setting of a DDP, as shown below.

The utility provides a basis for using the fitted regression model to do a decision analysis that links the trial's results to medical practice. This provides a rational basis for choosing a treatment X for a new patient with prognostic covariates $\boldsymbol{Z}$ by maximizing the posterior predictive distribution of the utility, computed under the fitted DDP model. Uncertainty associated with the decision-making procedures is quantified through the posterior distributions.

Let Y_i denote PFS time, defined as the time to death or progression (failure) for patient $i = 1,\ldots,n$. Let Y_i^o denote the time to failure or administrative right censoring, with indicator $\delta_i = 1$ if PFS time is observed, $Y_i^o = Y_i$, and $\delta_i = 0$ if censored, $Y_i > Y_i^o$. Censoring is assumed to be independent of Y_i, toxicity occurrences, and covariates. Denote the ordinal variable $W_{i,k} \in \{0,1,\ldots,J-1\}$ representing the maximum grade that patient i experienced for toxicity type $k = 1,\ldots,K$, with $\boldsymbol{W}_i = (W_{i,1},\ldots,W_{i,K})$. Denote the observed data from n patients by

$$\mathscr{D}_n = \{(Y_i^o, \delta_i, \boldsymbol{W}_i, X_i, \boldsymbol{Z}_i),\ \ i = 1,\ldots,n\}.$$

The breast cancer dataset includes $n = 340$ patients, $K = 21$ toxicity categories, and $J = 6$ severity grades, from grade 0 = no occurrence to grade 5 = regimen related death. The $p = 3$ prognostic covariates are Z_1 = Age, an indicator Z_2 of measurable disease, and an indicator Z_3 of whether the patient's disease-free interval prior to trial entry was > 24 months. The treatment indicator is $X = 1$ for L+B and $X = 0$ for B.

To account for heterogeneity between patients not explained by $\boldsymbol{Z} = (Z_1, Z_2, Z_3)$, the BNP regression model includes latent $(K+1)-$dimensional real-valued normal patient frailty vectors, $\boldsymbol{s}_i = (s_{i,0}, s_{i,1}, \ldots, s_{i,K})'$ for $i = 1,\cdots,n$, with assumed mulivariate normal prior distribution $\boldsymbol{s}_i \sim$ *iid* $\mathrm{N}_{K+1}(\boldsymbol{0}, \Omega)$ and $\Omega \sim$ Inv-Wishart(a_s, Ω^0).

The following device was used to deal with the 21 ordinal toxicity grade variables, to obtain a joint distribution and facilitate computation. A multinomial probit model for each $\boldsymbol{W}_i$ is constructed by introducing a second set of unobserved real-valued latent variables, $\tilde{\boldsymbol{L}}_i = (\tilde{L}_{i,1},\cdots,\tilde{L}_{i,K})$. For each i, this is used to define the observed outcomes

$$W_{i,k} = j \text{ if and only if } c_{k,j} < \tilde{L}_{i,k} \leq c_{k,j+1},$$

where $c_{k,0} < c_{k,1} < \cdots < c_{k,J}$ denote toxicity type-specific cutoffs for each k. This is a common modeling strategy in which real-valued unobserved latent variables are used to induce a tractable multivariate distribution on a vector of observed ordinal categorical variables. See, for example, Chib and Greenberg (1998). Denoting the logarithm transformed PFS time by $\tilde{Y} = \log(Y)$, the latent variable representation of $\boldsymbol{W}$ in terms of $\tilde{\boldsymbol{L}}$ allows a joint regression model to be specified for the vector $(\tilde{Y}, \tilde{L})$, which greatly facilitates computation. The joint density function of this $K+1$ dimensional outcome vector is denoted $h(\tilde{Y}_i, \tilde{\boldsymbol{L}}_i \mid X_i, \boldsymbol{Z}_i, \boldsymbol{s}_i)$.

To model this conditional distribution, the following linear DDP was assumed, with the family of random probability distributions indexed by $(X, \boldsymbol{Z})$. The linear DDP prior induces covariate dependence through simple regression structures (De Iorio et al., 2004, 2009). Denote

$$\tilde{\boldsymbol{Z}}_i' = (1, X_i, \boldsymbol{Z}_i'),\ \boldsymbol{\beta} = (\beta_0, \beta_X, \beta_1, \ldots, \beta_p)',\ \boldsymbol{\alpha}_k = (\alpha_{k,0}, \alpha_{k,X}, \alpha_{k,1}, \ldots, \alpha_{k,p})',$$

and the parametric linear combinations $\eta_0(X_i, \boldsymbol{Z}_i) = \boldsymbol{\beta}'\tilde{\boldsymbol{Z}}_i$ for PFS time and $\eta_k(X_i, \boldsymbol{Z}_i) = \boldsymbol{\alpha}_k'\tilde{\boldsymbol{Z}}_i$ for toxicity type $k = 1,\ldots,K$. Denoting the vector of linear terms by

$$\boldsymbol{\eta}(X_i, \boldsymbol{Z}_i) = (\eta_0(X_i, \boldsymbol{Z}_i), \ldots, \eta_K(X_i, \boldsymbol{Z}_i))$$

the DDP model assumes

$$h(\tilde{Y}_i, \tilde{\boldsymbol{L}}_i \mid X_i, \boldsymbol{Z}_i, \boldsymbol{s}_i) = \int \phi_{K+1}(\tilde{Y}_i, \tilde{\boldsymbol{L}}_i \mid \boldsymbol{\eta}(X_i, \boldsymbol{Z}_i) + \boldsymbol{s}_i, \Sigma) dG(\boldsymbol{\beta}, \boldsymbol{\alpha}), \tag{12.1}$$

where $\phi_{K+1}(\cdot \mid \boldsymbol{a}, \boldsymbol{B})$ denotes the $(K+1)-$variate normal pdf with mean vector $\boldsymbol{a}$ and positive definite $(K+1) \times (K+1)$ covariance matrix $\boldsymbol{B}$. The structure of the mean vector in the multivariate normal kernel of the density in (12.1) shows that it is a sum of a vector $\boldsymbol{\eta}(X_i, \boldsymbol{Z}_i)$ of linear terms and a random effect vector $\boldsymbol{s}_i$ with prior mean $\underline{0}$. The DP prior for the random mixing distribution G in (12.1), gives the density h as a DP mixture of multivariate normal linear models,

$$h(\tilde{Y}_i, \tilde{\boldsymbol{L}}_i \mid X_i, \boldsymbol{Z}_i, \boldsymbol{s}_i) = \sum_{m=1}^{\infty} w_m \phi_{K+1}(\tilde{Y}_i, \tilde{\boldsymbol{L}}_i \mid \boldsymbol{\eta}_m(X_i, \boldsymbol{Z}_i) + \boldsymbol{s}_i, \Sigma). \tag{12.2}$$

The weights $\{w_m\}$ are constructed using Sethuraman's (1994) stick-breaking process by assuming $\upsilon_1, \upsilon_2, \cdots \overset{iid}{\sim} Be(1, \xi)$ and $w_m = \upsilon_m \prod_{r<m}(1 - \upsilon_r)$, for each $m = 1, 2, \ldots$, with fixed ξ. Priors assumed for the covariate and treatment effect parameters were

$$\boldsymbol{\beta}_m \mid \bar{\boldsymbol{\beta}}, \tilde{\sigma}_\beta^2 \overset{iid}{\sim} \text{MVN}_{p+2}(\bar{\boldsymbol{\beta}}, \tilde{\sigma}_\beta^2 \text{I}_{p+2}) \quad \text{and} \quad \boldsymbol{\alpha}_{k,m} \mid \bar{\boldsymbol{\alpha}}_k, \boldsymbol{V} \overset{iid}{\sim} \text{N}_{p+2}(\bar{\boldsymbol{\alpha}}_k, \boldsymbol{V}) \tag{12.3}$$

with $\bar{\boldsymbol{\beta}}$ and $\tilde{\sigma}_\beta^2$ fixed, where N_{p+2} denotes a $(p+2)$ dimensional multivariate normal distribution, and $\bar{\boldsymbol{\alpha}}_k = (\bar{\alpha}_{k,0}, \bar{\alpha}_{k,X}, \bar{\alpha}_{k,1}, \ldots, \bar{\alpha}_{k,p})$ is a $(p+2)-$dimensional mean vector and $\boldsymbol{V}$ = diag$[v_p^2]$ is a $(p+2) \times (p+2)$ matrix. Assume $\bar{\alpha}_{k,p} \overset{indep}{\sim} \text{N}(\bar{\bar{\alpha}}_p, v_\alpha^2)$ with fixed $\bar{\bar{\alpha}}_p$ and v_α^2, and $v_p^2 \overset{iid}{\sim} IG(a_v, b_v)$ with fixed a_v and b_v for $p = 0, \ldots, k+2$. The hierarchical structure in the model (12.3) for $\boldsymbol{\alpha}_{k,m}$ borrows information across the toxicity categories.

The model in (12.2) is a DDP prior that incorporates X_i and $\boldsymbol{Z}_i$ linearly in the mean of each normal summand. Because the distribution of $(\tilde{Y}_i, \tilde{\boldsymbol{L}}_i)$ is a weighted average of multivariate normal distributions, each with its own linear term, the model accounts for possible effects of X_i and $\boldsymbol{Z}_i$ on $(\tilde{Y}_i, \tilde{\boldsymbol{L}}_i)$ that can be quite complex, including complicated interactions between two or more variables in $(X_i, \boldsymbol{Z}_i)$. Assuming that $\Sigma = \text{diag}(\sigma_t^2, \sigma_z^2, \ldots, \sigma_z^2)$ implies conditional independence between $\tilde{Y}_i$ and $\tilde{Z}_{i,k}$ given $\boldsymbol{s}_i$. This also allows computation of marginal distributions from (12.2) as weighted averages of normal distributions. The marginal for log PFS time is

$$\begin{aligned} \tilde{Y}_i \mid X_i, \boldsymbol{Z}_i, s_{i,0} &\overset{indep}{\sim} f(\tilde{Y}_i \mid X_i, \boldsymbol{Z}_i, s_{i,0}) \\ &= \sum_{m=1}^{\infty} w_m \phi_1(\tilde{Y}_i \mid \eta_{m,0}(X_i, \boldsymbol{Z}_i) + s_{i,0}, \sigma_t^2) \end{aligned}$$

The marginal for the latent variable that determines the kth toxicity's grades is

$$\begin{aligned} \tilde{L}_{i,k} \mid X_i, \boldsymbol{Z}_i, s_{i,k} &\overset{indep}{\sim} g_k(\tilde{L}_{i,k} \mid X_i, \boldsymbol{Z}_i, s_{i,k}) \\ &= \sum_{m=1}^{\infty} w_m \phi_1(\tilde{L}_{i,k} \mid \eta_{m,k}(X_i, \boldsymbol{Z}_i) + s_{i,k}, \sigma_l^2). \end{aligned}$$

Thus, f and each $g_1, \cdots, g_K$ each is a linear DDP mixture distribution. The marginal distribution of the kth observed toxicity grade, $W_{i,k}$, is obtained by integrating over the kth latent variable distribution,

$$\mathrm{P}(W_{i,k} = j \mid X_i, \mathbf{Z}_i, s_{i,k}) = \sum_{m=1}^{\infty} w_m \int_{c_{k,j}}^{c_{k,j+1}} \phi_1(v \mid \eta_{m,k}(X_i, \mathbf{Z}_i) + s_{i,k}, \sigma_l^2) dv.$$

Marginalizing by averaging over $\boldsymbol{s}$ in (12.2), the resulting DP mixture model has covariance matrix $\Sigma + \Omega$. Thus, Ω induces dependence between PFS time and the toxicities within each patient.

To ensure identifiability of the multivariate ordinal regression model, σ_z^2 is fixed, $c_{k,1} = 0$ for all k, $c_{k,0} = -\infty$, $c_{k,J} = \infty$, $\mathrm{P}(Z_{i,k} < 0) = 0$ and $\mathrm{P}(Z_{i,k} \leq J-1) = 1$. The cut-offs $c_{k,j}$, for $j = 2, \ldots, J-1$ that define the toxicities are assumed to be random for flexibility, with the Markovian structure

$$c_{k,j} = c_{k,j-1} + e_{k,j-1}, \quad k = 1, \ldots, K,$$

with error terms $e_{k,j} \overset{iid}{\sim} \mathrm{Gamma}(a_e, b_e)$, for $j = 2, \ldots, J-1$. Lastly, $\sigma_t^2 \sim IG(a_t, b_t)$.

Collecting terms, the vector of all model parameters is

$$\boldsymbol{\theta} = (w_m, \boldsymbol{\beta}_m, \sigma_t^2, \boldsymbol{\alpha}_{k,m}, \mathbf{e}, \bar{\alpha}_{k,p}, \upsilon_p^2, \Omega)$$

and the vector of all fixed hyper-parameters is

$$\tilde{\boldsymbol{\theta}} = (M, \bar{\boldsymbol{\beta}}, a_t, b_t, X_p^2, \bar{\bar{\alpha}}_p, v_\alpha^2, a_v, b_v, a_s, \Omega^0).$$

Given $\tilde{\boldsymbol{\theta}}$ and data $\mathscr{D}$, the joint posterior of the parameter vector $\boldsymbol{\theta}$ and the patient specific random effect vectors $\boldsymbol{s} = \{\boldsymbol{s}_i, i = 1, \ldots, n\}$ is

$$p(\boldsymbol{\theta}, \boldsymbol{s} \mid \mathscr{D}, \tilde{\boldsymbol{\theta}}) \propto \left\{ \prod_{i=1}^{n} p(\tilde{Y}_i^o, \delta_i, \tilde{\boldsymbol{L}}_i \mid X_i, \mathbf{Z}_i, \boldsymbol{s}_i, \boldsymbol{\theta}\tilde{\boldsymbol{\theta}}) \times p(\boldsymbol{s}_i \mid \boldsymbol{\theta}, \tilde{\boldsymbol{\theta}}) \right\} p(\boldsymbol{\theta} \mid \tilde{\boldsymbol{\theta}}).$$

The joint likelihood for the ith patient is

$$\mathscr{L}(\tilde{Y}_i^o, \delta_i, \tilde{\boldsymbol{L}}_i, X_i, \mathbf{Z}_i, \boldsymbol{s}_i, \boldsymbol{\theta}, \tilde{\boldsymbol{\theta}}) = \prod_{k=1}^{K} p(\tilde{L}_{ik} \mid X_i, \mathbf{Z}_i, s_{i,k}, \boldsymbol{\alpha}_k, \boldsymbol{u}_k)$$

$$\times \{f(\tilde{Y}_i \mid X_i, \mathbf{Z}_i, s_{i,0}, \boldsymbol{\beta}, \sigma_t^2)\}^{\delta_i} \{1 - F(\tilde{Y}_i \mid X_i, \mathbf{Z}_i, s_{i,0}, \boldsymbol{\beta}, \sigma_t^2)\}^{1-\delta_i}.$$

MCMC simulation is used to generate posterior samples of the parameter and latent variable vectors, $(\boldsymbol{\theta}, \boldsymbol{s})$. When fitting the model, the DDP was approximated by truncating the infinite sum of mixture components of F and G_k to the finite value M, with the final weight set to $w_M = 1 - \sum_{m=1}^{M-1} w_m$ to ensure that F and each G_k are proper distributions. For sufficiently large L, the truncated sum produces inferences virtually identical to those with the infinite sum, as discussed by Ishwaran and James (2001)and Rodriguez and Dunson (2011). Assessment of the sensitivity of the model to M values for the breast cancer dataset led to using $M = 15$ mixture components in the BNP regression model for the breast cancer data analysis.

12.4 A Precision Utility Function

Denoting Q = TTB for brevity, a practical advantage of reducing the K-dimensional toxicity vector $\boldsymbol{W}$ to the 1-dimensional summary statistic Q is that it provides a readily interpretable statistic for constructing a utility as a function of the pair (Y,Q). Compared to previously defined outcome utilities as in, for example, Thall et al. (2011), Thall and Nguyen (2012), or Lee et al. (2019), a refinement is that here a *precision utility* U is constructed so that it varies with the patient's covariates $\mathbf{Z}$, as well as the outcome pair (Y,Q). This is qualitatively similar to the subgroup-specific utility used by Lee et al. (2021) in the context of a phase I-II dose-finding design. The decision analysis here will use $U(Y,Q,\mathbf{Z})$, and the fitted regression model, to provide physicians with a formal basis for choosing between L+B (X^{new} = 1) and L (X^{new} = 0) for a future patient with covariates $\mathbf{Z}^{new}$. The set of all $U(Y,Q,\mathbf{Z})$ as $\mathbf{Z}$ varies over the space of possible covariate vectors is a PUFF, defined here for data analysis and decision-making rather than trial design.

To quantify trade-offs between Y and Q, for any $\mathbf{Z}$, a utility function must satisfy the admissibility constraints

$$U(Y,Q,\mathbf{Z}) > U(Y,Q',\mathbf{Z}) \text{ if } Q < Q' \text{ for any } Y$$

and

$$U(Y,Q,\mathbf{Z}) < U(Y',Q,\mathbf{Z}) \text{ if } Y < Y' \text{ for any } Q.$$

These inequalities formalize the requirement that, when each outcome is considered individually with the other outcome fixed, smaller TTB and longer PFS each is more desirable. That is, the utilities must make sense.

The particular form of the utility function given here, and the specific numerical values that it takes on, were determined by two medical oncologists, one of whom is a breast cancer subspecialist. The first step of the construction was to specify the overall precision utility function as the product

$$U(Y,Q,\mathbf{Z}) = U_{PFS}(Y,\mathbf{Z}) \times U_{TTB}(Q,\mathbf{Z}), \tag{12.4}$$

with $0 < U_{TTB}(q,\mathbf{Z}) \leq 1$. For any $\mathbf{Z}$, the above admissibility requirements imply that $U_{PFS}(Y,\mathbf{Z})$ must increase with Y since longer PFS is better, and $U_{TTB}(Q,\mathbf{Z})$ must decrease with Q since more toxicity is worse. These requirements are formalized by the constraints

$$0 < U_{PFS}(Y,\mathbf{Z}) < U_{\max} \quad \text{and} \quad 0 < U_{TTB}(Q,\mathbf{Z}) \leq 1.$$

The idea underlying this construction begins with the assumption that, for given $\mathbf{Z}$, the PFS time utility component $U_{PFS}(Y,\mathbf{Z})$ would be the patient's payoff if the patient had no toxicity. The TTB utility component $U_{TTB}(Q,\mathbf{Z})$ on the domain [0, 1] acts as a multiplicative penalty term that decreases $U_{PFS}(Y,\mathbf{Z})$, with larger Q leading to a smaller value of $U_{TTB}(Q,\mathbf{Z})$ and thus a smaller overall utility $U(Y,Q,\mathbf{Z})$.

To apply this structure, a particular functional form of (12.4) was constructed to reflect patent preferences in the treatment setting for hormone receptor positive advanced breast cancer. Denote the prognostic covariates Z_1 = Age, and indicators Z_2 = 1 for measurable disease, and Z_3 = 1 if the patient's disease free interval prior to trial entry was > 24 months. These three covariates are included, along with the treatment indicator X, in the regression model for PFS and $\boldsymbol{W}$. While the randomization for the breast cancer trial was stratified by Z_2 and Z_3 to improve precision, based on clinical experience it was decided that neither Z_2 nor Z_3 should have any effect on the utility function, whereas Age was included in U_{TTB}. This was a subjective decision based on the observation that younger advanced breast cancer patients tend to care more about extending PFS than controlling TTB, while older patients have shorter expected survival time and thus they tend to care more about maintaining good QOL, quantified by a smaller Q. This motivated using only Age in the utility function, with $U(Y,Q,Age)$ constructed so that, for any PFS time Y, a given TTB value Q decreases the utility for an older patient more than for a younger patient. It also was assumed that $U_{PFS}(Y,Age)$ = $U_{PFS}(Y)$, that is, the PFS time payoff does not vary with age, although in general this assumption is not strictly necessary., Setting $U_{\max} = 100$ for convenience because the domain (0, 100) is easy to interpret, the construction requires that the multiplier $U_{TTB}(0,Age) = 1$ for any *Age*. Thus, if the patient has no toxicity, there is no multiplicative penalty, formally $U(Y,0,Age)$ = $U_{PFS}(Y)$. For example, if $Q = 0.50$, $U_{PFS}(36)$ = 80, and $U_{TTB}(0.50,60)$ = 0.50 then $U(36,0.50,60)$ = 40 for a 60-year-old patient, whereas the toxicity penalty $U_{TTB}(0.50,40)$ = 0.70 gives the much higher utility $U(36,0.50,40)$ = 56 for a 40-year-old patient.

For $U_{PFS}(Y)$, the clinicians specified the numerical values $U_{PFS}(24) = 50$ and $U_{PFS}(48) = 95$, and required the limiting value $\lim_{t\to\infty} U_{PFS}(t) = 100$ for patients of any *Age*. Given these values, the following parametric function was specified :

$$U_{PFS}(Y) = \begin{cases} U_0\left(\frac{Y}{t_0}\right)^a & \text{if } Y < y_0 \\ \\ \frac{U_{\max}}{1+\exp(-b_1 Y)} & \text{if } Y \geq y_0, \end{cases} \tag{12.5}$$

setting y_0=48 months, U_0=95, and $U_{\max}$=100. This function increases in T up to 48 months, with a small additional increase for $Y > 48$ months. Since $U_{PFS}(24)$ = 50, the equation $95 \times (24/48)^a = 50$ gives $a = 0.926$. Similarly, since $U_{PFS}(48)$ = $100/\{1+\exp(-b_1 48)\} = 95$, this gives $b_1 = 0.061$. The resulting function $U_{PFS}(Y)$ is plotted in Figure 12.2(a).

To reflect the preferences of older patients with advanced breast cancer, $U_{TTB}(Q,Age)$ was defined to vary with *Age* so that it decreases at a faster rate for older *Age*. Numerical values of U_{TTB} in [0, 1] were established for each pair of (Q,Age) values specified on a grid, and a parametric function was constructed to closely approximate the elicited TTB utilities by exploring various functional forms. This fitting process led to the exponential function

$$U_{TTB}(Q,Age) = \exp\left\{\frac{-Q^2}{2\exp\{2(c_0+c_1 Age)\}}\right\}, \quad \text{for } 0 \leq Q \leq 1. \tag{12.6}$$

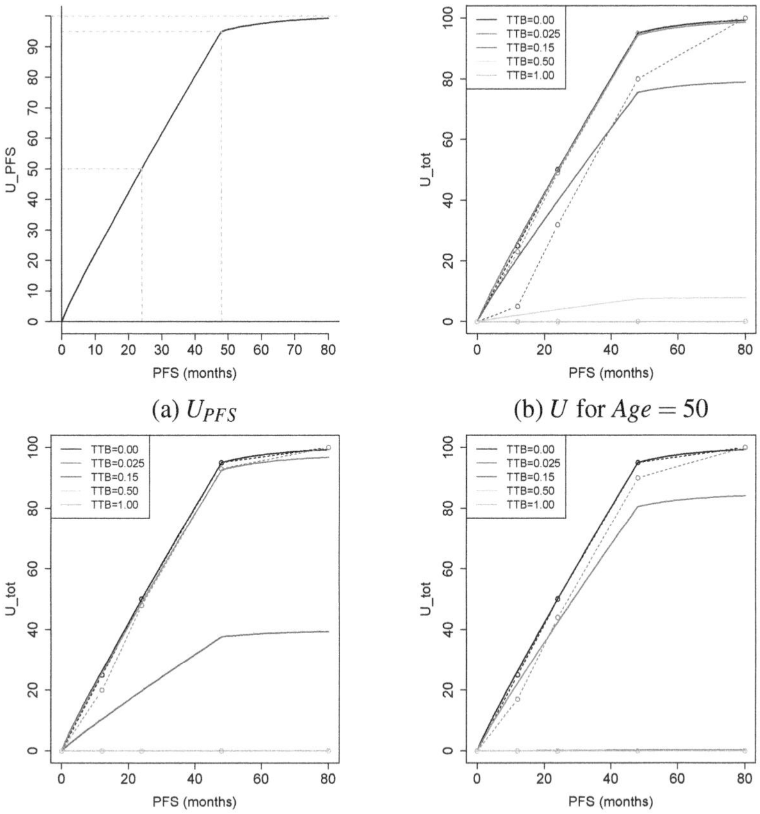

(a) U_{PFS} (b) U for $Age = 50$

(c) U for $Age = 65$ (d) U for $Age = 85$

FIGURE 12.2
Illustration of $U_{PFS}(Y)$ and overall utility $U(Y,Q,Age)$ for Age = 50, 65, 0r 85, where Y = PFS time and Q = TTB.

As Q increases, the overall utility given in (12.5) decreases by a multiplicative factor of the above exponential function of Q^2, and U_{PFS} is penalized in U using this exponential value. If no toxicity occurs, then $U_{TTB}(0,Age) = 1$ and $U(Y,Q,Age) = U_{PFS}(Y)$ for any Age. The function $U_{TTB}(Q,Age)$ has an inflection point at $Q = \exp\{c_0 + c_1 Age\}$, and for any $Q > \exp\{c_0 + c_1 Age\}$ U decreases to 0 very quickly, with $U_{TTB}(Q,Age)$ decreasing in Q faster for larger values of Age. This is obtained by restricting $c_1 < 0$ and calibrating the values of c_0 and c_1 using the elicited numerical utilities. The values $c_0 = 0.823$ and $c_1 = -0.05$ yielded a good approximation. Figure 12.2(b)–(c) compares $U(Y,Q,Age)$ (solid line) to the elicited values (dots connected by dotted lines) for Age = 50, 65, and 85. In each plot, different colors represent different values of Q. The figure illustrates how $U(Y,Q,Age)$ decreases with Q, and how the magnitude of decrease changes with Age.

To use this structure for individualized treatment selection, the fitted Bayesian regression model allows one to compute the joint posterior predictive (PP) distribution of the future PFS time Y^* and toxicity outcome vector $\boldsymbol{W}^*$ for a new patient with prognostic covariates $\boldsymbol{Z}^* = (Z_1^*, Z_2^*, Z_3^*)$. If treatment X is given to the patient, the PP is

$$p(Y^*, \boldsymbol{W}^* \mid X, \boldsymbol{Z}^*, \mathscr{D}) = \int_{\boldsymbol{\theta}} \int_{\boldsymbol{s}} p(Y^*, \boldsymbol{W}^* \mid \boldsymbol{\theta}, \boldsymbol{s}, \boldsymbol{Z}^*, X) p(\boldsymbol{s}, \boldsymbol{\theta} \mid \mathscr{D}_n) d\boldsymbol{s} d\boldsymbol{\theta}. \quad (12.7)$$

By averaging the likelihood of the new patient's future outcomes $(Y^*, \boldsymbol{W}^*)$ over the joint distribution of $(\boldsymbol{\theta}, \boldsymbol{s})$ given $\mathscr{D}$, the PP distribution in (12.7) provides a fully model-based criterion for making inferences to compare treatments. The PP distribution of the utility, $p(U(Y^*, Q, Age^*) \mid X, \boldsymbol{Z}^*, \mathscr{D})$, for the new patient with prognostic covariates $\boldsymbol{Z}^*$ is computed from (12.7).

Comparing the predictive mean total utilities for the two treatments may be used as a basis for optimal treatment selection, as follows, For $X = 0$ corresponding to L and $X = 1$ for L+B, the predictive mean total utility for a new patient with covariates $\boldsymbol{Z}^*$ is

$$\bar{u}(X, \boldsymbol{Z}^*) = \sum_{z_1=0}^{J-1} \cdots \sum_{z_K=1}^{J-1} \int_{\mathbb{R}} U(y, q, Age^*) p(y, \boldsymbol{z} \mid X, \mathscr{D}, \boldsymbol{Z}^*) dy. \quad (12.8)$$

One may choose the treatment X having larger $\bar{u}(X, \boldsymbol{Z}^*)$ for the new patient.

12.5 Choosing a Best Personalized Treatment

To use the fitted BNP regression model for the breast cancer trial data as a basis for personalized treatment selection, a decision analysis is needed. Estimates of the posterior predictive distributions of Y, Q, and U are shown in Figure 12.3 for a future patient with $\boldsymbol{Z}^* = (Age^*, 0, 0)$. The top, middle, and bottom rows of the figure correspond to $Age^* = 55$, 65, and 75 years, with treatment L ($X = 0$) represented by red and L+B ($X = 1$) represented by blue. PP estimates of the survival functions

$$S(y \mid X, \boldsymbol{Z}^*, \mathscr{D}_n) = \Pr(Y > y \mid X, \boldsymbol{Z}^*, \mathscr{D}_n),$$

with 95% pointwise credible intervals, are given in the left column. The middle column gives estimates of the PP distributions, $p(Q \mid X, \boldsymbol{Z}^*, \mathscr{D}_n)$, of TTB. Estimates of the survival probabilities $S(y \mid X, \boldsymbol{Z}^*, \mathscr{D}_n)$ for L+B are slightly above those for L at all ages, showing a small overall improvement in PFS with L+B compared to L. In contrast, estimates of $p(Q \mid X, \boldsymbol{Z}^*, \mathscr{D}_n)$ for L+B have much longer and thicker right tails, showing a substantially increased risk of toxicity with L+B compared to L. From the figures, effects of Age on PFS and TTB are small. These Bayesian inferences for the outcomes PFS and TTB, considered separately, agree qualitatively with the results reported by Dickler et al. (2016), with the important distinction that the comparisons given here are based on posterior quantities, rather than p-values.

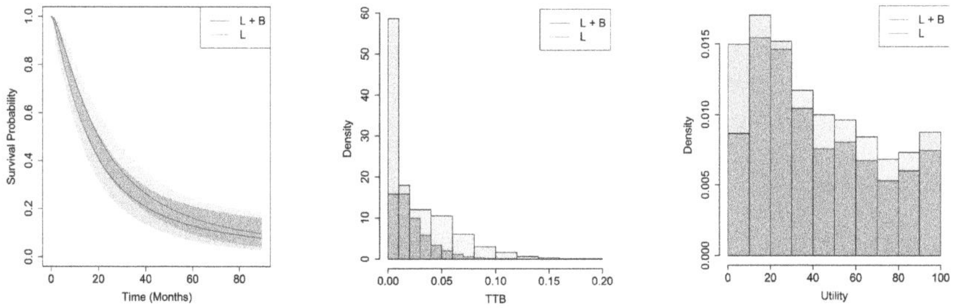

(a) $S(t \mid \mathscr{D})$ at age 55 (b) $p(q \mid \mathscr{D})$ at age 55 (c) $p(U_{\text{tot}} \mid \mathscr{D})$ at age 55

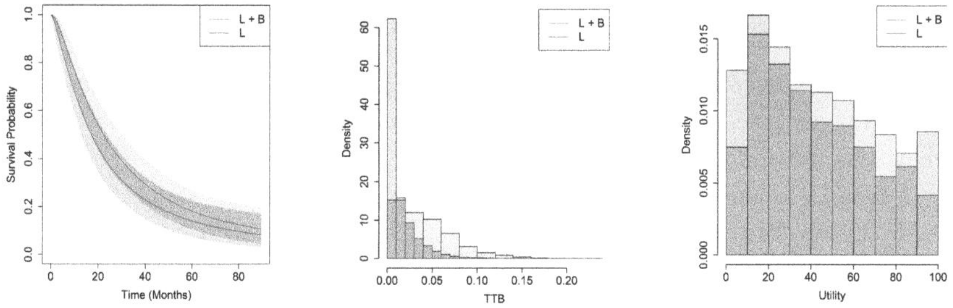

(d) $S(t \mid \mathscr{D})$ at age 65 (e) $p(q \mid \mathscr{D})$ at age 65 (f) $p(U_{\text{tot}} \mid \mathscr{D})$ at age 65

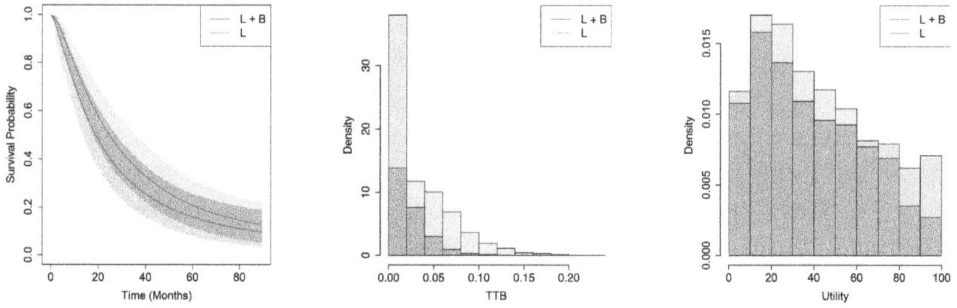

(g) $S(t \mid \mathscr{D})$ at age 75 (h) $p(q \mid \mathscr{D})$ at age 75 (i) $p(U_{\text{tot}} \mid \mathscr{D})$ at age 75

FIGURE 12.3
Estimated posterior predictive survival functions are in panels (a), (d), and (g). Estimates of posterior predictive distributions of TTB (q) are in panels (b), (e), and (h). Posterior predictive distribution estimates of U_{tot} are in panels (c), (e) and (i). The top, middle, and bottom rows correspond to ages $Z_1^* = 55$, 65, and 75 years, respectively, with $(Z_2^*, Z_3^*) = (0,0)$ fixed. In each panel, red represents treatment L and blue represents treatment $L + B$.

Recall that this analysis was motivated by the problem that treatment comparisons based on TTB or PFS considered separately lead in opposite directions. Consequently, the fitted DDP regression model does not provide a clear basis for choosing between L+B and L, either overall or for individual patients. Considering T and Q together using the joint utility function, which accounts for *Age*, provides a tool for resolving this problem. A criterion for a utility-based treatment comparison that accounts for Age uses the predictive distributions for the outcomes Y_X^* and Q_X^* of a future patient with Age^* treated with either $X = L$ or $X = L+B$. The utilities are $U(Y_X^*, Q_X^*, Age^*)$, and the PP probabilities are computed conditional on the data $\mathscr{D}$ using Z_1^* = Age*. For a new patient, this decision criterion is the PP probability

$$\Delta(L, L+B, Age^*) = \Pr\{U(Y_L^*, Q_L^*, Age^*) < U(Y_{L+B}^*, Q_{L+B}^*, Age^*) \mid \mathscr{D}_n\}. \quad (12.9)$$

This is the PP that $L+B$ will give a higher utility than L, so values above .50 are in favor of $L+B$ and values below .50 are in favor of L. An important property of the posterior predictive criterion (12.9) is that it is based on the utility difference

$$U(Y_{L+B}^*, Q_{L+B}^*, Age^*) - U(Y_L^*, Q_L^*, Age^*). \quad (12.10)$$

This is in accordance with the general fact that, from a causal viewpoint, randomization ensures unbiased comparisons between pairs of potential outcomes for a future patient, which could be the causal effect $Y_{L+B}^* - Y_L^*$ for PFS, or $Q_{L+B}^* - Q_L^*$ for TTB. In this case, it is the difference (12.10) between the combined statistics obtained from the utility function, for two versions of a future patient with Age* given either treatment $L+B$ or L.

The plots in the rightmost column of Figure 12.3 compare PP distributions of U for the two treatments as functions of *Age*, showing that the utility advantage of *L+B* over L diminishes with increasing *Age*. Posterior estimates $\hat{\Delta}(\mathbf{Z}^*, L, L+B)$ of the probabilities that *L+B* has greater utility than L for 55, 65, and 75-year-old patients are, respectively, 0.56, 0.54, and 0.46. This implies that, in terms of the predicted overall utility accounting for PFS, TTB, and Age, decisions based on $\hat{\Delta}(\mathbf{Z}^*, L, L+B)$ would be to give *L+B* to patients with $Age^* < 70$, and give L to patients with $Age^* \geq 70$. This illustrates how precision medicine may be guided by a Bayesian decision analysis, in which the decision criterion is based on the posterior distribution of a covariate-specific utility function, estimated using a robust Bayesian regression model fit to the trial data.

Figure 12.4 illustrates $\hat{\Delta}(\mathbf{Z}^*, L, L+B)$ on a grid of $Age^* = Z_1^*$ values for the combinations (Z_2^*, Z_3^*) =(0, 0), (0, 1), (1, 0) and (1, 1). This figure provides a clear basis for making Age-specific treatment decisions. It shows that Z_2 and Z_3 have very little effect on the decision criterion, but $\hat{\Delta}(\mathbf{Z}^*, L, L+B)$ varies substantively with Z_1 = Age. *L+B* is more desirable for younger patients, while L is more desirable for older patients. Although PFS is greater for *L+B* than L for patients of all ages, the increases in TTB with both *L+B* and Age make *L+B* much less desirable for older patients. For example, $\hat{\Delta}(\mathbf{Z}^*, L, L+B) < 0.50$ for a patient with Age = 70 or older, $\mathbf{Z}^* = (70, 0, 1)$ (dashed line), implying that L is a more desirable treatment option for this patient. Effects of the other two covariates on $\hat{\Delta}(\mathbf{Z}^*, L, L+B)$ are negligible since differences in $\hat{\Delta}$ for varying (Z_2^*, Z_3^*) values are small, and the values of Z_2 and Z_3 do not change any decisions significantly. Posterior expected utilities are computed for a grid of different Ages for each treatment option, shown in panels (b) and (c) of the figure. The figure in panel (c) shows that the expected utility decreases rapidly with Age for *L+B*, while in contrast it increases slightly for L.

As new targeted agents emerge from preclinical development, there will continue to be a pressing need for clinical trials to evaluate their safety and efficacy.

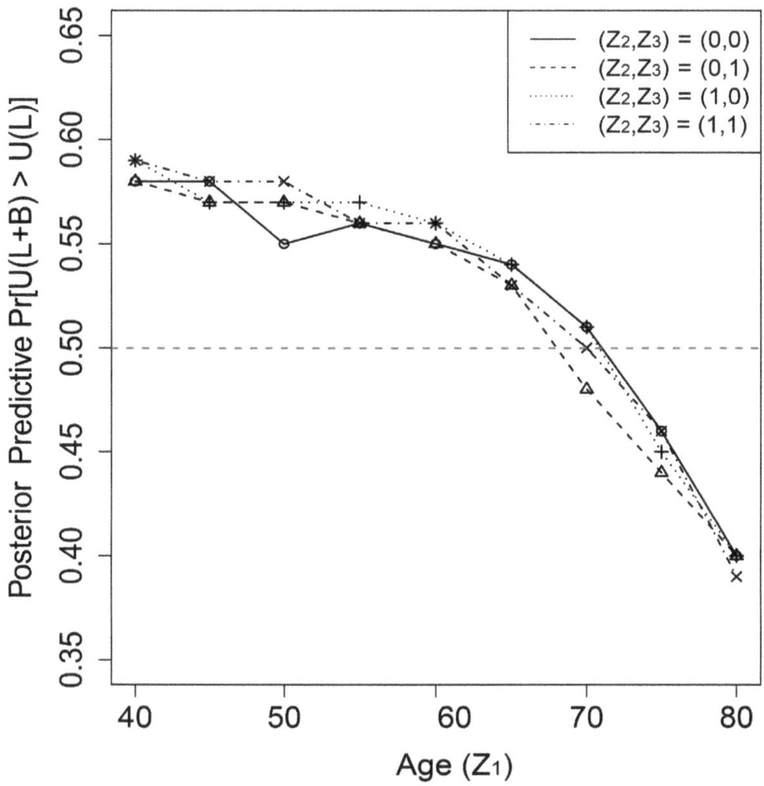

FIGURE 12.4
Posterior predictive probability $\hat{\Delta}(\mathbf{Z}, L, L+B)$ that treatment L+B has larger utility than treatment L for each of several different cases of $\mathbf{Z} = (Z_1, Z_2, Z_3)$, where Z_1 = age, Z_2 = I(measurable disease), and Z_3 = I(disease-free interval prior to trial entry $>$ 24 months).

As noted above, to provide a tool for physicians to choose treatments for individual patients, simply presenting a statistical data analysis is not enough. A decision analysis is needed to bridge the gap between statistical inference and an explicit rule for choosing each patient's treatment using their covariates. Certainly, the breast cancer application is very elaborate. It includes specifying and fitting a highly structured Bayesian nonparametric regression model, constructing a covariate-dependent utility function that accurately reflects patient preferences, and computing posterior estimates of utility payoffs for each treatment as a function of patient covariates. The fact that many subjective decisions and assumptions were made during the process of developing this statistical machinery may be regarded as a strength, rather than a weakness, since they were guided by statistical and medical considerations. The advanced breast cancer decision analysis may be regarded as a prototype for any medical setting where the risk-benefit trade-off between desirable and undesirable outcomes is a key element guiding treatment choice, and the outcomes and trade-off may vary with individual patient characteristics. Since a utility function is a natural tool for characterizing the trade-off between clinical outcomes, this approach may be tailored to particular settings where the goal is to develop rules for personalized decision-making by practicing physicians.

Computer Software A computer program 'utility-analysis' for fitting the model given by Lee et al. (2022) is available from *https://users.soe.ucsc.edu/juheelee/*.

Bibliography

Aalen, O. O. (1988). Heterogeneity in survival analysis. *Statistics in Medicine*, 7:1121–1137.

Abrahams, E. and Silver, M. (2009). The case for personalized medicine. *Journal of Diabetes Science and Technology*, 3:680–684.

Adashek, J. J., Genovese, G., Tannir, N. M., and Msaouel, P. (2020). Recent advancements in the treatment of metastatic clear cell renal cell carcinoma: A review of the evidence using second-generation p-values. *Cancer Treatment and Research Communications*, 23:100166.

Al Sayed, M., Ruckstuhl, C., and Hilmenyuk, T., et al. (2017). CD70 reverse signaling enhances nk cell function and immunosurveillance in CD27-expressing B-cell malignancies. *Blood*, 130:297–309.

Albert, J. H. and Chib, S. (1993). Bayesian analysis of binary and polychotomous response data. 88:669–679.

Ali, M., Prieto-Alhambra, D. and Lopes, L., et al. (2019). Propensity score methods in health technology assessment: principles, extended applications, and recent advances. *Frontiers in Pharmacology*, 10:973.

Almirall, D., Nahum-Shani, I., Sherwood, N., and Murphy, S. (2014). An evaluation of phase I cancer clinical trial designs. *Translational Behavioral Medicine*, 4:260–274.

Anderson, J., Cain, K., and Gelber, R. (1983). Analysis of survival by tumor response. *Journal of Clinical Oncology*, 1:710–719.

Andersson, B., Thall, P., Valdez, B., Milton, D., Al-Atrash, G., Chen, J., Gulbis, A., Chu, D., Martinez, C., Parmar, S., Popat, U., Nieto, Y., Kebriaei, P., Alousi, de Lima, M., Rondon, G., Meng, Q., Myers, A., Kawedia, J., Worth, L., Fernandez-Vina, M., Madden, T., Shpall, E., Jones, R., and Champlin, R. (2017). Fludarabine with pharmacokinetically guided iv busulfan is superior to fixed-dose delivery in pretransplant conditioning of aml/mds patients. *Bone Marrow Transplantation*, 52:580–587.

Andersson, B. S., Madden, T., Tran, H. T., Hu, W. W., Blume, K. G., Chow, D. S.-L., Champlin, R. E., and Vaughan, W. P. (2000). Acute safety and pharmacokinetics of intravenous busulfan when used with oral busulfan and cyclophosphamide as pretransplantation conditioning therapy: a phase i study. *Biology of Blood and Marrow Transplantation*, 6:548–554.

Andersson, B. S., Thall, P. F., Madden, T., Couriel, D., Wang, X., Tran, H. T., Anderlini, P., De Lima, M., Gajewski, J., and Champlin, R. E. (2002). Busulfan systemic exposure relative to regimen-related toxicity and acute graft-versus-host disease: defining a therapeutic window for iv bucy2 in chronic myelogenous leukemia. *Biology of Blood and Marrow Transplantation*, 8:477–485.

Ashford, J. R. and Sowden, R. R. (1970). Multi-variate probit analysis. *Biometrics*, 26(3):535–546.

Austin, P. and Stuart, E. (2015). Moving towards best practice when using inverse probability of treatment weighting (iptw) using the propensity score to estimate causal treatment effects in observational studies. *Statistics in Medicine*, 35: 3662–3679.

Bagočius, V., Zavadskas, E. K., and Turskis, Z. (2014). Multi-person selection of the best wind turbine based on the multi-criteria integrated additive-multiplicative utility function. *Journal of civil engineering and management*, 20(4):590–599.

Bai, X., Tsiatis, A. A., Lu, W., and Song, R. (2017). Optimal treatment regimes for survival endpoints using a locally-efficient doubly-robust estimator from a classification perspective. *Lifetime Data Analysis*, 23:585–604.

Barrientos, A. F., Jara, A., and Quintana, F. A. (2012). On the support of maceachern's dependent dirichlet processes and extensions. *Bayesian Analysis*, 7:277–310.

Bartelink, I. H., Lalmohamed, A., van Reij, E. M., Dvorak, C. C., Savic, R. M., Zwaveling, J., Bredius, R. G., Egberts, A. C., Bierings, M., Kletzel, M., et al. (2016). Association of busulfan exposure with survival and toxicity after haemopoietic cell transplantation in children and young adults: a multicentre, retrospective cohort analysis. *The Lancet Haematology*, 3(11):e526–e536.

Bechhofer, R., Santner, T., and Goldsman, D. (1995). *Design and Analysis of Experiments for Statistical Selection, Screening and Multiple Comparisons*. John Wiley & Sons, New York.

Bekele, B. N., Ji, Y., Shen, Y., and Thall, P. F. (2008). Monitoring late-onset toxicities in Phase I trials using predicted risks. *Biostatistics*, 9:442–457.

Bekele, B. N. and Thall, P. F. (2004). Dose-finding based on multiple toxicities in a soft tissue sarcoma trial. *Journal of the American Statistical Association*, 99: 26–35.

Benet, L. and Sheiner, L. (1985). *The Pharmacological Basis of Therapeutics,* 7th Ed, Pharmacokinetics: The dynamics of drug absorption, distribution and eliminationy, pages 3–34. MacMillan Publishing Company, New York, NY.

Blackwell, D. and MacQueen, J. (1973). Ferguson distributions via polya urn schemes. *Annals of Statistics*, 1:352–355.

Brannath, W., Zuber, E., Branson, M., Bretz, F., Gallo, P., Posch, M., and Racine-Poon, A. (2009). Confirmatory adaptive designs with bayesian decision tools for a targeted therapy in oncology. *Statistics in medicine*, 28(10):1445–1463.

Bredeson, C., LeRademacher, J., Kato, K., DiPersio, J. F., Agura, E., Devine, S. M., Appelbaum, F. R., Tomblyn, M. R., Laport, G. G., Zhu, X., et al. (2013). Prospective cohort study comparing intravenous busulfan to total body irradiation in hematopoietic cell transplantation. *Blood*, 122:3871–3878.

Brentjens, R. J., Davila, M. L., Riviere, I., Park, J., Wang, X., Cowell, L. G., Bartido, S., Stefanski, J., Taylor, C., Olszewska, M., et al. (2013). Cd19-targeted t cells rapidly induce molecular remissions in adults with chemotherapy-refractory acute lymphoblastic leukemia. *Science translational medicine*, 5(177):177ra38–177ra38.

Brooks, S., Gelman, A., Jones, G., and Meng, X.-L. (2011). *Handbook of Markov Chain Monte Carlo*. Chapman & Hall/CRC, Boca Raton, FL.

Brumback, B. (2021). *Fundamentals of Causal Inference*. Chapman & Hall/CRC, New York.

Buyse, M. and Piebois, P. (1996). On the relationship between response to treatment and survival time. *Statistics in Medicine*, 15(15):2797–2812.

Chapple, A. and Thall, P. (2018). Subgroup-specific dose finding in Phase I clinical trials based on time to toxicity allowing adaptive subgroup combination. *Pharmaceutical Statistics*, 17:734–749.

Chen, W., Wiecek, M. M., and Zhang, J. (1998). Quality utility: a compromise programming approach to robust design. In *International Design Engineering Technical Conferences and Computers and Information in Engineering Conference*, volume 80326, page V002T02A032. American Society of Mechanical Engineers.

Cheung, Y. K. (2011). *Dose Finding by the Continual Reassessment Method*. Chapman & Hall/CRC, New York.

Cheung, Y. K. and Chappell, R. (2000). Sequential designs for Phase I clinical trials with late-onset toxicities. *Biometrics*, 56:1177–1182.

Chib, S. and Greenberg, E. (1998). Analysis of multivariate probit models. *Biometrika*, 85(2):347–361.

Chipman, H., George, E., and McCulloch, R. (1998). Bayesian cart model search. *Journal of the American Statistical Association*, 93:935–948.

Chipman, H., George, E., and McCulloch, R. (2010a). Bart: Bayesian additive regression trees. *The Annals of Applied Statistics*, 4:266–298.

Chipman, H., George, E., and McCulloch, R. (2010b). Bart: Bayesian additive regression trees. *The Annals of Applied Statistics*, 4:266–298.

Chu, Y. and Yuan, Y. (2018a). A Bayesian basket trial design using a calibrated Bayesian hierarchical model. *Clinical Trials*, 15:149–158.

Chu, Y. and Yuan, Y. (2018b). Blast: Bayesian latent subtype design for basket trials accounting for patient heterogeneity. *Journal of the Royal Statistical Society, Series C*, 67:723–740.

Clavien, P. A., Sanabria, J. R., and Strasberg, S. M. (1992). Proposed classification of complications of surgery with examples of utility in cholecystectomy. *Annals of Surgery*, 111:518–526.

Cox, D. R. (1972). Regression models and life-tables (with discussion). *Journal of the Royal Statistical Society*, 34:187–220.

Cunanan, K. M., Iasonos, A., Shen, R., Begg, C. B., and Gonen, M. (2017). An efficient basket trial design. *Statistics in Medicine*, 36:1568–1579.

Dagogo-Jack, I. and Shaw, A. (2018). Tumour heterogeneity and resistance to cancer therapies. *Nature Reviews Clinical Oncology*, 15:81–94.

Davila, M. L., Riviere, I., Wang, X., Bartido, S., Park, J., Curran, K., Chung, S. S., Stefanski, J., Borquez-Ojeda, O., Olszewska, M., et al. (2014). Efficacy and toxicity management of 19-28z CAR T cell therapy in B cell acute lymphoblastic leukemia. *Science Translational Medicine*, 6(224):224ra25–224ra25.

Dazzi, F., Szydlo, R., Cross, N., et al. (2000). Durability of responses following donor lymphocyte infusions for patients who relapse after allogeneic stem cell transplantation for chronic myeloid leukemia. *Blood*, 15:2712–2716.

De Iorio, M., Johnson, W. O., Müller, P., and Rosner, G. L. (2009). Bayesian nonparametric nonproportional hazards survival modeling. *Biometrics*, 65(3):762–771.

De Iorio, M., Müller, P., Rosner, G. L., and MacEachern, S. N. (2004). An anova model for dependent random measures. *Journal of the American Statistical Association*, 99(465):205–215.

DeIorio, M., Müller, P., Rosner, G. L., and MacEachern, S. (2004). An anova model for dependent random measures. *Journal of the American Statistical Association*, 99:205–215.

Denison, D., Mallick, B., and Smith, A. (1998). A Bayesian cart algorithm. *Biometrika*, 85:363–377.

Dey, D., Ghosh, S., and Mallick, B. K. (2000). *Generalized Linear Models: A Bayesian Perspective*. Marcel Dekker, New York.

Dickler, M. N., Barry, W. T., Cirrincione, C. T., Ellis, M. J., Moynahan, M. E., Innocenti, F., Hurria, A., Rugo, H. S., Lake, D. E., Hahn, O., et al. (2016). Phase III trial evaluating letrozole as first-line endocrine therapy with or without bevacizumab for the treatment of postmenopausal women with hormone receptor–positive advanced-stage breast cancer: Calgb 40503 (alliance). *Journal of Clinical Oncology*, 34(22):2602–2609.

Dindo, D., Demartines, N., and Clavien, P. (2004). Classification of surgical complications: a new proposal with evaluation in a cohort of 6336 patients and results of a survey. *Annals of Surgery*, 240:205–213.

Dixon, D. O. and Simon, R. (1991). Bayesian subset analysis. *Biometrics*, 47: 871–881.

Dixon, D. O. and Simon, R. (1992). Bayesian subset analysis in a colorectal cancer clinical trial. *Statistics in Medicine*, 11:13–22.

Dobson, A. and Barnett, A. (2008). *An Introduction to Generalized Linear Models,* 3rd Edition. CRC Press.

Douillard, J., Siena, S., Cassidy, J., et al. (2010). Randomized, Phase III trial of panitumumab with infusional fluorouracil, leucovorin, and oxaliplatin (folfox4) versus folfox4 alone as first-line treatment in patients with previously untreated metastatic colorectal cancer: the prime study. *J Clinical Oncology*, 28:4697–4705.

Escobar, M. and West, H. (1995). Bayesian density estimation and inference using mixtures. *Journal of the Amercian Statistical Association*, 90:577–588.

Estey, E. H., Thall, P. F., Pierce, S., Cortes, J., Beran, M., Kantarjian, H., Keating, M., Andreff, M., and Freireich, E. (1999). Randomized Phase II study of fludarabine+ cytosine arabinoside+ idarubicin $\pm$ all-trans retinoic acid $\pm$ granulocyte colony-stimulating factor in poor prognosis newly diagnosed acute myeloid leukemia and myelodysplastic syndrome. *Blood*, 93:2478–2484.

Eubank, R. (1999). *Nonparametric Regression and Spline Smoothing,* 2nd Edition. CRC Press.

Ferguson, T. S. (1973). A Bayesian analysis of some nonparametric problems. *The Annals of Statistics*, 1(2): 209–230.

Ferguson, T. S. (1983). Bayesian density estimation by mixtures of normal distributions. *Recent Advances in Statistics: Papers in Honor of Herman Chernoff on His Sixtieth Birthday* (Siegmund, Rustage, Rizvi, eds), pages 287–302.

Fisher, R. A. (1925). *Statistical Methods for Research Workers.* Oliver and Boyd, Edinburgh.

Fleming, T. and Powers, J. (2012). Biomarkers and surrogate endpoints in clinical trials. *Statistics in Medicine*, 31:2973–2984.

Freidlin, B., Jiang, W., and Simon, R. (2010). The cross-validated adaptive signature design. *Clinical Cancer Research*, 16:691–698.

Freidlin, B., Korn, E., and Gray, R. (2014). Marker sequential test (mast) design. *Clinical Trials*.

Freidlin B, Korn EL, and Marker G. R. (2014). Sequential Test (MaST) design. *Clin Trials,* 11(1):19-27. doi:10.1177/1740774513503739.

Gammerman, D. and Lopes, H. (2006). *Markov Chain Monte Carlo,* 2nd Edition. Chapman & Hall/CRC, Boca Raton, FL.

Garralda, E., Dienstmann, R., Piris-Gimenez, A., Brana, I., Rodon, J., and Tabernero, J. (2019). New clinical trial designs in the era of precision medicine. *Molecular Oncology*, 13:549–557.

Gauthier, J., Thall, P., and Yuan, Y. (2019). Bayesian phase 1/2 trial designs and cellular immunotherapies: a practical primer. *Cell and Gene Therapy Insights*, 5:1483–1494.

Gehan, E. A. and Freireich, E. (1974). Non-randomized controls in cancer clinical trials. *New England Journal of Medicine*, 290:198–203.

George, E. I. and McCulloch, R. E. (1993). Variable selection via gibbs sampling. *Journal of the American Statistical Association*, 88(423):881–889.

Gershman, S. and Blei, D. (2012). A tutorial on Bayesian nonparametric models. *Journal of Mathematical Psychology*, 56:1–12.

Green, P. (1995). Reversible jump Markov chain Monte Carlo computation and Bayesian model determination. *Biometrika*, 82:711–732.

Greenland, S., Senn, S., Rothman, K., Carlin, J., Poole, C., Goodman, S., and Altman, D. (2016). Statistical tests, p values, confidence intervals, and power: a guide to misinterpretations. *European Journal of Epidemiology*, 31:337–350.

Grupp, S. A., Kalos, M., Barrett, D., Aplenc, R., Porter, D. L., Rheingold, S. R., Teachey, D. T., Chew, A., Hauck, B., Wright, J. F., et al. (2013). Chimeric antigen receptor–modified t cells for acute lymphoid leukemia. *New England Journal of Medicine*, 368(16):1509–1518.

Guo, W., Ji, Y., and Catenucci, D. (2017). A subgroup cluster-based bayesian adaptive design for precision medicine. *Biometrics*, 73:367–377.

Gyurkocza, B. and Sandmaier, B. (2014). Conditioning regimens for hematopoietic cell transplantation: one size does not fit all. *Blood*, 124:591–600.

Hanson, T. and Johnson, W. O. (2002). Modeling regression error with a mixture of polya trees. *Journal of the American Statistical Association*, 97:1020–1033.

Harrell, F. (2001). *Regression Modeling Strategies*. Springer-Verlag, New York.

Heng, D. Y., Xie, W., Regan, M. M., Harshman, L. C., Bjarnason, G. A., Vaishampayan, U. N., Mackenzie, M., Wood, L., Donskov, F., Tan, M.-H., et al. (2013). External validation and comparison with other models of the international metastatic renal-cell carcinoma database consortium prognostic model: a population-based study. *The Lancet Oncology*, 14(2):141–148.

Hernan, M., Brumback, B., and Robins, J. M. (2000). Marginal structural models to estimate the causal effect of zidovudine on the survival of HIV-positive men. *Epidemiology*, 11(4):561–570.

Hey, S. and Kimmelman, J. (2014). The questionable use of unequal allocation in confirmatory trials. *Neeurology*, 82:77–79.

Hill, J. L. (2011). Bayesian nonparametric modeling for causal inference. *Journal of Computational and Graphical Statistics*, 20(1).

Hjort, N., Holmes, C., Müller, P., and Walker, S.G., et al. (2010). *Bayesian Nonparametrics*. Cambridge Series in Statistical and Probabilistic Mathematics, Cambridge.

Holland, P. (1986). Statistics and causal inference. *Journal of the American Statistical Association*, 81:945–960.

Horvitz, D. and Thompson, D. (1952). A generalization of sampling without replacement from a finite universe. *Journal of the American Statistical Association*, 47:663–685.

Hsieh, L.-L., Erl, T.-K., Chen, C.-C., Hsieh, J.-S., Chang, J.-G., and Liu, T.-C. (2012). Characteristics and prevalence of KRAS, BRAS, and PIK3CA mutations in colorectal cancer by high-resolution melting analysis in taiwanese population. *Clinica Chimica Acta*, 413:1605–678.

Ibrahim, J. G., Chen, M.-H., and Sinha, D. (2001). *Bayesian Survival Analysis*. Springer, New York.

Ibrahim, J. G., Chen, M.-H., and Sinha, D. (2005). *Bayeasian Survival Analysis*. Wiley Online Library.

Imbens, G. and Rubin, D. (2015). *Causal Inference for Statistics, Social, and Biomedical Sciences: An Introduction*. Cambridge University Press, Cambridge, UK.

Isakoff, S. J. (2010). Triple negative breast cancer: role of specific chemotherapy agents. *Cancer Journal (Sudbury, Mass.)*, 16(1):53.

Ishwaran, H. and James, L. F. (2001). Gibbs sampling methods for stick-breaking priors. *Journal of the American Statistical Association*, 96(453):161–173.

Ishwaran, H., Rao, J. S., et al. (2005). Spike and slab variable selection: frequentist and bayesian strategies. *The Annals of Statistics*, 33(2):730–773.

James, G., Witten, D., Hastie, T., and Tibshirani, R. (2017). *An Introduction to Statistical Learning: with Applications in R,7th printing*. Springer Texts in Statistics, New York.

James, N., Sydes, M., Clarke, N., Mason, M., Dearnaley, D., Anderson, J., Popert, R., Sanders, K., Morgan, R., Stansfeld, J., Dwyer, J., Masters, J., and Parmar, M. (2009). Systemic therapy for advancing or metastatic prostate cancer (stampede): a multi-arm, multistage randomized controlled trial. *BJU International*, 103: 464–469.

Jenkins, M., Stone, A., and Jennison, C. (2011). An adaptive seamless Phase II/III design for oncology trials with subpopulation selection using correlated survival endpoints. *Pharmaceutical Statistics*, 10(4):347–356.

Jennison, C. and Turnbull, B. (2007). Adaptive seamless designs: selection and prospective testing of hypotheses. *Journal of Biopharmaceutical Statistics*, 17:1135–1161.

Jennison, C. and Turnbull, B. W. (1999). *Group Sequential Methods with Applications to Clinical Trials*. CRC Press.

Jiang, L., Thall, P., Yan, F., Kopetz, S., and Yuan, Y. (2023). Bayesian treatment screening and selection using subgroup-specific utilities of response and toxicity. Manuscript in press.

Jiang, W., Freidlin, B., and Simon, R. (2007). Biomarker-adaptive threshold design: a procedure for evaluating treatment with possible biomarker-defined subset effect. *Journal of the National Cancer Institute*, 99(13):1036–1043.

Jin, I. H., Liu, S., Thall, P. F., and Yuan, Y. (2014). Using data augmentation to facilitate conduct of Phase I–II clinical trials with delayed outcomes. *Journal of the American Statistical Association*, 109(506):525–536.

Joffee, M. and Rosembaum, P. (1999). Invited commentary: propensity scores. *American Journal of Epidemiology*, 150:327–333.

Kalbfleisch, J. D. and Prentice, R. L. (1981). Estimation of the average hazard ratio. *Biometrika*, 68:105–112.

Kaplan, E. L. and Meier, P. (1958). Nonparametric estimation from incomplete observations. *Journal of the American Statistical Association*, 53:457–481.

Karabatsos, G. (2016). A menu-driven software package of Bayesian nonparametric (and parametric) mixed models for regression analysis and density estimation. *Behavior Research Methods*, 49:335–362.

Kashyap, A., Wingard, J., Cagnoni, P., Roy, J., Tarantolo, S., Hu, W., Blume, K., Niland, J., Palmer, J. M., Vaughan, W., Fernandez, H., Champlin, R., Forman, S., & Andersson, B. S. (2002). Intravenous versus oral busulfan as part of a busulfan/cyclophosphamide preparative regimen for allogeneic hematopoietic stem cell transplantation: decreased incidence of hepatic venoocclusive disease (HVOD), HVOD-related mortality, and overall 100-day mortality. *Biology of blood and marrow transplantation: journal of the American Society for Blood and Marrow Transplantation*, 8(9), 493–500. https://doi.org/10.1053/bbmt.2002.v8.pm12374454.

Keil, A., Daza, E., Engel, S., Buckley, J., and Edwards, J. (2018). A Bayesian approach to the g-formula. *Statistical Methods in Medical Research*, 27:3183–3204.

Kim, E., Herbst, R., Wistuba, I., and et al. (2011). The battle trial: personalizing therapy for lung cancer. *Cancer Discovery*, 1:44–53.

Kim, S., Chen, M., Dey, D. K., and Gamerman, D. (2007). Bayesian dynamic models for survival data with a cure fraction. *Lifetime Data Analysis*, 13:17–35.

Kimani, P. K., Todd, S., and Stallard, N. (2015). Estimation after subpopulation selection in adaptive seamless trials. *Statistics in Medicine*, 34(18):2581–2601.

Kopetz, S., Guthrie, K., Morris, V., et al. (2021). Randomized trial of irinotecan and cetuximab with or without vemurafenib in braf-mutant metastatic colorectal cancer (swog s1406). *Journal of Clinical Oncology*, 39:285–294.

Lan, K. K. G. and DeMets, D. L. (1983). Discrete sequential boundaries for clinical trials. *Biometrika*, 70(3):659–663.

Lee, J., Thall, P. F., Ji, Y., and Müller, P. (2016). A decision-theoretic Phase I-II design for ordinal outcomes in two cycles. *Biostatistics*, 2016:304–319.

Lee, J., Thall, P. F., Lim, B., and Msaouel, P. (2022). Utility based Bayesian personalized treatment selection for advanced breast cancer. *Journal of the Royal Statistical Society, Series C*, 71:1605–1622.

Lee, J., Thall, P. F., and Msaouel, P. (2021). Precision Bayesian Phase I–II dose-finding based on utilities tailored to prognostic subgroups. *Statistics in Medicine*, 40:123–456.

Lee, J., Thall, P. F., and Msaouel, P. (2023). Bayesian treatment screening and selection using subgroup-specific utilities of response and toxicity. *Biometrics*, 79:2458–2473.

Lee, J., Thall, P. F., and Rezvani, K. (2019). Optimizing natural killer cell doses for heterogeneous cancer patients based on multiple event times. *J Royal Statistical Society, Series C*, 68:461–474.

Lin, J., Gamalo-Siebers, M., and Tiwari, R. (2018). Propensity score matched augmented controls in randomized clinical trials: a case study. *Pharmaceutical Statistics*, 17:629–647.

Lin, R., Thall, P., and Yuan, Y. (2021a). Bags: a Bayesian adaptive group sequential trial design with subgroup-specific survival comparison. *J American Statistical Assoc*, 116:322–334.

Lin, R., Thall, P., and Yuan, Y. (2021b). A Bayesian adaptive group sequential trial design with subgroup-specific survival comparisons. *J Amercian Statistical Assoc*, 116:322–344.

Lin, R., Thall, P., and Yuan, Y. (2021c). A Phase I–II basket trial design to optimize dose-schedule regimes based on delayed outcomes. *Bayesian Analysis*, 16: 179–202.

Liu, C., Ma, J., and Amos, C. I. (2015). Bayesian variable selection for hierarchical gene–environment and gene–gene interactions. *Human genetics*, 134(1):23–36.

Liu, E., Marin, D., Macapinlac, H., Thompson, P., Basar, R., Kerbauy, L., Overman, B., Thall, P., Kaplan, M., Nandivada, V., Kaur, I., Cortes, A., Cao, K., Daher, M., Hosing, C., Cohen, E., Kebriaei, P., Mehta, R., Neelapu, S., Nieto, Y., Wang, M., Wierda, W., Keating, M., Champlin, R., Shpall, E., and Rezvani, K. (2020). Il-15 armored car-transduced nk cells against cd19 positive b cell tumors. *New England J Medicine.*, 382:345–353.

Loewenstein, G. F., Thompson, L., and Bazerman, M. H. (1989). Social utility and decision making in interpersonal contexts. *Journal of Personality and Social psychology*, 57(3):426.

Luger, S., Ringden, O., and Zhang, M. (2012). Siimilar outcomes using myeloablative versus reduced intensity regimens for allogeneic transplants for aml or mds. *Bone Marrow Transplantation*, 47:203–211.

Lunceford, J. K. and Davidian, M. (2004). Stratification and weighting via the propensity score in estimation of causal treatment effects: a comparative study. *Statistics in Medicine*, 23:2937–2960.

MacEachern, S. N. (1999). Dependent nonparametric processes. In *ASA proceedings of the Section on Bayesian Statistical Science*, pages 50–55. American Statistical Association, pp. 50–55, Alexandria, VA.

MacEachern, S. N. and Müller, P. (1998). Estimating mixture of Dirichlet process models. *Journal of Computational and Graphical Statistics*, 7(2):223–238.

Madigan, D. and Raftery, A. (1994). Model selection and accounting for model uncertainty in graphical models using occam's window. *Journal of the American Statistical Association*, 89:1535–1546.

Magnusson, B. and Turnbull, B. (2013). Group sequential enrichment design incorporating subgroup selection. *Statisticss in Medicine*, 32:2695–2714.

Malfuson, J., Etienne, A., Turlure, P., et al. (2008). Risk factors and decision criteria for intensive chemotherapy in older patients with acute myeloid leukemia. *Haematologica*, 93:1806–1818.

Maurer, W. and Bretz, F. (2013). Multiple testing in group sequential trials using graphical approaches. *Statistics in Biopharmaceutical Research*, 5:311–332.

Mc Grayne, S. B. (2011). *The Theory That Would Not Die*. Yale University Press, New Haven & London.

McCullagh, P. (1980). Regression models for ordinal data (with discussion). *Journal of the Royal Statistical Society*, 42:109–142.

McGranahan, N. and Swanton, C. (2017). Clonal heterogeneity and tumor evolution: past, present, and the future. *Cell*, 168:613–628.

McKeague, I. W. and Tighiouart, M. (2000). Bayesian estimators for conditional hazard functions. *Biometrics*, 56(4):1007–1015.

Mehta, C., Schäfer, H., Daniel, H., and Irle, S. (2014). Biomarker driven population enrichment for adaptive oncology trials with time to event endpoints. *Statistics in Medicine*, 33(26):4515–4531.

Merino, M., Kasamon, Y., Theoret, M., et al. (2023). Irreconcilable differences: the divorce between response rates, progression-free survival, and overall survival. *Journal of Clinical Oncology*, 41:2706–2711.

Mitchell, T. J. and Beauchamp, J. J. (1988). Bayesian variable selection in linear regression. *Journal of the American Statistical Association*, 83(404):1023–1032.

Mitra, R. and Müller, P. e. (2015). *Nonparametric Bayesian Inference in Biostatistics*. Springer-Verlag, New York.

Mok, T.S., Wu, Y-L., Thongprasert, S. et al. (2009). Gefitinib or carboplatin–paclitaxel in pulmonary adenocarcinoma. *New England Journal of Medicine*, 361:947–957.

Moodie, E. E. M., Richardson, T. S., and Stephens, D. A. (2007). Demystifying optimal dynamic treatment regimes. *Biometrics*, 63(2):447–455.

Motzer, R., Alekseev, B., Rha, S.-Y., Porta, C., Eto, M., Powles, T., Grünwald, V., Hutson, T. E., Kopyltsov, E., Méndez-Vidal, M. J., et al. (2021). Lenvatinib plus pembrolizumab or everolimus for advanced renal cell carcinoma. *New England Journal of Medicine*, 384(14):1289–1300.

Motzer, R. J., Hutson, T. E., Cella, D., Reeves, J., Hawkins, R., Guo, J., Nathan, P., Staehler, M., de Souza, P., Merchan, J. R., et al. (2013). Pazopanib versus sunitinib in metastatic renal-cell carcinoma. *New England Journal of Medicine*, 369(8):722–731.

Motzer, R. J., Penkov, K., Haanen, J., Rini, B., Albiges, L., Campbell, M. T., Venugopal, B., Kollmannsberger, C., Negrier, S., Uemura, M., et al. (2019a). Avelumab plus axitinib versus sunitinib for advanced renal-cell carcinoma. *New England Journal of Medicine*, 380(12):1103–1115.

Motzer, R. J., Rini, B. I., McDermott, D. F., Frontera, O. A., Hammers, H. J., Carducci, M. A., Salman, P., Escudier, B., Beuselinck, B., Amin, A., et al. (2019b). Nivolumab plus ipilimumab versus sunitinib in first-line treatment for advanced renal cell carcinoma: extended follow-up of efficacy and safety results from a randomised, controlled, phase 3 trial. *The Lancet Oncology*, 20(10):1370–1385.

Msaouel, P., Lee, J., and Karam J.A. Thall, P. F. (2022). A causal framework for making individualized treatment decisions in oncology. *Cancers*, 14:3923.

Msaouel, P., Lee, J., and Thall, P. F. (2021). Making patient-specific treatment decisions using prognostic variables and utilities of clinical outcomes. *Cancers*, 13:2741.

Müller, P. and Mitra, R. (2013). Bayesian nonparametric inference: why and how. *Bayesian Anaysis*, 8:269–302.

Müller, P., Quintana, F., and Rosner, G. (2011). A product partition model with regression on covariates. *Journal of Computational and Graphical Statistics*, 20:260–278.

Müller, P. and Rodriguez, A. (2013). Nonparametric Bayesian inference. *IMS-CBMS Lecture Notes. IMS*, 270.

Müller, P., Xu, Y., and Jara, A. (2016). A short tutorial on Bayesian nonparametrics. *Journal of Statistical Research*, 48-50:1–19.

Murphy, S. (2005a). An experimental design for the development of adaptive treatment strategies. *Statistics in Medicine*, 24:1455–1481.

Murphy, S. (2005b). Generalization error for q-learning. *Journal of Machine Learning*, 6:1073–1097.

Murphy, S. A. (2003). Optimal dynamic treatment regimes. *Journal of the Royal Statistical Society: Series B (Statistical Methodology)*, 65(2):331–355.

Murphy, S. A. and Bingham, D. (2009). Screening experiments for developing dynamic treatment regimes. *Journal of the American Statistical Association*, 104(485):391–408.

Murray, T., Thall, P., Yuan, Y., McAvoy, S., and Gomez, D. (2017). Robust treatment comparison based on utilities of semi-competing risks in non-small-cell lung cancer. *Journal of the American Statiastical Association*, 112:11–23.

Murray, T., Yuan, Y., Thall, P., Elizondo, J., and Hoffstetter, W. (2018). A utility based design for randomized comparative trials with ordinal outcomes and prognostic subgroups. *Biometrics*, 74:1095–1103.

Neyman, J. (1923). Sur les applications de la theorie des probabilites aux experiences agricoles: Essai des principes. *Masters Thesis, excerpts reprinted in English, Statistical Science, 5: 463–472.*

Neyman, J. (1934). On the two different aspects of the representative method: the method of stratified sampling and the method of purposive selection. *Journal of the Royal Statistical Society, Ser. A*, 97:558–606.

Nieto-Barajas, L. E. and Walker, S. (2002). Markov beta and gamma processes for modelling hazard rates. *Scandinavian Journal of Statistics*, 29:413–424.

O'Brien, P. and Fleming, T. (1979). A multiple testing procedure for clinical trials. *Biometrics*, 35:549–556.

Ondra, T., Joibjornsson, S., Beckman, R., Burman, C.-F., Konig, F., Stallard, N., and Posch, M. (2019). Optimized adaptive enrichment designs. *Statistical Methods in Medical Research*, 28:2096-2111.

O'Quigley, J., Pepe, M., and Fisher, L. (1990). Continual reassessment method: a practical design for phase I clinical trials in cancer. *Biometrics*, 46:33–48.

Park, Y., Liu, S., Thall, P., and Yuan, Y. (2022). Bayesian group sequential enrichment designs based on adaptive regression of response and survival time on baseline biomarkers. *Biometrics*, 100:1204–1214.

Parmigiani, G. (2002). *Modeling in Medical Decision Making: A Bayesian Approach.* Wiley, New York.

Pearl, J. and Bareinboim, E. (2011). Transportability of causal and statistical relations: a formal approach. pages 247–254. AAAI Press.

Pennings, J. M. and Smidts, A. (2003). The shape of utility functions and organizational behavior. *Management Science*, 49(9):1251–1263.

Pocock, S. J., Assmann, S. E., Enos, L. E., & Kasten, L. E. (2002). Subgroup analysis, covariate adjustment and baseline comparisons in clinical trial reporting: current practice and problems. *Statistics in medicine*, 21(19), 2917–2930. https://doi.org/10.1002/sim.1296

Pocock, S. and Simon, R. (1975). Sequential treatment assignment with balancing for prognostic factors in the controlled clinical trial. *Biometrics*, 31:101–115.

Porter, D. L., Levine, B. L., Kalos, M., Bagg, A., and June, C. H. (2011). Chimeric antigen receptor–modified T cells in chronic lymphoid leukemia. *New England Journal of Medicine*, 365(8):725–733.

Prasetyant, P. and Madema, J. (2017). Intra-tumor heterogeneity from a cancer stem cell perspective. *Molecular Cancer*, 16:41:DOI 10.1186/s12943–017–0600–4.

Proschan, M. and Evans, S. (2020). Resist the temptation of response-adaptive randomization. *Clinical Infectious Diseases*, 71:3002–3004.

Psioda, M., Xu, J., Jiang, Q., Ke, C., Yang, Z., and Ibrahim, J. (2019). Bayesian adaptive basket trial design using model averaging. *Biostatistics*, 22:19–36.

Quintana, F., Müller, P., Jara, A., and MacEachern, S. (2022). The dependent Dirichlet process and related models. *Statistical Science*, 37:24–41.

Rauch, G., Brannath, W., Brückner, M., and Kieser, M. (2018). The average hazard ratio–a good effect measure for time-to-event endpoints when the proportional hazard assumption is violated? *Methods of Information in Medicine*, 57:89–100.

Redig, A. J. and Jänne, P. A. (2015). Basket trials and the evolution of clinical trial design in an era of genomic medicine. *Journal of Clinical Oncology*, 33:975–977.

Reich, B. and Fuentes, M. (2007). A multivariate semiparametric Bayesian spatial modeling framework for hurricane surface wind fields. *Annals of Applied Statistics*, 1:249–264.

Rezvani, K. and Rouce, R. H. (2015). The application of natural killer cell immunotherapy for the treatment of cancer. *Frontiers in immunology*, 6:578.

Rini, B. I., Plimack, E. R., Stus, V., Gafanov, R., Hawkins, R., Nosov, D., Pouliot, F., Alekseev, B., Soulières, D., Melichar, B., et al. (2019). Pembrolizumab plus axitinib versus sunitinib for advanced renal-cell carcinoma. *New England Journal of Medicine*, 380(12):1116–1127.

Rittmeyer, A., Barlesi, F., Waterkamp, D., Park, K., et al. (2017). Atezolizumab versus docetaxel in patients with previously treated non-small-cell lung cancer (oak): a phase 3, open-label, multicentre randomised controlled trial. *The Lancet*, 389:255–265.

Robert, C. P. and Cassella, G. (1999). *Monte Carlo Statistical Methods*. Springer, New York.

Robins, J. (1986a). A new approach to causal inference in mortality studies with a sustained exposure period–application to control of the healthy worker survivor effect. *Mathematical Modelling*, 7(9):1393–1512.

Robins, J., Orellana, L., and Rotnitzky, A. (2008). Estimation and extrapolation of optimal treatment and testing strategies. *Statistics in Medicine*, 27(23):4678–4721.

Robins, J. M. (1986b). A new approach to causal inference in mortality studies with a sustained exposure period–application to control of the healthy worker survivor effect. *Mathematical Modeling*, 7(9):1393–1512.

Robins, J. M., Hernán, M. Á., and Brumback, B. (2000). Marginal structural models and causal inference in epidemiology. *Epidemiology*, 11(5):550–560.

Rodriguez, A. and Dunson, D. B. (2011). Nonparametric Bayesian models through probit stick-breaking processes. *Bayesian analysis (Online)*, 6(1).

Rosenbaum, P. (2017). *Observation and Experiment: An Introduction to Causal Inference*. Harvard University Press, Cambridge.

Rosenberger, W. and Lachin, J. (2004). *Randomization in Clinical Trials: Theory and Practice*. John Wiley & Sons.

Rosenblum, M., Luber, B., Thompson, R. E., and Hanley, D. (2016). Group sequential designs with prospectively planned rules for subpopulation enrichment. *Statistics in Medicine*, 35(21):3776–3791.

Roy, S. K., Maity, G., and Weber, G.-W. (2017). Multi-objective two-stage grey transportation problem using utility function with goals. *Central European Journal of Operations Research*, 25(2):417–439.

Rubin, D. (1974). Estimatng causal effects of treatments in randomized and nonrandomized studies. *Journal of Educational Psychology*, 66:688–701.

Rubin, D. (1978). Bayesian infererence for causal effects: the role of randomization. *Annals of Statisttics*, 6:34–58.

Rubin, D. (2005). Causal inference using potential outcomes. *Journal of the American Statisttical Association*, 100:322–331.

Schaid, D. J., Wieand, S., and Therneau, T. M. (1990). Optimal two stage screening designs for survival comparisons. *Biometrika*, 77:507–513.

Schumacher, M., Olschewski, M., and Schmoor, C. (1987). The impact of heterogeneity on the comparison of survival times. *Statistics in Medicine*, 6:773–784.

Sethuraman, J. (1994). A constructive definition of Dirichlet priors. *Statistica Sinica*, 4:639–650.

Sharma, S., Bell, D., Settleman, J., and Haber, D. (2007). Epidermal growth factor receptor mutations in lung cancer. *Nature Reviews*, 7:169–181.

Simon, N. and Simon, R. (2013). Adaptive enrichment designs for clinical trials. *Biostatistics*, 14(4):613–625.

Simon, R. (2010). Clinical trials for predictive medicine: new challenges and paradigms. *Clinical Trials*, 7:516–524.

Simon, R., Geyer, S., Subramanian, J., and Roychowdhury, S. (2016). The Bayesian basket design for genomic variant driven Phase II trials. *Nature Reviews Drug Discovery*, 14: 613–625.

Simon, R. and Makuch, R. (1984). A non-parametric graphical representation of the relationship between survival and the occurrence of an event: application to responder versus non-responder bias. *Statistics in Medicine*, 3:35–44.

Simon, R. and Roychowdhury, S. (2013). Implementing personalized cancer genomics in clinical trials. *Nature Reviews Drug Discovery*, 112:358–5369.

Simon, R. M. (1989). Optimal two-stage designs for Phase II clinical trials. *Controlled Clinical Trials*, 10:1–10.

Sinha, D., Chen, M.-H., and Ghosh, S. K. (1999). Bayesian analysis and model selection for interval-censored survival data. *Biometrics*, 55(2):585–590.

Snapinn, S. and Jiang, Q. (2011). On the clinical meaningfulness of a treatment's effect on a time-to-event variable. *Statistics in Medicine*, 30(19):2341–2348.

Spiegelhalter, D. J., Best, N., Carlin, B., and van der Linde, A. (2002). Bayesian measures of model complexity and fit (with discussion). *J. R. Statist. Soc. B*, 64:583–639.

Stallard, N. and Todd, S. (2003). Sequential designs for Phase III clinical trials incorporating treatment selection. *Statistics in Medicine*, 22:689–703.

Sutton, R. S. and Barto, A. G. (1998). *Reinforcement Learning: An Introduction.* MIT Press, Cambridge, MA.

Sverdlov, O., editor (2015). *Modern Adaptive Randomized Clinical Trials: Statistical and Practical Aspects.* CRC Press, Taylor and Francis, Boca Raton.

Tannir, N. M., Msaouel, P., Ross, J. A., Devine, C. E., Chandramohan, A., Gonzalez, G. M. N., Wang, X., Wang, J., Corn, P. G., Lim, Z. D., et al. (2020). Temsirolimus versus pazopanib (tempa) in patients with advanced clear-cell renal cell carcinoma and poor-risk features: a randomized phase II trial. *European Urology Oncology*, 3(5):687–694.

Teh, Y., Jordan, M., Beal, M., and Blei, D. (2004). Sharing clusters among related groups: hierarchical dirichlet processes. pages 1385–1392. https://proceedings.neurips.cc/paper/2004/file/fb4ab556bc42d6f0ee0f9e24ec4d1af0-Paper.pdf

Thall, P. (2020). *Statistical Remedies for Medical Researchers.* Springer Series in Pharmaceutical Statistics, Gewerbestrasse, Switzerland.

Thall, P. (2021). Adaptive enrichment designs in clinical trials. *Annual Review of Statistics and its Application*, 8:393–411.

Thall, P., Fox, P., and Wathen, J. (2015). Statistical controversies in clinical research: scientific and ethical problems with adaptive randomization in comparative clinical trials. *Annals of Oncology*, 26(8):1621–1628.

Thall, P., Wathen, J. K., Bekele, B. N., Champlin, R. E., Baker, L. H., and Benjamin, R. S. (2003). Hierarchical Bayesian approaches to Phase II trials in diseases with multiple subtypes. *Statistics in Medicine*, 22(5):763–780.

Thall, P. F. and Cook, J. D. (2004). Dose-finding based on efficacy-toxicity trade-offs. *Biometrics*, 60:684–693.

Thall, P. F., Herrick, R., Nguyen, H., Venier, J., and Norris, J. (2014). Using effective sample size for prior calibration in Bayesian Phase I–II dose-finding. *Clinical Trials*, 11:657–666.

Thall, P. F., Lee, J. J., Tseng, C. H., and Estey, E. H. (1999). Accrual strategies for Phase I trials with delayed patient outcome. *Statistics in Medicine*, 18:1155–1169.

Thall, P. F., Millikan, R. E., and Sung, H.-G. (2000). Evaluating multiple treatment courses in clinical trials. *Statistics in Medicine*, 19(8):1011–1028.

Thall, P. F., Mueller, P., Xu, Y., and Guindani, M. (2017). Bayesian nonparametric statistics: a new toolkit for discovery in cancer research. *Pharmaceutical Statistics*, 16:414–423.

Thall, P. F. and Nguyen, H. Q. (2012). Adaptive randomization to improve utility-based dose-finding with bivariate ordinal outcomes. *Journal of Biopharmaceutical Statistics*, 22(4):785–801.

Thall, P. F., Simon, R., and Ellenberg, S. S. (1988). Two-stage selection and testing designs for comparative clinical trials. *Biometrika*, 75(2):303–310.

Thall, P. F. and Simon, R. M. (1994). Practical Bayesian guidelines for Phase IIB clinical trials. *Biometrics*, 50:337–349.

Thall, P. F. and Sung, H.-G. (1998). Some extensions and applications of a Bayesian strategy for monitoring multiple outcomes in clinical trials. *Statistics in Medicine*, 17:1563–1580.

Thall, P. F., Sung, H.-G., and Estey, E. H. (2002). Selecting therapeutic strategies based on efficacy and death in multicourse clinical trials. *Journal of the American Statistical Association*, 97(457):29–39.

Thall, P. F., Szabo, A., Nguyen, H. Q., Amlie-Lefond, C. M., and Zaidat, O. O. (2011). Optimizing the concentration and bolus of a drug delivered by continuous infusion. *Biometrics*, 67(4):1638–1646.

Thall, P. F. and Wathen, J. K. (2005). Covariate-adjusted adaptive randomization in a sarcoma trial with multistage treatments. *Statistics in Medicine*, 24(13): 1947–1964.

Thall, P. F. and Wathen, J. K. (2007). Practical Bayesian adaptive randomisation in clinical trials. *European Journal of Cancer*, 43(5):859–866.

Thall, P. F., Zhang, Y., and Yuan, Y. (2023). Generalized Phase I–II designs to increase long term therapeutic success rate. *Pharmaceutical Statistics*, 22:111–123.

Thompson, W. (1933). On the likelihood that one unknown probability exceeds another in view of the evidence of the two samples. *Biometrika*, 25:285–294.

Thorlun, K., Dron, L, ., Park, J., and et al. (2020). Synthetic and external controls in clinical trials–a primer for researchers. *Clinical Epidemiology*, 12:467–467.

Trippa, L. and Alexander, B. (2017). Bayesian baskets: a novel design for biomarker-based clinical trials. *Journal of Clinical Oncology*, 35:681–687.

Tseng, Y., Chen, Y.-H., Catalano, P., and Ng, A. (2015). Rates and durability of response to salvage radiation therapy among patients with refractory or relapsed aggressive non-hodgkin lymphoma. *International J Radiation Oncology, Biology, Physics*, 91:223–231.

Tsiatis, A. (2006). *Semiparametric Theory and Missing Data.* Springer, New York.

Tsiatis, A., Davidian, M., Shannon, T., and Laber, E. (2021). *Dynamic Treatment Regimes: Statistical Methods for Precision Medicine.* Chapman and Hall/CRC, New York.

Uozumi, R. and Hamada, C. (2017). Interim decision-making strategies in adaptive designs for population selection using time-to-event endpoints. *Journal of Biopharmaceutical Statistics*, 27(1):84–100.

VanderWeele, T. and Onyebuchi, A. (2011). Bias formulas for sensitivity analysis of unmeasured confounding for general outcomes, treatments, and confounders. *Epidemiology*, 22:42–52.

Wahed, A. S. and Thall, P. F. (2013). Evaluating joint effects of induction–salvage treatment regimes on overall survival in acute leukaemia. *Journal of the Royal Statistical Society: Series C (Applied Statistics)*, 62(1):67–83.

Walsh, W. E., Tesauro, G., Kephart, J. O., and Das, R. (2004). Utility functions in autonomic systems. In *International Conference on Autonomic Computing, 2004. Proceedings.*, pages 70–77. IEEE.

Walter, R., Oths, M., Borthakur, G., et al. (2011). Prediction of early death after induction therapy for mewly diagnosed acute myeloid leukemia withppretreatment risk scores: a novel paradigm for treatment assignment. *Journal of Clnical Oncology*, 29:4417–4423.

Wang, R., Lagakos, S., Ware, J., Hunter, D., and Drazen, J. (2007). Statistics in medicine — reporting of subgroup analyses in clinical trials. *New England Journal of Medicine*, 357:129–133.

Wathen, J. K. and Thall, P. F. (2008). Bayesian adaptive model selection for optimizing group sequential clinical trials. *Statistics in Medicine*, 27:5586–5604.

Wathen, J. K. and Thall, P. F. (2017). A simulation study of outcome adaptive randomization in multi-arm clinical trials. *Clinical Trials*, 14:432–440.

Wathen, J. K., Thall, P. F., Cook, J. D., and Estey, E. H. (2008). Accounting for patient heterogeneity in Phase II clinical trials. *Statistics in Medicine*, 27(15):2802–2815.

Westreich, D. and Greenland, S. (2013). The Table 2 fallacy: presenting and interpreting confounder and modifier coefficients. *American Journal of Epidemiology*, 177:292–298.

Xu, Y., Muller, P., Wahed, A. S., and Thall, P. F. (2016a). Bayesian nonparametric estimation for dynamic treatment regimes with sequential transition times. *Journal of the American Statistical Association*, 111:921–950.

Xu, Y., Thall, P., Hua, W., and Andersson, B. (2019). Bayesian nonparametric survival regression for optimizing precision dosing of intravenous busulfan in allogeneic stem cell transplantation. *Journal of the Royal Statistical Society, Series C*, 68:809–828.

Xu, Y., Thall, P., Mueller, P., and Mehran, R. (2017). A decision-theoretic comparison of treatments to resolve air leaks after lung surgery based on nonparametric modeling. *Bayesian Analysis*, 12:639–652.

Xu, Y., Trippa, L., Mueller, P., and Ji, Y. (2016b). Subgroup-based adaptive (SUBA) designs for multi-arm biomarker trials. *Statistics in Biosciences*, 8:159–180.

Yan, F., Thall, P. F., Lu, K., Gilbert, M., and Yuan, Y. (2018). Phase I-II clinical trial design: a state-of-the-art paradigm for dose finding with novel agents. *Annals of Oncology*, 29:694–699.

Ye, Y., Li, A., Liu, L., and Yao, B. (2013). A group sequential holm procedure with multiple primary endpoints. *Statistics in Medicine*, 32:1112–1124.

Yuan, Y., Nguyen, H., and Thall, P. (2016). *Bayesian Designs for Phase I–II Clinical Trials*. CRC Press, Boca Raton, FL.

Zhang, T. and George, D. J. (2020). Choosing the best approach for patients with favorable-risk metastatic renal cell carcinoma. *Clinical Advances in Hematology & Oncology: H&O*, 18(4):204–207.

Zhang, Z., Li, M., Soon, G., Greene, T., and Shen, C. (2017). Subgroup selection in adaptive signature designs of confirmatory clinical trials. *Journal of the Royal Statistical Society, Series C*, 66:345–361.

Zhao, Y.-Q., Zeng, D., Laber, E. B., Song, R., Yuan, M., and Kosorok, M. R. (2015). Doubly robust learning for estimating individualized treatment with censored data. *Biometrika*, 102(1):151–168.

Zhou, H. and Hanson, T. (2018). A unified framework for fitting bayesian semiparametric models to arbitrarily censored survival data, including spatially-referenced data. *Journal of the Statistical Association*, 113:571–581.

Zhou, Y., Lee, J., and Yuan, Y. (2019). A utility-based bayesian optimal interval (u-boin) phase I/II design to identify the optimal biological dose for targeted and immune therapies. *Statistics in Medicine*, 33:5299–5316.

Index

Note: Page numbers in *italics* and **bold** refer to figures and tables, respectively.

A

Accelerated failure time (AFT) models, 267
ACGSD, 222–224
Acute ischemic stroke, 55
Acute leukemia, 51
Acute lymphocytic leukemia (ALL), 82
Acute myelogenous leukemia (AML), 27, 177, 242
Acute promyelocytic leukemia (APL), 243
Acute respiratory distress syndrome, 177
Adaptive cut-off function, 169
Adaptive decisions, 218
  making, 208
  rules, 45
Adaptive enrichment design (AED), 212–213, 225
Adaptive interim decisions, 134
Adaptively learning, 190
Adaptive matching, 132–133
  simulated covariate distributions, 134–137
Adaptive randomization (AR), 98, 101
Adaptive signature designs (ASD), 193–195
  biomarker subgroups as random partitions, 204–211
  hybrid utility-based enrichment design, 198–202
  model, 193
  one biomarker, 195–198
  phase II-III enrichment design, 202–204
  two regressions on biomarkers, 211–225
AED, *see* Adaptive enrichment design
ALL, *see* Acute lymphocytic leukemia
Allogeneic stem cell transplant, 27
Allogeneic stem cell transplantation, 259–263
Allosct, 259
All-trans retinoic acid (ATRA), 243
AML, *see* Acute myelogenous leukemia
AML/MDS, 243
Anastomotic leak, 154
Anemia, 71, 138
Anti-disease effects, 71, 119, 211, 272
Anti-EGFR monoclonal antibody, 35
Anti-hypertension treatment, 18
APL, *see* Acute promyelocytic leukemia
Apoptosis, 176
Apoptosism, 111
Ara-C, 243
ASD, *see* Adaptive signature designs
Atezolizumab, 160, *161,* 175
  arm, 172
Atrial fibrillation, 154
Augmented IPTW (AIPTW), 250
Average hazard ratio (AHR), 168
Average treatment effect (ATE), 40
  among the treated (ATT), 40
  among the untreated (ATU), 40

B

Baseline survival function, 165
Basket trials

with Bayesian model averaging, 121–126
early basket trial, 118–121
monitoring response and a longitudinal biomarker, 126–131
patient heterogeneity, 111–113
phase II design for non-exchangeable subgroups, 113–118
time-to-event outcome, 111
Bayesian adaptive design, 211
Bayesian adaptive GS (BAGS) design, 159–160, 166–170, 179, 238
Bayesian adaptive regression tree (BART), 44
Bayesian adaptive synthetic control (BASIC), 132–134, 136
Bayesian analysis of covariance, 112
Bayesian and frequentist statistical methods, 20
Bayesian basket trial design, 113
Bayesian computation, 53
Bayesian dependent Dirichlet process, 44
Bayesian designs, 70
Bayesian futility monitoring, 115
Bayesian G Computation, 52–54
Bayesian GS testing, 169
Bayesian hierarchical latent variable approach, 163
Bayesian hierarchical Markov-gamma process, 163
Bayesian hierarchical model (BHM), 112–113, **130,** 131, 164
Bayesian hierarchical piecewise model, 171
Bayesian hierarchical probability model, 206
Bayesian inferences, 24, 26
Bayesian latent subgroup trial (BLAST), 126–127, 129, **130**
Bayesian Learning, 25
Bayesian methods, 56, 201
Bayesian model, 3, 21, 26, 102, 113, 156, 182, 213
averaging (BMA), 121–124
selection problem, 167
Bayesian nonparametric models, 21
dependent Dirichlet processes, 238–241
Dirichlet process priors, 229–238
regression analysis in medical research, 226–229
Bayesian nonparametric (BNP) regression model, 52, 229–230, 259, 277, 287
Bayesian partition model, 204
Bayesian phase II design, 123
Bayesian piecewise exponential regression model, **27**
Bayesian posterior
computation, 216
decision, 218
predictive distribution, 52
probabilities, 134
Bayesian probability models, 230
Bayesian regression models, 112, 114, 238
Bayesian statistics, 20–28, 230
analysis, 277
methodology, 76
regression analysis, 263
Bayesian subgroup adaptive enrichment design, 204
Bayesian test statistics, 168
Bayesian utility-based criteria, 198
Bayes' law, 22–23, 25, 124
Bayes' theorem, 24
Bernoulli distribution, 217, 220
Beta-binomial computation, 26
Binary biomarkers, 37, 112
Binomial-beta model, 231
Binomial distribution, 3, 42, 124
Biomarker, 1, 179, 196
negative patients, 1, 10, 36, 113, 184–187, 200
positive patients, 1, 10, 17, 113, 177, 184–188, 200, **202**

positive subgroup, 187
selection, 213
subgroups, 160
trajectory, 129
vector values, 205, 211
Biomarker adaptive threshold design (BATD), 195
Bivariate
joint distribution, 139
normal distribution, 139
Bladder cancer, 127
Bonferroni test, 199
BRAF mutation, 187
BRAFV600E, 132
Brain tumors, 127
BRCA-mutant ovarian cancer, 214
Breast cancer decision analysis, 287
Breast cancer treatment
Bayesian regression analysis, 277–280
Kaplan-Meier estimates, 274–275
limitation of medical literature, 271–273
personalized treatment, 284–288
precision utility function, 281–284
targeted agents, 273–277
Breast cancer tumor, 33
Busulfan, 260

C

Cancer cells, 79
apoptosis, 111
killing of, 111
proliferation, 17
Cancer immunotherapy, 79
Cancer-killing immune cells, 34
Candidate dose set, 101
Carboplatin + paclitaxel, 36
Cardiac complications, 154
CATE, *see* Conditional average treatment effect
Causal diagrams, *38*
of relationships, 249
of treatment, *19*
of treatment and predictive variable effects, *35*
Causal models, 29
Causal treatment effect, 39
CDKN2A deletions, 127
Cell lung cancer, 36
Cellular proliferation, 214
Cellular therapy, 4
Cerebral bleeding, 55
Chimeric antigen receptor (CAR), 79–81
Chinese restaurant franchise, 237
Chinese restaurant process, 237
Chi-square test, 134
Chromophobe, 35
Chromosomal abnormalities, 242
Chronic lymphocytic leukemia (CLL), 18, 82
CIHM-based designs, 125
Clavien-Dindo post operative morbidity (POM) scores, **155,** *155*
Clinical treatment evaluation
estimation and testing, 1–9
outcomes, early and late, 18–20
practical Bayesian statistics, 20–28
Simpson's paradox, 9–18
Clinical trial design, 3–4
Cluster membership indicator, 74
Colorectal cancer (CRC), 132, 186
Complete remission (CR), 242
Complete response (CR), 70, 119, 138
Comprehensive Cancer Network (NCCN) guidelines, 138
Conditional average treatment effect (CATE), 40
Conditionally independent hierarchical model (CIHM), 119
Confidence intervals, 134
Consistency, 41
Consolidation therapy, 242
Counterfactual outcome, 39
Covariance kernel, 240
Cox proportional hazards regression model, 228, 244–245, 271, 277

Credible intervals, 134
Cross-validated ASD (CV-ASD), 194
Cutoff function, 169
Cut-off probability, 77
Cytosine arabinoside, 177

D

Data dredging, 37
DDP-GP model-based approach, 256
Decision analysis, 277
De Moivre, Abraham, 7
De Moivre-Laplace theorem, 7
Dependent Dirichlet process (DDP), 238–240
Developmental therapeutics, 17
Deviance information criterion (DIC), 128
Directed acyclic graph (DAG), 18
Dirichlet distribution, 230–233
Dirichlet-multinomial model, 102
Dirichlet prior, 104
Dirichlet process, 257
  mixture model, *235*
Discrimination function, 178, 196
Disease heterogeneity and targeted agents, 176–178
Disease recurrence, 260
Dispersion hyper-parameters, 77
DNA sequencing, 176
Docetaxel, 160, *161,* 173, 175
Dose limiting toxicity (DLT), 55
Dose selection, **107–108**
Dose-toxicity
  functions, 59
  hazard functions, 58
  relationships, 58
Doubly incorrect statistical test, 7
"Doubly robust" augmented IPTW (AIPTW), 256
DP mixture (DPM) distribution, 233
Dynamic treatment regimes (DTRs), 45, 245

E

Edema, 55
EGFR gene mutation, 36
Elicitation, 120
Engraftment, 259
Enrichment, 213
  disease heterogeneity and targeted agents, 176–178
  estimation bias, 190–192
  outcome-adaptive randomization (OAR), 181–184
  treatment-biomarker interactions, 184–190
Epidermal growth factor receptor (EGFR) tyrosine kinase inhibitor, 36
E-sensitive patients, 177–179, 188, 193–195, 198–199, 203, 211
E-sensitive subgroup, 222
Esophageal cancer, 127
  patients, 154
Estimation and testing, 1–9
Estimation bias, 190–192
Estrogen therapy (ET), 273
Evidence-based medicine, 76
Exchangeability, 41
External confounders, 14, 32

F

Factual outcome, 39
Familywise type I error rate (FWER), 202, 204
Federal regulatory agency (FRAG), 90
Fisher, Ronald, 4
Fludarabine (Flu), 262
Fludarabine + ara-C + idarubicin (FAI), 243
Fludarabine + Busulfan (FB), 27
Fludarabine + Clofarabine + Busulfan (FCB), 27, **28**
FOLFOX4, 35
Fundamental problem of causal inference, 39
FWER, *see* Familywise type I error rate

G

Gamma function, 177
Gastric cancer, 127

Gastric dumping syndrome, 154
Gaussian kernel, 253, 265
Gaussian mixture model, 141
Gaussian process, 240, 253, 257
Gefitinib, 36
Generalized computation methods, 43
  Bayesian G computation, 52–54
  casual effects, 48–49
  longitudinal data, 51
  multistage treatment, 51
  single-stage data, 44
  two-stage data, 46–47, 50–51
  two-stage strategies, 49–51
Generalized phase I-II (Gen I-II) design, 96–110
  dose outcome curves, *109*
  dose selection, **107–108**
Generalized power (GP), 180
Gen I-II design, 96, 104
G-estimation, 245, 250
Graft-*versus*-host disease (GVHD), 80, 259
Granulocyte colony-stimulating factor (GCSF), 243
Group sequential (GS) design, 191
GSED, 202–203
GS enrichment design, 221
GVHD, *see* Graft-*versus*-host disease

H

Hazard function, 165
Hazard ratio (HR), 228
Hematological malignancy, 9
HER2/neu gene, 17, 33
Heterogeneity, types, 176
Hierarchical Bayesian model, 121, **121**
Hierarchical Markov gamma process, 166
High-dose ara-C (HDAC), 243
Homo-logrank design, 173
House Party algorithm, 60–61, 74
Human immunodeficiency virus (HIV), 23, 37
Humanized antiprogrammed death-ligand 1 (PD-L1) pathway, 160
Human leukocyte antigen (HLA), 259
Hypercalcemia, 71, 138
Hypermean, 120

I

IMDC score, 73
Immature hematopoietic cells, 263
Immune checkpoint inhibitors (ICIs), 71
Immune disorders, 177
Immunotherapy, 260
Inequality, 21
Inhibition of apoptosis, 17
International Metastatic Renal-Cell Carcinoma Database Consortium (IMDC), 138
  prognostic risk score, 71
Interquartile range (IQR), 257
Intra-operative air leaks (IALs), 234
Intravenous Bu (IV-Bu), 260, 262
Inverse probability of treatment weighting (IPTW), 38–42
Ipilimumab, 71, 76–77
IPTW, *see* Inverse probability of treatment weighting

J

Joint probability model, 215

K

Kaplan–Meier estimator, 57, *161,* 252
  of survival time distributions, 264, *264*
Kaplan–Meier plots, 244, *244,* 271
Karnofsky performance status, 138

L

Laplace, Pierre-Simon, 7, 23
Late-onset Efficacy-Toxicity (LO-ET) design, 58
Late onset toxicities, 56
Law of Reverse Probability, 21
Law of total probability, 11–14, 20–22, 43

Lebesgue integral, 40, 50–51
Letrozole, 273
Leukemia, 259
Bayesian regression analysis, 253–258
multistage therapy, 245–253
Likelihood
contribution, 57
evaluation, 142
function, 24
Longitudinal biomarker process, 128
Look ahead method, 56
Low bulk disease (HBD), 82–84
Lung cancer, 160
Lurking variable, 14
Lying with statistics, 14
Lymphatic fluid leaks, 154
Lymphoma, 259

M
Marginal probabilities, 217
Marker sequential test (MaST) design, 179
Markov chain Monte Carlo (MCMC) methods, 24, 27
posterior sampling, 164
proposal steps, 61
Markovian structure, 280
Maslow, Abraham, 227
Maximum a posteriori (MAP) estimation, 167
Maximum tolerated dose (MTD), 98
Melanoma, 127
Metastatic clear cell renal cell carcinoma (ccRCC), 71
Metastatic renal cell carcinoma (mRCC), 66, 138
Misclassification rate (MCR), 170
Model-based clustering, 140
Monte Carlo sample, 145
Mouse-to-human transportability, 4
Multinomial distribution, 146
Multivariate normal distributions, 128
Myelodysplastic syndrome (MDS), 27, 242, 259

N
Natural killer (NK) cells, 79
Neoadjuvant systemic therapy (NST), 207
Neutrophilia, 71, 138
Nivolumab, 71, 76–77
plus ipilimumab, 138
NK cell, 83
Non-Hodgkin lymphomas (NHLs), 18, 82
Non-proportional odds (NPO) model, 156
Non-randomized single-arm phase II trials, 15
Non-small-cell lung cancer, 160
NSCLC tumor cell proliferation, 36
Null hypothesis, 15, 197, 203
NuPrehab, 156–157, **159**
Nutritional prehabilitation (N), 154

O
O'Brien-Fleming boundaries, 173
Olaparib, 214
Oncology, 176
Operating characteristics (OCs), 101
Oracle-BAGS design, 173
Outcome-adaptive randomization (OAR), 181–184
Overall survival (OS) time, 248

P
Pancreatic cancer, 127
Panitumumab, 35
Panitumumab + FOLFOX4, 35–36
Parametric Bayesian models, 24
Parsimonious parametric function, 57
Parsimonious parametric model, 104
Pavlov, Ivan, 32
Pembrolizumab, 34–35
Permutation distribution, 196
Personalized benefit index (PBI), 213, 218
Personalized medicines, 1
Pharmacokinetic (PK) model, 260
Phase II basket trial designs, 122

Phase II sarcoma trial design, 122
Phase II trials, 9
Piecewise exponential distribution, 76
Piecewise exponential hazard model, 215
Play-the-winner (PTW) rule, 181
Pneumonia, 154
Poisson process, 125
Poly(ADP-ribose) polymerase (PARP), 214
Polya urn scheme, 236–237
Polymerase inhibitor, 207
PO-model-based design, 156
Positivity, 41
Posterior distributions, 145, 278
Posterior predictive (PP) mean utilities, 53, 77
Posterior predictive probability, 147
Posteriors, 104, 124
Post operative morbidity (POM), 154
Precision dose optimization
  choosing estimands, 65–69
  dose finding for natural killer cells, 79–95
  generalized Phase I-II designs, 95–110
  time to toxicity, 55–59
    MCMC proposal steps, 61–65
    subgroup clustering model and algorithm, 59–61
  utility functions, 70–73
Precision medicines, 1, 160
Precision randomized phase III designs
  nutritional prehabilitation and post-operative morbidity, 154–159
  precision confirmatory survival time comparisons, 159–175
Precision treatment screening and selection, 137
  decision criteria and trial conduct, 142–146
  prior specification and posterior computation, 141–142
  screen-and-select design, 137–141
Precision utility function family (PUFF), 67
Predictive biomarker, 17
Predictive covariate, 177
Predictive probabilities of toxicity, 56
Presumptive enrichment, 189
Probability, 181
Progression-free survival (PFS), 27, 95
Progressive disease (PD), 70, 119, 138
Propensity score, 42
  matching, 133
  models, 43
Proportional hazards (PH), 228
Proportional odds (PO) model, 156

Q

Q functions, 52
Q model, 44

R

Radiation therapy trials, 55
Random function, 240
Randomization, 15, 29, 33, *34*, 39, 48, 134, 145, 227, 245
  confirmatory study, 165
Randomization controlled trial (RCT), 132
Random probability measures, 238
RCC, *see* Renal cell carcinoma
RCT, *see* Randomization controlled trial (RCT)
Real-valued random variables, 231
Regression analyses, 29–30, 228, 263
Regression coefficient vector, 216
Regression model-based estimates, 134
Regression models, 39, 57, 84, 226, 228
Remission duration (RD), 96
Renal cell carcinoma (RCC), 34, 70–71
Response durability, 96
Response probabilities, 99, 185, 205
Rich get richer property, 237
Right-censored time-to-toxicity, 58
Right censoring time, 100
Risk-benefit trade-offs, 272–273, 276
Robustness, 196
R programming language, 26, 159

S
Safety monitoring, 90
Salvage
chemo, 242
therapy, 66, 246
treatments, 249
Sampling-based method, 45
Secondary covariates, 37
Sequential clustering, 62
Sequentially ignorable treatment assignment, 49
Sequential multiply randomized adaptive randomized trial (SMART), 48
Severity scores, 274
Simon design, 125
Simon's two-stage design, 125
Simpson's paradox, 9–18, **10, 12,** 31
Single-arm clinical trials, 15
Single-arm design (SA), 135, *136*
with matched controls (SC), *136*
Single-arm phase II design, 112, 132
Single-arm phase II trials, 114
Single atoms DDP, 240
Single solid tumor, 176
Single-stage frequentist designs, 199
Single-stage observational data, 43
Single-strand DNA breaks, 214
Smoking cigarettes, 227
S-NTI design, **118**
Spike-and-slab priors, 141
SSD, 222
STAMPEDE trial, 191
Standard normal, 7
Statistical estimates, 134
Statistical thinking, 1
Statistics and causality
causal relationships, 29–38
generalized computation methods, 43–44
Bayesian computation, 52–54
causal effects, 48–49
longitudinal data, 51–52
single-stage data, 44–46
two-stage data, 46–47, 50–51
two-stage strategies, 49–50
inverse probability of treatment weighting, 38–43
Stem cell transplantation
allogeneic, 259–263
Allosct data, 268–270
DDP with refined gaussian process prior, 263–265
personalized dose intervals, 266
simulation study, 266–268
Stickiness, 182
S-TI design, **118**
Stochastic ordering, 140
Strong hierarchy interaction constraint, 216
Strong ignorability, 41
SUBA, 204–205, 209, **209**
Subgroup-specific
decision-making, 141
decisions, 169
dose acceptability rule, 62
futility, 114
response probabilities, 5, **5,** 111
stopping probabilities, 124
utility function, 68, 147
Subgroup-treatment interactions, 147
Substantial randomness, 32
SubTiTE, 58, 61, 63, 74, 160
Survival function, 72
Survival probabilities, 67
and hazards, 66
for treatment, 12, 19
S value, 244
Synthetic control, 133

T
Table 2 Fallacy, 37
Targeted agents, 176
Targeted biomarker, 126
Targeted immunotherapy (TI), **12**
T-cells, 177
Temporality, 32
Thrombocytosis, 71, 138

Time-to-event continual reassessment method (TiTE-CRM), 56–58, 63
computer software, 64–65
Time-to-event (TTE) variables, 55
survival function, 57
Total hazard ratio, 168
Total toxicity burden (TTB), 274
Toxicity, 227
and efficacy, 77
event time distribution, 72
probabilities, 57, 61, 125
Toxicity-*versus*-no toxicity utility, 68
Transformed Bernstein polynomials (TBP), 267
Treatment-biomarker interactions, 184–190
Treatment safety, 2
Treatment–subgroup interactions, 117
Trial BATTLE, 183
Triply incorrect statistical test, 8
T-test, 134
Tumor-infiltrating immune cells, *161*
Two-stage basket trial design, 113
Two-stage phase II designs
adaptive matching, 132–133
simulated covariate distributions, 134–137
singe-arm to RCT decision, 133–134
singe-arm to singe-arm decision, 133
precision treatment screening and selection, 137
decision criteria and trial conduct, 142–146
prior specification and posterior computation, 141–142
screen-and-select design, 137–141
steps for trial conduct, 146–153
Type I error probability, 157, 193, 195
Tyrosine kinase inhibitor (TKI), 71

U
Uniformly unacceptably low efficacy, 208
Utility-based approach, 99
Utility-based Gen I-II trial, 100
Utility elicitation process, 68

V
Validation cohorts, 194
Vascular endothelial growth factor (VEGF), 176, 273
Vector of biomarkers, 178–181
Vemurafenib + irinotecan + cetuximab, 132
Veno-occlusive disease (VOD), 260–261
Viral infections, 177

W
Weibull distribution, 76, 91, 104, 177, **223**
Working likelihood function, 57

Z
Zidovudine (AZT), 45
Z-negative patients, 190
Z-score statistics, 199

For Product Safety Concerns and Information please contact our EU
representative GPSR@taylorandfrancis.com
Taylor & Francis Verlag GmbH, Kaufingerstraße 24, 80331 München, Germany

www.ingramcontent.com/pod-product-compliance
Ingram Content Group UK Ltd.
Pitfield, Milton Keynes, MK11 3LW, UK
UKHW021110180425
457613UK00001B/19
9781032754468